PLANT BREEDING AND BIOTECHNOLOGY

Societal Context and the Future of Agriculture

This accessible survey of modern plant breeding traces its history from the earliest experiments at the dawn of the scientific revolution in the seventeenth century to the present day and the existence of high-tech agribusiness. Denis Murphy tells the story from the perspective of a scientist working in this field, offering a rational and evidence-based insight into its development. Crop improvement is examined from both a scientific and socio-economic perspective, and the ways in which these factors interact and impact on agricultural development are discussed. In conclusion, some concerns over the future of plant breeding are highlighted, as well as potential options to enable us to meet the challenges of feeding the world in the twenty-first century. This thoroughly interdisciplinary and balanced account will serve as an essential resource for everyone involved with plant breeding research, policy and funding, as well as those wishing to engage with current debates about agriculture and its future.

DENIS J. MURPHY is Professor of Biotechnology at the University of Glamorgan, UK. His career in plant biotechnology research spans three decades, including ten years on the management team of the John Innes Centre, arguably Europe's premier research centre in plant science. He is currently highly involved with the ongoing debate on genetically modified food and crops, both locally and internationally, providing expertise and advice to numerous organisations and government agencies, as well as engaging with the general public and the media.

Plant Breeding and Biotechnology

Societal Context and the Future
of Agriculture

DENIS J. MURPHY

University of Glamorgan

CAMBRIDGE
UNIVERSITY PRESS

Shaftesbury Road, Cambridge CB2 8EA, United Kingdom

One Liberty Plaza, 20th Floor, New York, NY 10006, USA

477 Williamstown Road, Port Melbourne, VIC 3207, Australia

314–321, 3rd Floor, Plot 3, Splendor Forum, Jasola District Centre, New Delhi – 110025, India

103 Penang Road, #05–06/07, Visioncrest Commercial, Singapore 238467

Cambridge University Press is part of Cambridge University Press & Assessment,
a department of the University of Cambridge.

We share the University's mission to contribute to society through the pursuit of
education, learning and research at the highest international levels of excellence.

www.cambridge.org
Information on this title: www.cambridge.org/9780521530880

First published 2007

A catalogue record for this publication is available from the British Library

ISBN   978-0-521-82389-0   Hardback
ISBN   978-0-521-53088-0   Paperback

In the frozen midwinter of 1941–1942, the great metropolis of Leningrad began its epic 900-day siege by the encircling German army. Throughout the ruined city, thousands of people were dying of cold, starvation, and shellfire. In the world's oldest seed bank, at the Institute of Plant Industry, a dedicated team of breeders and curators sought to guard and preserve their priceless samples for posterity. This collection of over 160 000 plant varieties had been set up in the 1920s by Nikolai Vavilov, the doyen of twentieth century plant breeding. Hardly any food reached the biologists as they maintained their protective vigil. One by one, they succumbed to starvation, surrounded by bags of edible seeds and tubers. The oats curator L. M. Rodina died, as did rice curator D. S. Ivanov, and peanut curator A. G. Shtchukin, and seven more of their heroic colleagues, one of whom even expired at his desk, working until the end. When the city was eventually liberated in January 1944, the entire collection was intact. It has since been used to supply new edible plant varieties to millions of people around the world.

This book is dedicated to all the many heroes of plant breeding, both past and present, including: Norman Borlaug, Robert Carsky, Charles Darwin, Thomas Fairchild, Jack Harlan, Monkombu Swaminathan, Nikolai Vavilov, those brave workers from Leningrad and from other more recently threatened seed banks in Asia and Africa; and, of course, the untold generations of anonymous farmer-breeders, most of whom were women.

It is to you that we truly owe our daily bread.

Contents

Preface

This book is an evidence-based and, in places, a personal account of the development of scientific plant breeding over the past two hundred years. The work is informed by my background and experiences as a biologist who, while largely trained in the UK, has also worked extensively in the USA, Germany, Australia and, more recently, in the Far East. It is the story of how breeding evolved from an empirical endeavour, practised for millennia by farmers and amateur enthusiasts, to become the globalised corporate agribusiness enterprise of today. I was moved to write this account after spending over two decades working at the interface of academic plant science and its practical application in crop breeding. During this time I have witnessed the steady erosion of plant breeding as a worthwhile and respected aspect of plant science, especially in the public sector. One of my principal motives in writing the book is to raise the profile of plant breeding as a valued and useful profession. I also wish to highlight some of the many imbalances that now bedevil our approach to breeding, some of which have coloured today's often contentious discourse on agriculture and crop improvement in general.

There are many misapprehensions, among scientists and the general public alike, about the way that plant breeders go about their business. In particular, the supposedly revolutionary nature of the 'new' (actually now more than two decades old) technologies of genetic engineering has been exaggerated by virtually everybody involved in the debate, whether they be researchers, politicians, agbiotech companies or anti-GM (genetic manipulation) campaigners. The current fixation on this at times overhyped phenomenon is coupled with a worrying dearth of knowledge and understanding about the many other (non-transgenic) forms of plant breeding which, as I will show, can in principle be subjected to many of the same objections that are levelled against GM technologies. So, why is it that this particular aspect of plant breeding is deemed so threatening that it can elicit violence and disorder among often idealistic and well-intentioned anti-GM activists, while the same people know little of the rudiments of plant breeding in the wider sense? And it is not just anti-GM campaigners who have little knowledge of the broader socio-scientific dimensions of plant breeding; a similar charge can be levelled against many people in the broader realms of science, politics, the media etc.

In Part I, we will begin to address these issues by discussing the basic scientific background of plant breeding. These three chapters are the only ones that focus on science per se. The remaining five parts of the book are devoted to examining the interface between science/technology and society, and the manner in which these forces have mutually influenced each other in the case of plant breeding and the production of improved crops. In this analysis, I will take it as read that science and technology are deeply embedded in the wider socio-economic milieux from which they both arise. This is particularly true in the case of plant science and its applications via various technologies to effect improvements in crop performance. As we will see, an 'improvement' is normally so defined by the improver; hence one person's improvement might even be another's curse. Science and society are respectively made up of many players, all of whom interact with and affect one another, often in subtle ways that are not always obvious to the casual observer. In the case of GM crops, the peculiar, and unusually contentious, trajectory of this technology has been determined by interactions between a host of factors including scientific discoveries (how to transfer genes), legislative measures (patenting plants), the economic environment (privatisation), political opportunism (policy based on pressure groups), ideology (policy based on belief systems), and so on. I will show how it was a particular conjunction of circumstances in the 1980s and 1990s that has led to the current, and arguably inappropriate, domination of the agbiotech/GM phenomenon in both scientific and public discourse. A key message that I wish to convey is the need to rediscover a sense of perspective in our attitude to crop improvement and to raise our gaze beyond the narrow confines of the GM debate, so that we can behold the real challenges and opportunities that confront international agriculture in the twenty-first century.

Acknowledgements

I am indebted to the many friends and colleagues who have, wittingly or not, inspired and assisted me in various ways during the writing of this book. This includes those colleagues at the John Innes Centre and University of Glamorgan with whom I have had numerous fruitful discussions over the past 15 years. Special thanks are due to Eddie Arthur and Ray Matthias with whom I tried (and failed) to interest funding agencies in the domestication of new crops – rest assured, our time will come. As an initially somewhat reductionist molecular biologist, I was challenged in the early 1990s by Colin Law who remained a true believer in traditional plant breeding, and a sceptic of the many chiliastic claims of agbiotech. I guess some of his sentiments must have eventually rubbed off on me, as this book shows. Other colleagues, including Ian Bartle, Gerry Roberts and Colette Murphy provided valuable feedback on various drafts of the manuscript, as did various anonymous referees. Finally, many thanks to Katrina Halliday and the staff at Cambridge University Press for their patience and encouragement during the gestation of this project.

Denis J. Murphy
Glamorgan, Wales, April 2006

Using this book

This book is aimed at several audiences, from botanists to economists, and from business people to agronomists. Each group of readers will have different technical backgrounds and different types of expertise. The book is therefore written on three levels, namely the main text, a series of more than 900 endnotes, and a bibliography. For the general reader or specialist in other areas, the main text of each chapter should suffice to convey its key message. However, for those wishing to follow up points in more detail, the endnotes provide additional information that is in turn linked to a comprehensive bibliography of over 750 citations, mostly from the peer-reviewed primary literature. Wherever possible, I have also provided web links to the many articles that are now available online. Many of the better scientific journals now make their articles freely available on the Internet within a year of publication, and such web links tend to be relatively stable. These primary research articles can often be surprisingly accessible, even to the interested layperson, and I recommend interested readers to consult at least one or two examples.

Secondary literature sources, e.g. scholarly reviews, government reports, conference papers etc., are also often available on the Internet and can frequently be useful resources, especially for the technical specialist from a different field. Such articles generally give a broader perspective than primary research papers, but may not necessarily be peer reviewed. 'Tertiary' sources, including newspaper and magazine articles, are rarely peer reviewed. Such articles tend to be more 'fresh' and accessible in their content, but can also be less factually reliable, and rarely provide a broad overview of the topic in question. Tertiary sources also tend to be more ephemeral in their Internet locations (hence, *caveat lector* and my apologies in advance if some of the web links no longer work). Nevertheless, newspaper and magazine articles often add a welcome degree of colour and immediacy to a discussion that contrasts favourably with the more sober and restrained tone of most mainstream scientific literature.

Nomenclature and terminology

Measurements and dates

The metric system is used throughout for all physical measurements. All prices are given in US dollars ($) and, unless stated otherwise, they relate to the period in question, i.e. prices are not corrected to present-day values. Some historical dates are expressed as BCE (before common era) or CE (common era). All unqualified dates refer to CE.

Initials and acronyms

I have tried to forebear, as much as possible, from using unfamiliar initials and acronyms in the main text. Where this is impractical, I give the full version of each term in the text when it is first used. A full list of such terms, plus some additional explanation of their significance, is also given in the Abbreviations and glossary section (see below).

Industrial and developing countries

In describing the major global economic blocs, the term 'industrial' is used to describe those countries that have already completed a thorough industrialising process. In many cases such countries are now at a post-industrial stage of development. Included here are the major globalised economies of North America, Europe, Australasia and Japan. I use the term 'developing' to describe those countries in which industrialisation is still proceeding, albeit often at an advanced stage. This group includes the Asian giants India and China, as well as most of the remainder of Asia, Africa and Latin America. This terminology is by no means perfect and there will always be exceptions. Neither is it meant as a value judgement since all countries are always in the process of some sort of development. But it remains, withal, a useful shorthand.

xix

Abbreviations and glossary

ADAS   Agricultural Development and Advisory Service (UK) – The former public sector agricultural advisory service for England and Wales. ADAS was gradually privatised during the 1990s until it became a private company in 1997.

AEBC   Agriculture and Environment Biotechnology Commission (UK) – the UK government strategic advisory body on biotechnology issues affecting agriculture and the environment. Established in 2000, the AEBC was wound up in 2005, following criticisms of its narrow remit and dissension among members.

BBSRC   Biotechnology and Biological Sciences Research Council (UK) – the major UK public sector funding agency for research in biological sciences, with an annual budget of about $550 million.

BIOS   Biological Innovation for Open Society – an initiative of **CAMBIA** to 'foster democratic innovation in applications of biological technologies to sustainable development'.

Bt   *Bacillus thuringiensis* – a bacterium that produces a variety of insecticidal protein toxins. Bt sprays (containing bacterial spores and toxin crystals) are regularly used as insecticides by organic farmers, while the Bt toxin gene has been added to some crops to provide inbuilt insect protection.

CAMBIA   Center for the Application of Molecular Biology to International Agriculture – non-profit, Australian-based scientific organisation working for the development of new **open access technologies** for crop improvement across the world.

CGIAR   Consultative Group on International Agricultural Research – an alliance of countries, international and regional organisations, and private foundations that supports 15 Research Centres. The Centres work with national agricultural research systems and civil society organisations including the private sector and generate global public goods that are freely available to all. CGIAR research centres include **CIAT, CIFOR, CIMMYT, CIP, ICARDA, ICRAF, ICRISAT, IFPRI, IITA, ILRI, IPGRI, IRRI, IWMU, WARDA** and **WFC**.

CIAT   Centro Internacional de Agricultura Tropical (Columbia) – one of the **CGIAR** crop improvement centres.

CIFOR   Center for International Forestry Research (Indonesia) – one of the **CGIAR** crop improvement centres.

CIMMYT   Centro Internacional de Mejoramiento de Maiz y Trigo (International Maize and Wheat Improvement Center Mexico) – one of the **CGIAR** crop improvement centres.

CIP   Centro Internacional de la Papa (Peru) – one of the **CGIAR** crop improvement centres.

CSIRO   Commonwealth Scientific and Industrial Research Organisation – the principal public sector research organisation in Australia that covers agribusiness; information, manufacturing and minerals; and sustainable energy and environment. CSIRO manages several research centres that work on crop-related topics. One of the best known of these is CSIRO Plant Industry in Canberra, which is especially noted for its research on plant molecular and developmental biology. The annual budget is about $700 million.

Cultivar   a cultivated variety of a crop – such varieties have normally been selected by breeding and are adapted for a particular agricultural use or climatic region.

DEFRA   Department for Environment, Food and Rural Affairs (UK) – formerly known as **MAFF**, DEFRA was created in 2001 in the wake of the BSE scandal but lost its role in food safety to the new Food Standards Agency (**FSA**).

EMBRAPA   Empresa Brasileira de Pesquisa Agropecuária–Brazilian Agricultural Research Corporation.

EST   expressed sequence tag – a small portion of a gene that can be used to help identify unknown genes and to map their positions within a genome.

Ex situ conservation   the maintenance of biological specimens away from their normal habitat, normally under closely controlled conditions, such as in arboretums (trees), and botanical (plants) or zoological (animals) gardens. The term also refers to the keeping of stocks, such as seeds, cuttings, or other propagules in germplasm repositories.

FAO   Food and Agriculture Organisation – United Nations agency, set up in 1945, whose mandate is: 'to raise levels of nutrition, improve agricultural productivity, better the lives of rural populations and contribute to the growth of the world economy.' Its annual $750 million budget covers both ongoing programmes and emergency relief work.

Farm-scale evaluations   a $9 million research exercise in the UK to determine the on-farm effects to fauna and flora of growing and managing herbicide tolerant crops compared to non-tolerant varieties of the same crops.

FMD   foot and mouth disease – also known as hoof and mouth disease in the USA, this virulent viral disease spread across the UK in 2001. Following scientific advice that has since been questioned, the UK government implemented a drastic cull that resulted in the slaughter of 6 million animals, at an estimated cost to the economy of $15 billion.

FSA   Food Standards Agency (UK) – established in 2000 'to protect the public's health and consumer interests in relation to food.'

Germplasm   the genetic material, i.e. the DNA, of an organism. The term is often used in connection with the collection or conservation of seeds, cuttings, cell cultures, or other germplasm resources, in repositories such as gene banks.

GM   genetically modified or genetically manipulated – a term normally used to describe an organism into which DNA, containing one or more genes, has been transferred from elsewhere. The transferred DNA is never itself actually from another organism, but may be an exogenous copy of DNA (i.e. from a different species).

	Alternatively the transferred DNA may be an extra copy of an endogenous gene (i.e. from the same species). Finally, the transferred DNA may be completely synthetic and hence of non-biological origin. An organism containing any of these categories of introduced gene is called **transgenic**.
Heterosis	also called hybrid vigour, the phenomenon whereby a hybrid of genetically distinct (but often inbred) parents is sometimes much more vigorous than either parent. In crop terms, hybrids exhibiting heterosis can out-yield their parents by as much as 30–40%.
HRI	Horticulture Research Institute (UK) – also called Horticulture Research International, HRI is a former public sector plant science research centre that was transferred to ownership of the University of Warwick in 2005.
Hybrid	an organism resulting from a cross between parents of differing genotypes. Hybrids may be fertile or sterile, depending on qualitative and/or quantitative differences in the genomes of the two parents. Hybrids are most commonly formed by sexual cross-fertilisation between compatible organisms, but cell fusion and tissue culture techniques now allow their production from less related organisms.
ICARDA	International Center for Agricultural Research in the Dry Areas – this CGIAR-affiliated centre, established in 1977 with its head-quarters in Aleppo, Syria, has a mission 'to improve the welfare of poor people and alleviate poverty through research and training in dry areas of the developing world, by increasing the production, productivity and nutritional quality of food, while preserving and enhancing the natural resource base.'
ICRAF	World Agroforestry Centre (Kenya) – one of the **CGIAR** crop improvement centres.
ICRISAT	International Crops Research Institute for the Semi-Arid Tropics (India) – one of the **CGIAR** crop improvement centres.
IFPRI	International Food Policy Research Institute (USA) – one of the **CGIAR** agricultural improvement centres.

IGER
: Institute for Grassland and Environmental Research (UK – formed by a merger between the Welsh Plant Breeding Station and the Grassland Research Institute at Hurley.

IITA
: International Institute of Tropical Agriculture (Nigeria) – one of the **CGIAR** crop improvement centres.

ILRI
: International Livestock Research Institute (Kenya) – one of the **CGIAR** crop improvement centres.

Input trait
: a genetic character that affects how the crop is grown without changing the nature of the harvested product. For example, herbicide tolerance and insect resistance are agronomically useful input traits in the context of crop management, but they do not normally alter seed quality or other so-called **output traits** that are related to the useful product of the crop.

In situ conservation
: the maintenance of a species or population in its normal biological habitat. In the case of plants, this applies particularly to natural populations of crop species and/or their wild relatives that may be future sources of genetic variation, as well as to endangered species in general. *In situ* conservation is especially useful in the preservation of traditional crop **landraces**, many of which are under threat from the increasing use of higher yielding but more genetically uniform modern varieties in agriculture.

IPGRI
: International Plant Genetic Resources Institute (Italy) – an international **CGIAR**-affiliated research institute with a mandate to advance the conservation and use of genetic diversity for the well being of present and future generations.

IPR
: Intellectual Property Rights – as defined by the World Trade Organization: 'Intellectual property rights are the rights given to persons over the creations of their minds. They usually give the creator an exclusive right over the use of his/her creation for a certain period of time.' IPR covers literary and artistic works (via copyright) in addition to industrial inventions (via patents and trademarks) and typically lasts for about 20 years. IPR protection of living organisms, such as plant varieties, is a more recent and controversial development.

IRRI	International Rice Research Institute (Philippines) – an independent, non-profit agricultural research and training organisation and **CGIAR** centre that is focused on rice improvement. IRRI was established in 1960 by the Ford and Rockefeller foundations in cooperation with the Philippines government with its main site at Los Baños, near Manila
ISNAR	International Service for National Agricultural Research (USA) – assists developing countries in improving the performance of their national agricultural research systems and organisations by promoting appropriate agricultural research policies, sustainable research institutions, and improved research management.
IWMU	International Water Management Institute (Sri Lanka) – one of the **CGIAR** crop improvement centres.
JIC	John Innes Centre (UK) – a **BBSRC**-funded plant and microbial science research centre near Norwich.
Land Grant Universities	US network of agriculturally focused universities established by the Morill Act in 1862.
Landrace	a genetically diverse and dynamic population of a given crop produced by traditional breeding. Landraces largely fell out of favour in commercial farming during the twentieth century and many have died out. Landraces are often seen as potentially useful sources of novel genetic variation and efforts are under way to conserve the survivors.
MAFF	Ministry of Agriculture, Fisheries and Food (UK) – the government department responsible for oversight of UK agriculture, including the commissioning of some research areas. MAFF was reorganised as **DEFRA** in 2001.
NSDO	National Seed Development Organisation – commercial arm of **PBI** in the UK.
Nutraceutical	a neologism combining nutritional with pharmaceutical and meaning a food product that has been determined to have a specific physiological benefit for human health. The term has no regulatory definition and is primarily used in promotion and marketing.

Open access technology (OAT)	technology that may be protected but which is made available in the public domain. Some OATs might be freely available for unrestricted use, while others might only be available to members of a consortium who agree to use the technologies in a particular manner, e.g. solely as non-profit public goods. The best known OAT is the Linux computer operating system, but analogous OATs have recently been developed in the field of agbiotech, most notably by **CAMBIA**.
Output trait	a genetic character that alters the quality of the crop product itself, e.g. by altering its starch, protein, vitamin or oil composition.
PBI	Plant Breeding Institute – widely regarded as the premier centre of plant breeding research in the UK, based in Cambridge, PBI was privatised in 1989 and subsequently sold on to a series of multinational companies.
PBR	Plant breeders' rights – a form of intellectual property protection in the European Union (via **UPOV**) designed specifically for new varieties of plants.
PCR	Polymerase chain reaction – a technique for rapidly copying a particular piece of DNA in the test tube (rather than in living cells). PCR has made possible the detection of tiny amounts of specific DNA sequences in complex mixtures. It is now used for DNA fingerprinting in police work, in genetic testing and in plant and animal breeding.
PSIPRA	Public Sector Intellectual Property Resource for Agriculture (USA) – initiative of the Rockefeller and McKnight Foundations, in collaboration with ten of the major US **Land Grant Universities**. As with **CAMBIA**, this US initiative is designed to support plant biotechnology research in developing countries.
PVPA	Plant Variety Protection Act – legislation enacted in 1970 by the US Congress that extended **UPOV**-like legal protection to plant germplasm.
Quantitative genetics	the study of continuous traits (such as height or weight) and its underlying mechanisms.

Quantitative trait locus (QTL)	DNA region associated with a particular trait, such as plant height. While QTLs are not necessarily genes themselves, they are closely linked to the genes that regulate the trait in question. QTLs normally regulate so called complex or quantitative traits that vary continuously over a wide range. While a complex trait may be regulated by many QTLs, the majority of the variation in the trait can sometimes be traced to a few key genes.
RAE	Research Assessment Exercise – method used in the UK to rank university research on the basis of the perceived quality of a 'unit of assessment' that normally corresponds to a department. This ranking is then used to apportion funding selectively in favour of higher ranked departments.
SAES	State Agricultural Experiment Station (USA) – established by the Hatch Act of 1887, the nationwide network of SAESs works with **Land Grant Universities** to carry out a joint research/teaching/ extension mission.
Species	a group of organisms capable of interbreeding freely with each other but not with members of other species (this is a much simplified definition, the species concept is much more complex). A species can also be defined as a taxonomic rank below a genus, consisting of similar individuals capable of exchanging genes or interbreeding.
Teosinte	the original wild grass, native to Mexico, from which cultivated maize is derived; it is now classified as part of the same species as maize, *Zea mays*.
TILLING	Targeting Induced Local Lesions IN Genomes – the directed identification of random mutations controlling a wide range of plant characters. A more sophisticated DNA-based version of mutagenesis breeding, TILLING does not involve **transgenesis**.
Transgenesis	the process of creating a **transgenic** organism.
Transgenic	an organism into which DNA, normally containing one or more genes, has been transferred from elsewhere (see **GM**).
UPOV	Union for the Protection of New Varieties – established in 1960 by six European nations to extend legal ownership rights to plant germplasm.

USDA   United States Department of Agriculture (USA) – established by President Lincoln in 1862, USDA is the government department responsible for all matters pertaining to agriculture, including aspects of trade policy, food safety and the environment.

WARDA   Africa Rice Center (formerly called West Africa Rice Development Association) – one of the **CGIAR** crop improvement centres.

WFC   World Fish Center (Malaysia) – one of the **CGIAR** agricultural improvement centres.

Wide crossing   in plant breeding this refers to a genetic cross where one parent is from outside the immediate gene pool of the other, e.g. a wild relative from one species crossed with a modern crop cultivar of another species.

Wild relative   plant or animal species that is taxonomically related to crop or livestock species and serves as a potential source of genes for breeding new crops or livestock varieties.

Introduction

The purpose of this book is to examine the wider scientific and social contexts of modern plant breeding and agriculture. We will begin by examining the historical development of plant breeding over the past two centuries, before focusing on the dramatic changes of the last two decades. Perhaps the best-known recent development in plant breeding is the emergence of genetic engineering, with its attendant social and scientific controversies. But, as we shall see, GM crops and 'agbiotech' (agricultural biotechnology) are just one manifestation of a more extensive series of seismic changes that have profoundly altered the course of plant breeding since the 1980s. Today, in the middle of the first decade of the twenty-first century, plant breeding and crop improvement are at an historic crossroads. On one hand, are the tried and tested breeding methods that underpinned the Green Revolution and enabled us to feed the expanding world populations in the twentieth century. More recently, however, governments across the world have largely dismantled their applied research infrastructures and have greatly reduced the capacity for public-good applications of newly emerging breeding technologies, including transgenesis. Much of this institutional restructuring occurred as part of the ideologically driven privatisation of public assets in the 1980s and 1990s. The resulting depletion of public sector breeding has left a void that was filled by a few private sector companies who applied a new paradigm of crop improvement based on transgenesis – and from this, the agbiotech revolution was born.

As we confront the challenges of increasing populations, economic growth, rising affluence, the spread of environmental degradation, and the depletion of non-renewable resources, twenty-first century agriculture will need all the tools and scientific expertise that plant breeders can muster. Not to mention the appropriate crop management strategies, market freedoms, and social stability that will be necessary to translate the promise of the breeder into the reality of productive and profitable crops for the wellbeing of the farmer. We will see how research into plant science is becoming increasingly remote from its application for breeding. For a variety of different but linked reasons, public sector scientists are largely failing to provide the requisite leadership in the development of practical public-good technologies for crop improvement, especially in developing countries where the need is greatest. One of the main take-home messages of this book is that we must re-engage plant and agricultural science with the rest of society at a whole series of levels. These include better links between basic science and applied technologies, between scientific breeders and their farmer-customers, between the public sector and the private sector, between industrialised countries and the developing world, between inexpensive conventional breeding and the costliest high-tech methods, and between agronomists and managers, and the economists and politicians working in agriculturally related areas of their respective professions.

The book is divided into six parts that first introduce us to the science of plant breeding before describing its changing social organisation and evolution as a mixed public/private venture over the last two centuries. Part I includes a brief account of the origins of breeding and its transition from a farm-based empirical activity to the highly sophisticated scientific programmes of today. We will follow the increasingly successful efforts of plant scientists of the eighteenth and nineteenth centuries to harness their growing knowledge of plant reproduction and development for practical and profitable commercial application. We will see how agricultural innovators became ever more skilled in manipulating those twin pillars of breeding, namely genetic variation and selection. The rediscovery of the principles of Mendelian inheritance and their application to simple and complex genetic traits was the key scientific foundation of twentieth century crop breeding.

The practical application of genetic knowledge to crop improvement in the field was made feasible by the theoretical and statistical tools provided by quantitative genetics after 1918. In the 1920s, chemical and X-ray mutagenesis were first used to create new crop varieties, while the 1930s saw the beginnings of increasingly successful applications of tissue culture in breeding programmes. Soon, scientific breeders could create artificial hybrid combinations from different species, and even different genera. And it was not long before the first manmade crop species, a plant called triticale, was produced. By the 1950s, the technique of wide crossing, coupled with chemically induced chromosome manipulations, had enabled breeders to transfer chromosomes, or parts thereof, from plants that were normally much too distantly related to inter-breed. More effective types of radiation mutagenesis, using nuclear sources such as cobalt-60 or caesium-137, were effectively used after World War II to create more than 3000 new crop varieties.

In Part II, we will switch to consider the societal contexts of these scientific developments that led us from the farmer-breeder of the nineteenth century to today's multinational, high-tech agribusiness. During the nineteenth century, it was realised that the most effective method for applying scientific principles to crop improvement was to establish a professional body of trained plant breeders and researchers. In many of the newly industrialising countries, this was achieved by direct government action. Without a doubt, the most comprehensive, effective, and enduring crop improvement network is that of the USA, as originally established by the Morrill Act in 1862, during the depths of the Civil War. The British establishment, in contrast, took a distinctly more *laissez-faire* route to agricultural betterment. Here, there was a gradual evolution of a disparate group of mostly privately funded research centres during the late nineteenth and early twentieth centuries. It was in some of these British research centres that the application of the newly rediscovered principles of Mendelian genetics first propelled crop science into a new era. In the USA, the huge potential of hybrid crops, in terms of both yield and profitability, began to be realised during the 1920s with the introduction of the high-yielding maize varieties that eventually spread across the continent and beyond. For most of the twentieth century, plant breeding and crop science research were very much concentrated in the public sector, with major contributions from universities and specialised crop-focused research centres.

The success of this public sector based paradigm became ever more apparent as increasingly sophisticated breeding technologies were developed. These technologies, developed by public sector plant researchers as free public goods, were called upon to resolve the worsening food crisis as populations in developing countries expanded rapidly during the 1960s. The Green Revolution of the 1960s and 1970s was largely the result of the focused application of such public-good plant breeding, assisted by some US-based philanthropic foundations. Thanks to the work of a few groups of dedicated plant breeders, new high-yielding varieties of wheat and rice were developed, just in time to head off the spectre of mass hunger that haunted the Indian subcontinent and much of Eastern Asia. The spectacular success of the Green Revolution in much (but not all) of the developing world led to the establishment of an international network of plant research and breeding centres, including such vital resources as seed and germplasm banks.

In Part III, we move on to consider the turbulent events of the late twentieth century and the surprisingly rapid unravelling of the hitherto successful public/ private paradigm of plant breeding research and development. The 1970s and early 1980s marked the apogee of public sector and public-good international plant breeding. Within a few years, governments around the world began to dismantle their public sector plant science infrastructures, in line with the new privatisation agenda that emanated largely from the UK. Meanwhile, the private sector emerged from the shadows as an increasingly dominant force in the enterprise of crop modification and improvement. Two additional factors facilitated the growth of the private sector: the shift to a more benign regulatory environment for the legal protection of new plant varieties; and the invention of a new set of plant manipulation technologies that would allow the patenting of transgenic (GM) crop varieties. We will go on to follow the fate of some of the rationalised, reduced, or terminated public breeding programmes across the world and the resulting retreat of the vast majority of public sector researchers into more academic studies.

The topic of Part IV is agbiotech. The decade from 1985–1995 witnessed a fundamental shift in the world of plant breeding, as the private sector became the more dominant partner and transgenic technologies were increasingly promoted as the way forward for crop improvement in general. We will analyse the consequences of these important developments for the future of agriculture. I will present the case that it was not so much genetic engineering (transgenesis) itself that has been the root cause of the many public controversies about agricultural biotechnology (agbiotech). Rather, it is the context in which the technology was created, promoted, and then applied to crop manipulation, which was radically different to previous forms of high-tech scientific crop improvement. After World War II, highly intrusive and 'artificial' methods of crop genetic modification had already been developed in the public domain with little or no fanfare or public controversy. These technologies were used freely to create new crop varieties around the world and were especially widely applied in developing countries.

In contrast, transgenic technologies were largely developed and patented by the private sector. Some companies then used the new technologies for the manipulation of a few simple input traits in a few profitable commercial cash crops. In the meantime, however, these technologies had already been widely hailed, by public sector scientists and companies alike, as a radical and revolutionary breakthrough in plant breeding of almost unlimited potential for the future of agriculture. Subsequently, the fact that, notwithstanding the optimistic rhetoric, nothing of any matching public value has so far emerged from transgenesis, has engendered a mixture of public scepticism and distrust about the entire agbiotech enterprise. We will also see how the actions of a few agbiotech companies are currently in danger of sabotaging some rather promising future developments in transgene technology to produce cheap medicines via biopharming.

In Part V, we will discuss alternative methods of enhancing crop production, especially amongst the rapidly increasing populations of the developing world. I will show that there need not be any looming crisis in feeding the world population over the next fifty years. We already have the crops, the breeding expertise, and the organisational skills to achieve this task – providing it is managed properly. I will present the case for a judicious expansion of our use of arable land, especially in parts of South America, where a large amount of non-forested land is available for sustainable crop cultivation. Combined with re-use of fallow, abandoned, and set-aside land, these measures could significantly increase global food production over the next few decades. Other productivity enhancing measures include better on-farm management, improvement of physical and regulatory infrastructure (ports, roads, credit facilities, tax regimes etc.), and the ending of discriminatory tariffs and subsidies. Implementation of these exceedingly practical but relatively unglamorous measures, along with the prospect of continuing yield gains via plant breeding, should ensure that we will be able to 'feed the world' over the next fifty years, without recourse to more nebulous and uncertain 'magic bullet' solutions.

In Part VI, we will look forward to the future of plant breeding in the twenty-first century, whether in the public/private sectors, or in industrial/developing countries. We will discuss the uncertain situation of international organisations like CGIAR (Consultative Group on International Agricultural Research), our endangered global seed banks, and the often heroic, and largely unseen, efforts of breeders in countries from Iraq to Côte d'Ivoire in trying to maintain these precious resources against the depredations of warfare and civil strife and the more benign neglect of increasingly jaded funding bodies. We will then look forward to consider some new options that could allow a reinvigorated public sector to resume its place as a major partner in the global enterprise of crop improvement. The long-term success of international agriculture is dependent on a diverse, mixed ecology of public and private agents and agencies. We need strong, well-resourced public-good ventures, which in turn are balanced and complemented by appropriately regulated, for-profit, private sector ventures that are both innovative and truly competitive.

The current problems of plant breeding have not been helped by the fact that many public sector scientists have largely withdrawn from practical breeding and public debate, to the more secluded and serene realms of basic research. The latter are not so much ivory towers as ivory cloisters of an almost adamantine unworldliness. This withdrawal has left the public arena bereft of many of the voices that could bring some balance into the sterile and polarised discourse on transgenic crops that has plagued the debates of the past decade. It is only by regaining a sense of balance in each of these aspects of crop improvement that we can recapture public confidence, and move forward with a renewed sense of optimism to confront and resolve the many challenges of agriculture in the twenty-first century.

Part I

The science of plant breeding

Here Ceres' gifts in waving prospect stand,
And nodding tempt the joyful reaper's hand.
Alexander Pope (1688–1744)
Windsor Forest (l. 39)

1

Origins of plant breeding

For out of olde feldes, as men seith
Cometh al this newe corn fro yeer to yere;
And out of olde bokes, in good feith,
Cometh all this newe science that men lere.
Geoffrey Chaucer (c. 1382)
The Parlement of Foules

Introduction – the development of agriculture

For most of our history, we humans have been omnivores who enjoyed a varied plant and animal based diet that was derived from a hunter-gathering lifestyle. This special relationship has bound people and plants in mutual dependence for well over one hundred millennia. During this period, our Palaeolithic and Neolithic ancestors experimented with many different strategies of plant exploitation, especially during the last hundred millennia when climatic conditions changed repeatedly and other resources such as large animals often became progressively more difficult to obtain.[1] For tens of millennia before the start of formal agriculture, societies throughout the world were engaged in many types of relatively sophisticated management of their favoured food plants. For example, 23 000 years ago, people in the Jordan Valley were already harvesting and grinding wild cereal grains, and baking the flour into bread and cakes.[2] Discoveries of similar grinding implements dating back as far as 48 000 years ago might mean that the management and processing of cereals went on for well over 30 000 years before these plants were ever cultivated as crops.

During this period of informal plant management, our ancestors unwittingly began a process of plant selection that would lead to the domestication of a few genetically amenable species. These plants became the first successful crops, and their formal cultivation was already well under way in several regions of the world by 12 000 years ago, and possibly earlier. Following the development and spread of

9

agriculture, most of our ancestors came to rely increasingly on a much more restricted repertoire of domesticated plants for the majority of their food needs. Even today, most of the world depends on a small, carefully selected group of edible plants. We also use plants for a host of other purposes, such as clothing, shelter, medicines and tools. Although we now believe that the beginnings of the domestication process were probably non-intentional and unforeseen by Palaeolithic and Neolithic proto-farmers, these people soon learned how to improve their new crops by conscious forms of selection and breeding.[3]

Non-intentional selection

Selection can be said to be the backbone of crop breeding. There is little point in assembling or creating a group of genetic variants unless one has an effective mechanism to recognise and select the best adapted or most useful of these variants for further propagation. Such selection could have been either unintentional or deliberate on the part of the early farmers or would-be farmers. To a great extent, all living organisms act in concert with the abiotic (non-living) part of the environment as unintentional agents of selection. This sort of selection is normally negative, i.e. less fit individuals tend to be eliminated from the population. For example, simply by hunting for prey, a carnivore will tend to select those individuals that are easier to capture because they are less well protected, slower, more easily detected etc. As a result, the prey population is selected in favour of fitter individuals who, for example, may have adopted herding behaviour, are more fleet of foot, have acquired camouflage etc. Our ancestors started out as non-intentional agents of selection during the early stages of crop domestication. They then progressed to conscious selection of the relatively small number of favourable traits that were readily recognisable in a crop plant, e.g. seed size, vigour, or yield. In contrast, the scientific breeders of today have access to a battery of screening and selection strategies, many of them automated, that enable them to manipulate hundreds of often invisible traits in our major crop plants.

Most plants in non-agricultural ecosystems have been selected for traits such as indeterminate flowering and easy seed shedding from the parent plant. This minimises the chance of seed loss to herbivores. Several early human societies used techniques of plant management that had the by-product of selecting for a different set of traits. From recent genetic evidence, we know that these new traits enabled a few plant species to develop in the direction of domestication. The concept of non-intentional, or unconscious, selection by humans was first expounded by Darwin,[4] although he had no idea of the mechanism by which the favoured variants could transmit their variations to subsequent generations.[5] More recently, this mechanism of crop selection has been described in detail by Zohary.[6] Non-intentional selection probably led to those initial genetic changes that were the prerequisites to successful cultivation of plants as crops. The simplest form of traditional grain agriculture involves planting the seeds into tilled fields, harvesting the grain-bearing structures, and threshing out the seeds by mechanical agitation. Just by growing a crop in this way, a huge selection pressure is established that favours plants that do not shed their seeds before the harvester is ready.

In the wild, most seeds would be shed from the reproductive structures, but in a tilled field such shed seeds would fall to the ground and therefore would not be saved for re-planting. Gradually, the wild-type seed shedding trait would be lost from the population, which would instead be dominated by a new phenotype, i.e. non-shedding of seeds. This non-shedding phenotype would have been extremely maladaptive in the original ecosystem in which the plant had evolved, but it then became extremely useful under the new conditions of cultivation by humans. Similarly, seeds that germinated immediately upon planting would be automatically selected under the conditions created by cultivation. Wild-type varieties tend not to germinate straightaway. Instead, the seeds enter a period of dormancy that can be of variable duration, hence ensuring that they do not all germinate at the same time and compete with each other.[7] This type of seed dormancy trait is automatically selected against under cultivation and most seed crops have now lost their ability to delay germination.

Other important domestication-related traits that are automatically selected for by cultivation include an erect habit, synchronous flowering, thin seed coats, loss of camouflage colouration, and more numerous and larger seeds. This form of unconscious selection probably made possible the acquisition of most of the domestication-related traits in the early crops. The time taken for these evolutionary developments would have depended on the genetics of the individual crop. For example, if the variation for these traits were largely regulated by a small number of genes, and if these genetic loci were tightly linked, then new domestication-syndrome varieties could appear within a few dozen generations. It is likely that most of the early crops were genetically pre-adapted to cultivation, which means that the first farmers were at least spared some of the initial stages of trait selection. However, there are many other crop traits, the manipulation of which requires the deliberate intervention of humans. Such characteristics include many non-visible traits, such as those regulating seed quality or disease resistance.

Our most favoured plants were, therefore, the species that responded genetically to their prolonged association with human beings. In particular, a few plants were able to evolve some very specific traits that strongly encouraged their continued and more extensive use by their human guardians. These plants formed larger, more prominent seeds, which stayed on the main body of the adult plant and were therefore easier for people to see and collect. They germinated as soon as they were planted, ensuring a full crop each year. A few wild cereals hybridised with each other to produce many different types of starchy grain. This diversity of grains enabled people to select and grow a range of food types from a single family of crops. For example, people have learned to grow several different forms of wheat that are variously suitable for making bread, or cakes, or biscuits, or the many different types of pasta. Genetic evidence suggests that many changes involved in domestication occurred because of some very unusual arrangements of genes in the relatively few plant species that were successfully cultivated as staple crops. It seems that it was their peculiar genetic endowments that determined which plants would go on to become crops, rather than any conscious decision by human cultivators.[8]

In those ancient societies that persisted with agriculturally based plant exploitation, people soon started to manipulate their crops deliberately by a process of empirical breeding. As with the ancient craft of empirical biotechnology, which has given us such products as wine, beer, spirits and leavened bread, the empirical forms of plant breeding needed no knowledge of science to achieve far-reaching biological manipulations of crops. New varieties of wheat, rice and maize were developed, and farmers bred particular landraces of crops that were adapted for their own specific regions, soils and climates. This process resulted in a slow but steady increase in crop yields, and their adaptation to a host of new environments, as cultivation spread far and wide. But there were also many setbacks for farmers, as new diseases, warfare, and local climatic vagaries took their toll on food production. Hence, it is likely that the varieties of major crop staples being cultivated in late-medieval Europe were in many cases only marginally superior in yield to those grown by Neolithic farmers many millennia previously. However, all this was set to change after the sixteenth century, as a combination of scientific enquiry and entrepreneurial activity led to the transformation of agriculture in Northwestern Europe. This second phase of plant breeding occurred during the post-Enlightenment explosion in evidence-based knowledge of the past three hundred years, and we may therefore refer to this as 'scientific plant breeding'.

Variation and selection in breeding

Conceptually, at least, breeding is a fairly straightforward process. The two keys to the successful breeding of anything from a peony to a pony are *variation*[9] and *selection*.[10] In a nutshell, all that any breeder really needs is some degree of genetic variation between the individuals in a given population, plus a means of identifying and selecting the most suitable variants. These more useful variants are then mated or crossed with each other to produce a population that is now composed almost entirely of the newly selected genetic variety. This is how wolves were turned into

Chihuahuas, and wild Asian grasses became today's breadwheat and malting barley. All of these powerful genetic transformations were effected simply by the selection of variants in a given breeding population. Put like this, the process of deliberate selection of variants sounds like a pretty straightforward recipe for successful breeding. Unfortunately, as many a breeder would respond: 'would that things were so simple!' In the real world of plant breeding and cultivation, that key raw material of the breeder, namely genetic variation, is simultaneously a boon and a curse.

Despite the need for a degree of variation for crop breeding programmes, in practice most modern breeders tend to select field crops to be as uniform, and hence as invariant, as possible in most respects. Therefore, modern varieties are often highly inbred and hence genetically invariant, but even ancient farmers would have wished for a degree of uniformity in their crops. For a grain crop, the farmer normally wants each of his plants to flower and set seed at about the same time, so that the entire crop can be harvested simultaneously. Imagine the problems that farmers would face if they could only harvest their crop plants one by one, because each plant matured at a different time over a period of several weeks or even months. Clearly, therefore, uniformity of seed maturation time is very important for practical harvesting. We also want our crops to yield products of equal quality. In a field of wheat, we do not want the plants to produce a range of different types of grains with different starch compositions. If such a mixed grain crop were harvested and milled, the farmer would end up with a messy combination of many different starches that might well be useless for further processing into food. Once again, the farmer just wants the one variety of grain that he can easily process to make either bread, or cakes, or the various types of pasta, and so on. In this case, uniformity of grain composition is vital for food production. Naturally, the traditional farmer would also, if possible, select for a degree of variability in such characters as disease or pest resistance, as a hedge against future outbreaks. Hence traditional 'landraces' tend to be non-uniform in many respects, and are therefore fast becoming a rarity in modern breeding programmes, often to the point of extinction (see Chapter 8).

In the context of crop and livestock domestication, humans have gradually converted wild grasses into rice and wolves into poodles by means of a largely unconscious process of selection towards the production of populations that are relatively genetically uniform in comparison with their wild progenitors. The ever-present trend towards greater genetic uniformity in our major crops has become even more pronounced over the past century. Almost all commercial agriculture now consists of vast tracts of inbred monocultures in which all of the plants are virtually identical to one another in terms of their genetic endowment. The consequence is that we have, for very good practical reasons, more or less sacrificed genetic variation in most of our field crops. And yet, variation is still the sine qua non for any successful breeding programme. We still require a degree of variation in order to breed improved cultivars in the future, whether they are designed to give higher yields, have an increased resistance to newly evolving pests and diseases, or are able to respond to future environmental changes. In order to breed plants, therefore, it is still necessary to have a population that exhibits a sufficient degree of genetic variation between individuals to give the breeder some raw material to select. Despite all of our previous achievements, therefore, the simple fact remains that variation is the key to all crop improvement, as shown in Figure 1.

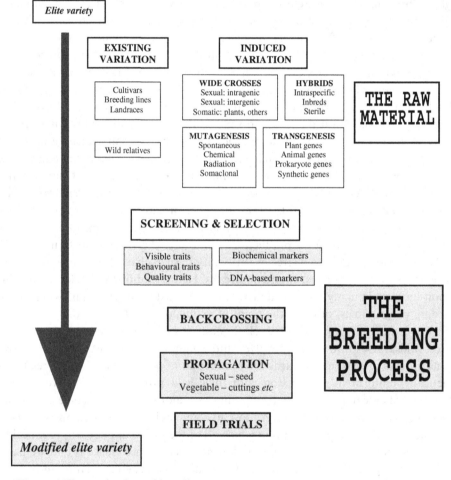

Figure 1 The mechanism of breeding.

This dichotomy, of both needing and abhorring variation at the same time, is at the core of modern crop breeding. We have been very successful in producing genetically uniform crops on a massive scale. Meanwhile, for almost three centuries, we have been searching for ever more radical, and ever more scientifically informed, ways in which to add back the 'lost' variation to our modern crops. The story of scientific plant breeding began with experiments such as 'Fairchild's mule', which was created by the eponymous entrepreneurial botanist in 1718. This was the first manmade hybrid of a carnation with the ornamental, Sweet William, to produce a new 'artificial' plant species. As with many breeding innovations, 'Fairchild's mule' was created solely for commercial purposes. The latest trick, genetic engineering or transgenesis, is merely another in a long series of such scientific innovations in crop breeding, such as cell fusion (1909), quantitative genetics (1918), chemical muta-genesis (1927), X-ray mutagenesis (1928), in vitro tissue culture and somaclonal mutagenesis (early 1940s), mass clonal propagation (1950s), and even the use of γ-rays from cobalt-60 sources to induce DNA damage (late 1940s).

Figure 1

Plant breeding can be summarised as the selection and propagation of distinctive variants from a population. Breeders rely on access to a range of genetically diverse plants to provide them with sufficient variation in agronomic traits to select suitable candidates for future programmes of improvement. Prior to scientific breeding, farmers only had access to the existing genetic variation in each crop species in a particular area. New variation was occasionally introduced into crops via fortuitous spontaneous mutation or hybridisation events, providing such variants were recognised and propagated by farmers. The deliberate creation of new variation began with sexual crosses between members of the same species, and gradually progressed to crosses between ever more distantly related plants from dif-ferent genera. Many wide crosses beyond the species barrier can only survive if plants are subject to techniques such as chromosome doubling or embryo rescue and tissue culture. Additional variation can be introduced by induced mutagenesis using radiation, chemical and somaclonal methods. The collection or generation of a population of variants gives rise to the *raw material* of plant breeding.

The raw material, sometimes consisting of tens of thousands of plants, is then screened, and favourable variants are selected by many diverse approaches. These can range from straightforward visual selection to a battery of increasingly sophisticated and often auto-mated genomic and metabolic techniques. This is the first stage in the *breeding process* itself. Following selection, suitable variants normally need to be backcrossed several times into existing elite breeding lines so that their favourable traits are retained, while unfavourable traits are eliminated. This produces an improved elite variety, which must be multiplied up and then propagated for external field trials. Several years of field trials are frequently necessary to ensure that the improved variety meets the DUS (distinctness, uniformity and stability) criteria required for registration as a new cultivar.

Note that transgenesis (genetic engineering) only involves the initial stages of breeding, i. e. it is simply another method of creating additional genetic variation. Even if it is successful in this regard, a transgenic variant will still need to undergo the full gamut of screening, selection, backcrossing, propagation and field trials as part of the normal breeding process that applies to all plants.

Pre-scientific empirical breeding

Empirical breeding involves the use of a kind of informed trial-and-error process to effect crop improvement. As we will see below, it was similar empirical techniques that were responsible for the various forms of traditional biotechnology that also flourished over the same period. The practitioners of such empirical methods had no under-standing of the scientific principles that underlay either fermentation or crop genetics, but this did not stop them employing their crafts to great effect. Indeed, one is tempted to comment that modern versions of beer, cheese, and certain other products of tra-ditional biotechnologies, are often decidedly inferior to many of the older versions that were produced by the non-scientific craft of the original ancient empiricists. Empirical plant breeding would have been possible as soon as the early farmers realised that they could manipulate the appearance and behaviour of their newly domesticated crops. Observant plant collectors and husbandmen would have already realised that the phenomenon of 'like begetting like' applied to the plant world just as much as it did to animals. For example, instead of saving a random sample of his harvested grain for replanting the next year, a farmer might decide to select only the larger, better-looking seeds. He would correctly expect progeny of the selected larger seeds to tend to pro-duce larger seeds. Alternatively, a farmer might decide to save seeds for replanting only from his best performing fields. In either case, we have the beginnings of conscious selection and breeding for particular characters or phenotypes.

An especially important breakthrough would have been the recognition that it was not only the external appearance of the crop that could be controlled by selection, but also traits like bread-making ability or taste that were invisible to the naked eye. There were other, even more subtle, traits that would have taken longer to recognise but would have been of great importance for the early farmers. An example would be the ability of a chosen crop to withstand attack by pests and diseases, especially if it were grown in newly cleared areas where new pathogens might lurk. This sort of trait could have been recognised by the farmer when a few, perhaps spontaneously mutated, plants survived an otherwise devastating disease. If an observant farmer propagated seeds only from these resistant plants, his crop would prosper more than those of his neighbours when that particular disease returned to the area. Gradually such improved varieties would have spread throughout a region, giving rise to a so-called 'landrace'. Much of empirical agriculture was, and in some cases still is, based on the selection of local landraces of crops, as we will discuss shortly.

Scientific breeding

Scientific breeding differs from empirical breeding in that it is based upon at least a partial understanding of the traits that regulate the agronomic performance of crops, coupled with some knowledge of how to manipulate such traits. There were several stages in the evolution of scientific breeding, over the past three centuries, as knowledge of plant biology and access to new forms of technology both improved. The early stages of this process, in the seventeenth and eighteenth centuries, involved a better understanding of how plants worked, and especially how they reproduced. During the latter part of the eighteenth and nineteenth centuries, there was an effort to utilise as much existing variation in crops as possible, e.g. by collecting a wide range of different landraces or wild forms of the crop, and then crossing them with each other to produce new, and possibly improved, cultivars. By the early twentieth century, it became possible to create mutants deliberately by various new technologies. Whereas hybrid breeding can be compared to reshuffling an existing deck of cards, mutagenesis involves adding completely new cards to the deck.

Beginnings of practical scientific breeding

Hybrids

It is often assumed that the application of scientific knowledge to plant breeding did not start until the rediscovery of Mendel's results on pea genetics in the early twentieth century. However, by this time, crop breeders had already been using scientific principles to inform their craft in a very practical sense for more than two centuries. The earliest steps towards an understanding of the microscopic structure of plants date back to the beginning of the application of modern scientific research methods by for example Hooke and Malpighi in the seventeenth century. At about the same time, in 1645, a manual by Richard Weston included one of the earliest recommendations in English on crop rotations, including the use of legumes or brassicas as break crops, instead of having fallow years during which the land lay unused and therefore unproductive.[11] Weston's book, which recommended either clover or turnips to replace fallow years in cereal rotations, was widely used in the Low Countries, which at this time were at the forefront of the new field of applied plant science. By the 1650s, Robert Child and his colleagues were performing some of the earliest experiments on the effect of nitrates on cereals, as well as working on the domestication of tree crops.[12]

In England, the Royal Society, founded in London in 1660, was especially concerned with agricultural improvement. Society luminaries such as Boyle and Newton studied and wrote about the mechanism of vegetative growth, and were acquainted with noted agrarian reformers including Weston and Child. The final quarter of the seventeenth century witnessed a resurgence of interest in agricultural improvement (i.e. enhanced profitability) in England, among both philosophers (natural and classical) and practical husbandmen.[13] Meanwhile, in society at large, there was an emerging appreciation of the potential of the new scientific knowledge to further the betterment of agriculture. Gradually, the long-standing British tradition of private initiative in the practice of husbandry began to incorporate government or quasi-government institutions, such as the Royal Society. The motivation was not just to improve existing staple crops, such as wheat and barley, but also to exploit the immense wealth of new crops now available from the newly seized colonies abroad.

By the end of the seventeenth century, Dutch breeders were undertaking the earliest systematic programme for the breeding of ornamental flowers, and the first hybrid hyacinth had been produced. At the same time, in Germany, Rudolph Camerarius demonstrated for the first time the existence of sexual reproduction in plants. Camerarius went on to suggest that cross breeding of different varieties, or even different species, could be used to create new and potentially more useful types of plant.[14] Another key achievement of this period was the creation, in 1718, of the first interspecific hybrid. This new plant was a cross between two species of the genus *Dianthus*, namely Sweet William and a carnation, which was bred by the English botanist, Thomas Fairchild. With these successes, the increasingly intrusive intervention of humans into plant reproduction was already well under way by the early eighteenth century. Interest in the deliberate production of hybrids and their potential uses in horticulture and agriculture continued throughout the eighteenth century. One of the most notable achievements was the discovery in 1761 of so-called 'hybrid vigour' by the German botanist, Josef Kölreuter, who also produced the first hybrid crop variety.[15]

In 1825, John Lorain in the USA described the possibility of growing maize as a hybrid crop.[16] Not only would such a crop yield more, it also would not be economically worthwhile for farmers to save any seed for replanting (see below). Therefore they would need to buy new seed each year from a specialised breeder who could recreate the vigorous hybrid for them. Lorain rationalised that the farmers would readily offset the extra cost of buying seed from a merchant if they could make greater profits from high-yielding hybrid crops. For the first time, it became apparent that there was the possibility of making a lot of money from the breeding of certain arable crops. This concept took another century to realise as a viable commercial proposition, but it eventually resulted in what is today's most profitable global crop, namely F_1 and F_2 inbred-hybrid maize, as we will see in the next chapter. Interestingly, this centuries-old strategy of preventing seed-saving by farmers and gardeners has recently been stigmatised by anti-GM campaigners as 'terminator technology', as if it were (a) something new, and (b) unique to transgenic crops.[17]

Back in the nineteenth century, the potential use of hybrids for crop improvement soon piqued the interest of several of the new scientific academies that had been established in Europe, following the earlier example of the Royal Society in England. In 1822, the Berlin Academy offered a cash prize for the elucidation of the mechanism of plant sexual reproduction. The winner was A. F. Wiegmann, an apothecary from Brunswick.[18] A few years later, in 1830, the ever pragmatic Dutch offered a similar prize, but stipulated that the research should explain how hybridisation could be exploited practically in plant breeding. This prize, from the Netherlands Academy of Sciences, was won in 1837 by the German botanist F. C. Gärtner. In his paper, he pointed to the occurrence of new trait combinations that could be used to create new varieties of cultivated plants. Gärtner also drew attention to the appearance of constant hybrid plants that were capable of propagating themselves as new species.[19] Finally, in 1860, the French Academy offered a further prize for an explanation of hybrid fertility, which was won by botanist C. Naudin for his 1863 paper on hybridisation in plants.[20] All of these competitions were international in scope and they illustrate the importance to the new scientific establishment of plant science research, both as a general public good and as a practical, and potentially profitable, tool for crop improvement.

Mutations

While hybrid strategies can give breeders access to considerable new genetic variation, mutations can be the source of much more radical changes of the sort that turned bushy, low-yielding teosinte plants into the erect, high-yielding crop that we now call maize. A mutation is a sudden change in chromosomal DNA that often leads to a change in gene function and a consequent alteration in the appearance, or phenotype, of an organism. There are many agents in the physical environment that cause mutations, and all living organisms on Earth are constantly exposed to such mutagenic agents. Some of the best-known mutagenic agents originate from solar radiation, especially ultraviolet, X-rays and γ-rays, which can cause DNA damage and cellular mutations. The other important category of mutagenic agent is the vast group of DNA-reactive chemicals that result from both biotic and abiotic processes. Some of these chemicals, such as benzene and mustard gas,[21] are manmade but many

others are by-products of processes like volcanism, or may result from normal cellular activity, for example during stress or ageing.[22] Organisms can undergo two
forms of mutation, i.e. somatic or germline. Somatic mutations occur in the normal
cells of the body (Greek: *soma*, meaning body) and may result in the death of a cell or
a malfunction in its development, e.g. leading to cancer. However, because all
somatic cells die with the body, such somatic mutations have no impact on the next
generation, which will be genetically 'normal'. Of far greater evolutionary consequence are the germline mutations, which affect the reproductive or germ cells, and
are therefore inherited by subsequent generations. In the normal course of events,
the appearance of a spontaneous germline mutation is relatively rare.

The rate of spontaneous mutation varies according to the species and to its
environmental exposure to mutagens. In complex multicellular eukaryotes, such as
plants or mammals, the rate is about one mutation per million cell divisions.
Exposure to environmental mutagens, including ultraviolet radiation from strong
sunlight, and certain chemicals, will increase the incidence of mutation to some
extent, depending on the nature of the mutagen and the duration of exposure.[23]
However, most crops have relatively massive genomes and numerous duplicated
genes, which gives them an impressive degree of genetic redundancy. This also means
that the likelihood of a useful spontaneous mutation suddenly appearing in any
given crop plant is extremely low. The existence of spontaneous mutants or 'sports'
had been known by farmers, and other observers of Nature, for many centuries, but
their significance in generating biological variation was not realised. This was largely
because most people across the world had long believed that all the original variations in living organisms had been produced by an external deity and did not require
further explanation. As a result, mutations were regarded as rare and aberrant
monsters that were generally considered abominations of the 'naturally created'
order of life.

Towards the end of the nineteenth century, several plant scientists were beginning
to realise the key role played by mutations in the creation of new variation, especially
in crops. One of the principal figures was the Dutch botanist Hugo de Vries, who was
studying the appearance of mutants, or sports, among evening primrose (*Oenothera
biennis*) plants. His observations led him to propose that new characteristics may
appear suddenly in a plant or animal and that these characteristics are sometimes
inherited by their offspring.[24] In other words, in contrast with classical Darwinian
principles, genes in germline cells could sometimes be altered during the lifetime of
an organism and any offspring from such a 'mutated' organism would also carry the
altered genes. At the same time as de Vries and others were demonstrating the
important role of mutations in generating new types of variation, other scientists
were beginning to realise that mutations could be deliberately manufactured by

human intervention. This was a radical step in the application of scientific knowledge and the use of newly available technology, such as X-ray sources, which had just been discovered at the turn of the twentieth century.

The ability to cause mutations deliberately in plants and animals enabled breeders to advance beyond a reliance on spontaneous genetic variation, and to begin the creation of a much wider pool of manmade variation. In the next chapter, we will look at some of the most effective and widely used techniques of induced mutagenesis and wide crossing that have been developed during the twentieth century. It was the use of these and other more intrusive approaches to plant breeding that allowed our food production to more than keep pace with the recent rapid expansion of the human population. By the early twentieth century, the combination of knowledge about the roles of both hybrids and mutations in enhancing variation, together with the rediscovery of Mendel's work on inheritance, set the scene for a dramatic leap forward in scientific plant breeding.

Quantitative genetics

Mendel's principles of heredity and the later rediscovery of Mendelian genetics were largely the work of plant breeders, whether theoretical (i.e. not immediately concerned with agriculture, like Mendel himself) or practical (crop breeders such as Biffen). Mendel's work dealt with individual traits that segregated simply and were controlled by a single or a very few genes. As we will see in Chapter 5, a few early scientific breeders, such as Biffen at Cambridge, used a Mendelian approach with great effect in selecting disease resistant varieties of wheat, including the very successful 'Little Joss', which was released in 1910. However, it was soon realised that the vast majority of agronomic characters in crops was not regulated by simple Mendelian inheritance. Important characters such as yield and stress tolerance often behave in a complex manner that is regulated by an often unequal combination of numerous genes, plus a variable environmental component.[25] The elucidation of the behaviour of such complex characters and the ways in which this knowledge might be harnessed for crop breeding became known as quantitative genetics. It is difficult to overestimate the importance of quantitative genetics for practical breeding. Indeed, it has been described in a recent review as "the intellectual cornerstone of plant breeding for close to 100 years."[26]

The foundations of quantitative genetics were laid by British geneticist Ronald Fisher in a seminal paper of 1918, in which he showed that continuous variation between members of a population could be as a result of Mendelian inheritance, albeit involving many genes, plus an environmental component.[27] Previous work by Mendelian geneticists such as William Bateson at Cambridge (see Chapter 5) suggested that Mendelian mechanisms only gave rise to large and discrete, or quantum, changes in phenotype. Fisher and others established a statistical framework that has underpinned the application of quantitative genetics in crop breeding for the past century.[28] In particular, quantitative genetics allowed breeders to develop reasonably robust mechanisms to predict phenotypic performance from their knowledge of a given genotype. It also enabled them to design manageable and affordable field trials that had a good likelihood of detecting useful characters without being too large or complex.[29]

There has been a tendency among molecular biologists to assume that modern genomics, and especially knowledge of the DNA sequences in a plant genome, will reduce the importance of quantitative genetics in the future. However, it seems more likely that existing genetic models built up from classical quantitative genetics could be adapted to provide the framework for handling the mass of both genetic and phenotypic information that is now accruing.[30] In the early twenty-first century, quantitative genetics remains the cornerstone of practical breeding and is only gradually being supplemented, but not replaced, by knowledge from genomics. As we will see in the next chapter, one of the important achievements of scientific breeding is the ability to create novel genetic variation using several different approaches. But it is important to remember that the various tools of quantitative genetics are still the major lens through which most breeders view their plant material and its agricultural potential.

2

Creating new genetic variation

Weep not that the world changes – did it keep
A stable, changeless state, it were cause indeed to weep
William Cullen Bryant (1794–1878)
Mutation

Introduction

For the past three hundred years, crop breeders have had access to a considerable amount of existing variation resulting from spontaneous mutation, and also from the various genetic manipulations, such as hybridisation, that they learned to perform as they improved their understanding of plant reproduction. By collecting varieties and landraces from around the world, breeders were also able to exploit a great deal of the variation present in the gene pool of the crop species itself. Breeders had learned how to force crops to hybridise with some of their wild relatives, and even with other more distantly related species. Not even the genus barrier could withstand their assault as the first experimental manmade intergenic hybrids were produced in the mid nineteenth century. At this stage, many of these achievements were only successful in glasshouses or field plots and had not yet resulted in the production of any new varieties of the major staple crops. To a great extent, farmers at the beginning of the twentieth century were still reliant on traditional crop landraces. Organised systems for the dissemination of improved seed stocks from the new scientific breeding programmes were just beginning to be established in a few countries, but were still far from effective. However, all of this was to change as the pace of scientific discovery and its application to agriculture accelerated steadily during the next few decades.

Two scientific developments revolutionised the ability of breeders to effect ever more radical genetic manipulations and to create more novel and useful variations in crop genomes. Firstly, the techniques of crop hybridisation were perfected in crops like maize, and used to create higher yielding varieties that opened up a whole new area of commercial opportunity for seed companies. In addition to the successes with

23

hybrid maize, hybridisation was used to create a totally new species of intergenic hybrid crop called triticale. Secondly, it became possible for the first time to produce mutations deliberately and thereby to increase the rate of mutagenesis in a population by many thousand-fold. This is the technique of induced mutagenesis, as introduced in the last chapter. In this chapter, we will look at the ways in which modern hybrid crops and induced mutant varieties have greatly improved yield and performance around the world.

Hybrid crops

Maize and other intraspecific hybrids

The commercial possibilities of F_1 hybrid crops had been recognised early in the nineteenth century, but the reduction of this concept to practice was beyond the capacity of the breeders of the time. An F_1 hybrid is created by using pollen from one parent to fertilise a genetically dissimilar parent. Although such hybrids are often more vigorous than either parent, they cannot normally be propagated as their progeny tend to be highly variable. Hence the F_1 hybrid must be recreated by a breeder every season. This provided a good commercial opportunity for breeders because purchasers of seed (whether crops or ornamental species) were obliged to return to the breeder each year for more seed. The development of F_1 hybrid crops was made possible by the new cadre of well-trained professional scientists that was emerging from the impressive system of public Land Grant Universities and agricultural experimental stations that had been set up in the USA during the nineteenth century, as we will discuss further in Chapter 4. Virtually all of the initial research and development that led to double-cross hybrid maize was done in such institutions. The first demonstrations of the true potential of the use of F_1 hybrids in commercial agriculture were made by W. J. Beale between 1878 and 1881.[31] Beale showed convincingly that inter-varietal hybrids of maize had greatly increased yields compared to conventional varieties. He also suggested their use on the farm, but was unable to generate large quantities of reproducible hybrid seed.

Early in the twentieth century, George Shull and Edward East independently began work on producing inbred maize lines that could be combined repeatedly to produce reproducible F_1 hybrids. Shull worked at the Station for Experimental Evolution at Cold Spring Harbour, New York, while East was at the Connecticut Agricultural Experiment Station in New Haven, Connecticut.[32] The first commercial 'crossed corn' was produced in 1917 and improved hybrids were subsequently developed at USDA research centres and at the various Corn Belt experiment stations.[33] In order to produce seed on a commercial scale, it is necessary to cross two different F_1 hybrids together, which results in an even more vigorous F_2, or double-cross, hybrid variety. Some of the experimental maize hybrids produced by Shull, East and colleagues exceeded the yields of conventional varieties by the impressive margin of 30%. However, despite this undoubted yield advantage, the new hybrids were far from being an instant success with US farmers. Most of these hardheaded and often impecunious folk remained suspicious of a crop that was impossible to use for re-sowing and hence required them to buy expensive new seed each year. Despite the obvious yield gains, farmers were sceptical that any increased returns would really make up for the extra expense incurred in buying the new seed. Then there was the loss of independence implied by being forced to purchase seed from a merchant, even in a bad year when ready cash might well be in short supply.

Despite repeated demonstrations of the virtues of hybrid maize and a series of promotion campaigns, it took several more decades to win over the wary Midwestern farmers. In 1919, magazine editor and soon-to-be breeder, Henry Wallace suggested that a new type of contest should be held at Midwestern agricultural shows, namely a yield test of hybrid versus conventional maize under controlled conditions.[34] He rationalised that this would show the sceptical farmers, before their very eyes, how much better the hybrid maize could perform. Iowa State University soon picked up on Wallace's idea and in 1920 started the Iowa yield tests.[35] Over the next few years, the results of these increasingly well-attended yield tests showed thousands of farmers just how good hybrid maize could be, compared with the conventional open-pollinating varieties that they were planting at the time. In 1924, one of Wallace's own hybrid varieties, called Copper Cross, won a coveted gold medal and he benefited greatly from the attendant publicity. Unfortunately for the boosters of the hybrid varieties, however, all their best efforts were then set back by the Great Depression of the late 1920s and early 1930s. This led to a drop in grain prices, leaving many farmers with little or no spare income for any fancy new seed, no matter how well it yielded.

Meanwhile, the commercial opportunities presented by the hybrid maize were gradually being explored by a clutch of newly formed private seed companies. An important prerequisite for the success of such ventures was the development of an effective and reliable supply and distribution system. Such a system would be necessary for the new crop to make any significant headway across the vast farming areas of the USA. One of the earliest seed companies to market the new maize hybrids was founded in 1926 by the same Henry Wallace who had originally suggested the idea of yield contests to promote the hybrids at agricultural shows. Wallace's firm, the Pioneer Hi-Bred Corn Company, soon emerged as the largest private seed breeder and merchant in the Midwest.[36] Pioneer and other seed companies provided a vital link for the commercial-scale multiplication and dissemination to farmers of hybrid seed that had been developed originally by the public sector institutions. These seed companies continued on to work closely with the public sector breeders over the succeeding decades to produce many dozens of new and improved hybrid varieties, a joint venture that continues to this day. In Chapter 6, we will discuss further the versatile Henry Wallace and his many roles in US and world agriculture.

Although hybrid maize was being vigorously promoted by both public and private sector boosters such as Wallace throughout the 1920s, the new varieties still only made up a paltry 1% of the US national maize crop by 1933. In addition to the financial worries caused to farmers by the Great Depression, there was now an additional problem for salesmen of the expensive new hybrid maize. By the early 1930s, the cheaper conventional open-pollinated maize had itself been much improved by breeders, and the hybrid maize was only outperforming it by 15%, rather than the 30% difference that had existed a decade or so earlier. Ironically, one of the principal factors that eventually tipped the balance in favour of the hybrids was not their yield advantage, but rather the greater susceptibility of conventional varieties to drought during the Dust Bowl years of the mid-1930s. This effectively removed the competition from open-pollinated varieties for a few crucial years, and the continued good performance of the hybrid varieties set the scene for the eventual mass adoption of hybrid maize across the USA. By the 1940s, hybrids dominated maize production, reaching 95% of the Midwestern Corn Belt crop by 1953 and almost 100% by 1965.[37]

Another, more controversial, view of this phenomenon is that the eventual success of hybrid maize occurred largely because most commercial and public sector breeders switched to using hybrids in their crop improvement programmes, at the expense of conventional open-pollinated varieties.[38] This meant that the latter were neglected while the hybrids benefited from the greatly increased attentions of the breeders. Perhaps, so the argument goes, hybrids would not have outperformed open-pollinated varieties so dramatically if both types of maize had received the same attention from breeders. In other words, the success of this new hybrid crop technology may not have been entirely due to its intrinsic superiority over other methods, but rather due to a switch to the new technology at the expense of the alternatives. This debate has clear parallels with many of the discussions about the use of transgenic crop technology today, as discussed further in Parts III to VI.

Whatever the reasons for their eventual adoption, it cannot be disputed that maize cultivation increasingly thrived as hybrid varieties spread across the USA. Following the almost universal adoption of hybrid varieties, US maize yields increased by a staggering 430%, from 1.8 tonnes/hectare in the 1920s to an impressive 7.8 tonnes/hectare in the 1990s. With yield gains of this magnitude, even the most sceptical farmer quickly lost his reluctance to purchase the relatively expensive hybrid seed from private suppliers such as Pioneer. Of course, this also meant that US farmers have now quietly abandoned their traditional independence in saving their farm-grown seed. The reward for this loss of their precious autonomy is that farmers of hybrid maize grow a bountiful crop that now acts as the food-bank of the world, a hedge against future crop failure, and a potent political tool for the use of the federal government. If this is, as has been claimed, a Faustian compact between scientists, companies and farmers, the devil has yet to claim any souls. Indeed, hybrids seem to be steadily spreading their reach in terms of both new crops and new countries across the world.

New types of hybrid crops are now increasingly grown in a wide range of com-mercial and subsistence agricultural systems. Originally bred for use in industrialised countries, hybrid crops are now being adapted for cultivation in developing coun-tries, where the potential for yield gains is much higher. For example, hybrid rice is grown extensively in China and increasingly in India.[39] The increasing impact of hybrid rice in Asian agriculture was recognised when the 2004 World Food Prize was jointly awarded to Yuan Longping, who discovered the genetic basis of heterosis in rice.[40] Yuan's first commercial hybrid variety, Nan-you No. 2, was originally released in 1974. With its 20% yield improvement, the new rice immediately started to transform Chinese agriculture towards self sufficiency in food production. Commercial hybrid rice varieties, derived from Yuan's early breeding lines, are now being cultivated in ten countries around the world, including the USA. Another widely grown hybrid crop is sorghum, *Sorghum bicolor*, a grain crop that is especially important in arid regions, which was developed in the USA in the 1950s and is now grown worldwide.[41] Hybrid sorghum varieties outperform conventional local vari-eties by at least 40% under favourable conditions, and give an even better yield advantage under drought stress. Globally, hybrids now account for 65% of the maize crop, 60% of sunflower, 48% of sorghum and 12% of rice.[42]

Hybrid wheat has been much more difficult to produce because of the architecture of the floral structures. Wheat flowers are very small and difficult to see, and often self-pollinate before opening. This has prevented controlled hybridisation on a large scale and has made wheat hybrids prohibitively expensive. In the 1980s and 1990s, the use of chemical hybridising agents, which cause male sterility without affecting female fertility, finally enabled hybrid wheat varieties to be produced on a com-mercial scale, although even today hybrids still only account for a tiny fraction of global wheat production.[43] Some other crops in which hybrid cultivars are increasingly being used include sugar beet, oilseed rape, several brassica vegetables, tomato, potato and onion. As with the commercial maize hybrids developed in the USA in the 1920s, all of these newer high-yielding hybrid crops make it impossible for farmers to save seed for replanting.[44] And likewise, as with maize, such Faustian compacts appear to have been entered into by farmers around the world, millions of whom have foregone their ability to save and replant seed in return for the obvious yield benefits of the new hybrid varieties.

Intergenic hybrids – triticale, a new manmade crop species

The vast majority of hybrid crops come from intraspecific or interspecific crosses. But the production of intergenic hybrids promises even greater opportunities to create useful variation, although it is technically a rather complex and time consuming process. The first commercially successful manmade crop species to be produced from such an intergenic hybrid is triticale. This completely new cereal species is the result of hybridising a member of the wheat genus (*Triticum* spp.) with rye, *Secale cereale*. Triticale combines several of the useful agronomic qualities of wheat and rye to give a hardier cereal crop that can also be used for making bread of a far higher quality than is possible with rye alone. Triticale can be grown under a wider range of climatic and soil conditions than wheat, including many of the marginal habitats that are often worked by some of the poorest farmers in developing countries. Triticale is also nutritionally superior to wheat in its protein content, in both overall quantity and essential amino acid profile. Depending on whether tetraploid (emmer) wheat or hexaploid (bread) wheat is used as the *Triticum* parent, the resultant triticale species can be either hexaploid or octoploid. With such divergent parents and complex genetic organisations, it is not surprising that early attempts to create this artificial species in the late nineteenth century only led to sterile plants that often had serious agronomic defects.[45]

Eventually, after many years of increasingly sophisticated manipulation by breeding crosses, a fertile version of triticale was produced.[46] However, it still took several more decades of experimentation with different wheat/rye combinations before a sufficiently good range of triticales was available for the new crop to be released on a large scale. Most of the successful triticale varieties grown today derive from a cross between tetraploid wheat and rye, and are therefore hexaploid. One of the key advances that enabled fertile triticale to be produced was the new chromosome doubling technology (see Chapter 3). Studies in Germany have shown that triticale can outperform winter wheat in yield, as well as using soil-borne nutrients and nitrate fertilisers more efficiently.[47] Hence triticale is both cheaper to grow and a more sustainable crop. By the mid-1980s, triticale was being grown on more than one million hectares throughout Europe and the Americas, and by the 1990s it had expanded to over 2.5 million hectares. Triticale is a new species resulting from hybridisation of two plant species from different genera, and is only viable thanks to the chemically induced modification of its genome. It is truly one of the ultimate products of genetic manipulation by modern plant breeding. However, since it is not transgenic, triticale is *not* classified as a GM crop. Indeed, triticale products can be found in many health food stores today and are especially popular with vegetarians and other dietarily conscious folk.[48]

Induced mutagenesis

As we saw in the previous chapter, the rate of spontaneous mutation due to environmental agents or electromagnetic radiation is very low. In practice, this means that the crop breeder must screen large numbers of plants for many years, and often decades or longer, before a useful mutation turns up. Even should such a happenstance arise, the new mutation could easily be missed in such a huge population, especially given the limited screening and selection technologies that were available in the early twentieth century. The first breakthrough came in the 1920s when it was realised that newly discovered sources of high-energy radiation and synthetic chemicals could create mutations much more efficiently than ambient environmental agents. In the case of irradiation, induced mutagenesis delivers the same sort of mutagenic agent as occurs in the open environment, but in a much more concentrated form. High-energy electromagnetic radiation is present in normal sunshine, but only in tiny amounts compared to that produced by radioactive isotopes, such as cobalt-60 or caesium-137. Induced mutation is now a core technology in plant breeding that has been responsible for the creation of thousands of new varieties of all the major crops.[49] There are three common methods of induced mutation currently used in plant breeding, namely radiation, chemical and somaclonal mutagenesis.

Radiation and chemical mutagenesis

In 1926, Hermann Muller discovered that X-rays could cause mutations in *Drosophila melanogaster* (fruit flies), a discovery for which he was awarded the 1946 Nobel Prize for Medicine. In 1927–1928, the earliest results of X-ray mutagenesis on tobacco, *Datura* (a solanaceous genus), and maize were published.[50] Since the late 1920s, X-ray and other forms of mutagenesis have been of immense value both for fundamental research in plant biology and for the production of improved crops. By the 1940s, researchers in Germany were using X-ray mutagenesis as a tool to study what was to become the model plant for genetic research, namely *Arabidopsis thaliana*.[51] Since the early days of radiation mutagenesis, the technology has been continually improved. In the 1940s, the emerging nuclear technologies enabled the use of more effective types of radiation, such as γ-rays and thermal neutrons, as mutagenic agents for crop improvement. The most commonly used chemical mutagens in plant breeding are alkylating agents that react directly with DNA bases to modify their structure. The best known of these alkylating agents are ethyl methane sulphonate and ethyl nitroso urea. Another useful mutagen is sodium azide, which is very effective in rice and barley but much less so in wheat. Therefore the decision about which chemical mutagen to use will vary according to the nature of the crop. Chemical mutagens are relatively cheap and readily available compared to radiation mutagenesis, which requires access to high-energy radiation sources, as well as involving all the safety considerations that arise when such powerful sources of radiation are used.

In the 1950s, the Food and Agriculture Organisation (FAO) of the United Nations began a long-term collaboration with the International Atomic Energy Authority (IAEA) in order to make irradiation technology more widely available, especially to developing countries.[52] New, portable γ-ray irradiators using cobalt-60 or caesium-137 sources have now been developed for use in regions that lack access to the more expensive large-scale mutagenesis facilities. During a mutagenesis programme, small batches of seeds of a crop plant will normally be exposed to a series of different radiation and/or chemical mutagens for various times in order to discover the best agent and conditions for that particular species. Although seeds are the most common part of the plant to be exposed to mutagens, cuttings, inflorescences, tissue cultures, and even whole plants may be exposed instead. Breeders often have to use huge numbers of seeds, sometimes in excess of 100 000 in order to produce a second, or M_2, generation of as many as 30 000–50 000 plants. Such large numbers of plants are necessary for selection of the tiny number of desirable mutations out of the thousands of random events, most of which will be deleterious to the plant. Even if we are just interested in one trait, a mutagenesis programme is a huge affair. For example, if a breeder wishes to generate a mutant population for semi-dwarf plants, it is necessary to grow and screen 10 000–15 000 M_2 plants. Those very few mutant plants that are eventually selected from the M_2 population often carry undesirable mutations as well as the desired trait and must be grown on, and often backcrossed to an elite cultivar, for many more generations before the new mutant variety is ready for commercial production.

Despite the huge numbers of plants required to give a reasonable chance of success, and the long time required to take a mutation-derived variety through to field release, the technology of induced mutagenesis has been very successful since it was first applied on a large scale in the 1950s. Public agencies, including FAO/IAEA and many universities have been effective proponents of mutagenesis technology. Most of the resultant crop varieties have been produced by and for developing countries, although there are also examples of profitable varieties being produced by private companies. For example, many commercial fruits, including apple, pear, grapefruit, apricot, peach and papaya have been improved by private companies using mutagenesis. Horticultural companies have also used this technology to produce hundreds of new types of ornamental plants for homes and gardens. The full list of mutation-bred crops is legion, including over 3000 varieties of all the major staple species grown in at least 59 countries, mostly in Asia.[53] Among the more successful mutagenised crop varieties are the following: Soghat bread wheat in Pakistan (sodium azide), Zhefu rice in Thailand (γ-rays), Shwewartun rice in Myanmar (γ-rays), Bajra pearl millet in India (γ-rays), Gold Nijisseiki pear in Japan (γ-rays), Rio Star grapefruit in the USA (neutrons), and Golden Promise malting barley in the UK (γ-rays), which has been used to make many popular types of beer.[54] Physical and chemical mutagenesis may be somewhat less well known than the more recent highly publicised transgenic methods of crop breeding but they are much more widely used across the world.[55]

Somaclonal mutagenesis

This technique arose from plant tissue culture, which will be discussed more fully in the next chapter. When researchers first started to grow plants, or parts of plants, in tissue culture, they often observed abnormalities in the cultured cells. In a few cases these abnormalities were inherited by the next generation of cells in the tissue culture, or were seen when the plants were regenerated, implying that a drastic genetic change had taken place and that this had affected most of the cells of the tissue. The phenomenon was first described by Alan Durrant at the University of Wales in 1962, and is now known to be due to changes in DNA that are induced during in vitro culture.[56] Such events are generally referred to as somaclonal variations. Somaclonal variation is normally regarded as an undesirable by-product of the stresses imposed on a plant by subjecting it to tissue culture. These stresses include physical factors such as cold, drought, or high salt concentrations, excess or dearth of nutrients, the effects of chemical growth regulators, and infections by pathogens.

The stresses of tissue culture can result in single-gene mutations, deletion or transposition of larger stretches of DNA, including chromosome segments, inappropriate methylation or de-methylation of genes, and even duplication or loss of entire chromosomes. Although some of these genetic changes are rather drastic, researchers soon realised that, provided they were carefully controlled, somaclonal changes in cultured plant cells could potentially provide a powerful new tool to generate variation for the crop breeder.[57] During the past few decades, many new crop varieties have been produced using somaclonal mutagenesis. Crops ranging from tomato and sugar cane to banana and poplar have been improved with respect to disease resistance, insect resistance, nutritional value, drought and salt tolerance and many more useful traits.

Mutagenesis – an acceptable technology for genetic manipulation?

Mutagenesis breeding has largely been developed and applied by public sector scientists. Such a public sector paradigm for mutagenesis is in marked contrast to the more recent recombinant DNA-based technologies of genetic engineering. Not only has mutagenesis per se always been regarded as a public good, its development within public research and breeding institutes meant that it was not exploited prematurely and was primarily focused on improving crops in the long term, rather than on a short-term quest for higher profits. This is not to say that private companies have not used mutagenesis in their own breeding programmes. After all, the technology is freely available to all. Numerous commercial varieties have been developed by mutagenesis breeding. However, unlike plants produced by genetic engineering, plants developed by mutagenesis methods cannot be patented per se by companies. Because mutagenesis (unlike genetic engineering) only manipulates existing genes and has poor patent protection, many companies have understandably tended to steer away from this technology and towards agbiotech.

We will explore the private sector agbiotech paradigm in greater detail in Part IV, but in the meantime it is interesting to reflect on the contrasting public treatment of the mutagenesis (ignorance/approval) versus contemporary agbiotech (disapproval). It would be very easy to mount a scare campaign against radiation or chemical mutagenesis using emotive terms like 'nuclear radiation', 'carcinogenic chemicals', 'unnatural', 'γ-rays', 'cobalt-60', 'DNA damage', 'abnormality' and so on. But nobody has done so. One is justified in asking, 'why not?' Partly, the lack of public concern about mutagenesis is due to ignorance. The technology was developed in the 1920s and has been getting more refined ever since. Prior to the 1990s there was far less public sensitivity to such issues than there is today. However, this is by no means the full story. Mutagenesis was developed by public sector scientists and the vast majority of the resulting plant varieties were, and still are, not produced for profit. Today, the technology is mostly used in developing countries for the benefit of poor farmers, who as a result get higher yielding and disease resistant seed from public breeding centres, e.g. the International Rice Research Institute (IRRI), and the International Maize and Wheat Improvement Center (CIMMYT). It would therefore be a rather brave (or foolish) eco-warrior who campaigned against mutagenised crops. We should bear these factors in mind when trying to explain the unprecedentedly violent reaction to genetically engineered crops, and the lack of such reaction to the products of the many alternative forms of sophisticated genetic manipulation practised by modern plant breeders.

Wide crossing

A wide cross is the process of crossing a crop cultivar with another more distantly related plant that is outside its immediate gene pool and therefore normally sexually incompatible. The normal purpose of wide crossing is not to produce true hybrids, i.e. progeny containing significant parts of both parental genomes, but rather to obtain a plant that is virtually identical to the original crop cultivar, except for a few genes contributed by the distant relative. In some cases, it may even be possible to use the wide crossing strategy to obtain a plant that is identical to an elite cultivar of a crop except for the presence of a single new gene that has been transferred from a different species. The production of a plant containing one or more genes from a different species is also what happens during transgenesis, also known as 'genetic modification', or 'genetic engineering', or simply 'GM'. However, crop varieties produced from wide crosses are not defined as 'GM', no matter how much intrusive modification of their genome has occurred and despite the fact that, to the crop breeder, the outcome from the two processes is, to all intents and purposes, exactly the same.

The strategy of obtaining useful genes from other species via wide crosses was only made possible by advances in plant tissue culture that had been made from the 1930s onwards. A particular challenge was to circumvent the biological mechanisms that normally prevent interspecific and intergenic crosses. For example, while it is often possible to force the pollen of one plant species to fertilise the eggs of a different species, the resulting embryos will almost invariably abort. The same thing normally happens when enforced fertilisation beyond a species or genus level occurs in animals. This spontaneous rejection of the hybrid embryo is part of the mechanism to ensure reproductive isolation of populations and to avoid the production of non-viable or debilitated hybrid progeny. Rejection, or abortion, of unsuitable embryos or foetuses is a major mechanism of reproductive quality control in all of the higher organisms, both plant and animal.[58] A broadly similar phenomenon occurs in plants, where a high proportion of wide-hybrid seeds may not develop to maturity or may not contain a living embryo. In the case of plants, it is often possible to remove the tiny embryos from the female parent before this normal abortion occurs. When wide crosses have been made between very distantly related species, embryo abortion occurs at an extremely early stage, which previously made it almost impossible to collect viable embryos. However, several ingeniously skilful new culture techniques now allow these minute embryos to be 'rescued'.[59] Mortality rates are often rather high, but enough embryos normally survive the rigours of removal, transfer, tissue culture and regeneration to produce a few adult hybrid plants.

First-generation wide-hybrid plants are rarely suitable for cultivation because they have only received half of their genes from the crop parent. From the other (non-crop) parent they would have received not only the one or two desirable genes wanted by the breeder but also many thousands of undesirable genes that must be removed by further manipulation. This is achieved by re-crossing the hybrid with the original crop plant. Another round of embryo rescue is then required to grow up the new hybrids. The resulting second-generation hybrid will now have three-quarters of its genes from the crop and only a quarter from the original non-crop parent. The breeder selects only the progeny that contain the small number of desirable genes from the non-crop parent and eliminates the remainder. This process, which is called backcrossing, is then repeated for about six generations (or sometimes more), by which time the breeder ends up with a plant that is 99.9% identical to the original crop cultivar, except that it now contains a small number of desirable genes from the non-crop parent plant. Virtually all of the unwanted genes from this plant, which is often a wild species, would have been removed during the lengthy backcrossing process. Wide-crossing programmes take many years, and often a decade or more, to bring to fruition. They involve thousands of plants, a great deal of scientific expertise and skilled labour, and success is never guaranteed. Nevertheless, wide crosses have been hugely successful in enabling breeders to access useful genetic variation that is beyond the normal reproductive barrier at the species/genus level. A few practical examples can illustrate this point.

Wide crosses in rice

Rice is a major crop of interest at the International Rice Research Institute (IRRI), in the Philippines, as is breeding for disease resistance (see Chapter 6). Like most crops, rice is susceptible to a host of bacterial, fungal and viral diseases that can regularly cause yield losses of 30–50% to the relatively impoverished subsistence farmers in this part of Asia. One of the most serious diseases is caused by a virulent tenuivirus, called the grassy stunt virus, to which cultivated rice has little genetic resistance. The pathogen is transmitted to the plant by a leaf-dwelling insect known as the brown planthopper, *Nilaparvata lugens*. By the 1960s and 1970s, the grassy stunt virus had become an endemic disease that was severely affecting the crop and threatening the food supply of the steadily increasing human population. There was a particularly severe disease outbreak in Indonesia during the mid-1970s. During an earlier collecting expedition to India, scientists from IRRI had found a tiny population of a wild Indian plant, called *Oryza nirvara*, that was resistant to the virus.

Normally, it would be impossible to cross these two rather different *Oryza* species, but IRRI scientists were able to use tissue culture to produce a crude wide hybrid of the Indian plant with Asian rice. Eventually, in 1974, after many years of repeatedly backcrossing this hybrid with local cultivars of Asian rice, three new virus resistant varieties of Asian rice were finally released, much to the delight of the hard-pressed farmers of the region.[60] Despite repeated searching, the original Indian population of virus resistant *Oryza nirvara* has never been found again and may well have been lost for good. Luckily, on this occasion at least, some of the useful *Oryza nirvara* genes have been saved by the IRRI scientists, although these genes are now located in the genomes of the three new varieties of Asian rice, *Oryza sativa*. However, many other potentially beneficial species or local populations of wild plants are being destroyed every year by habitat degradation, industrialisation and, ironically, agricultural expansion. This illustrates the need for an inventory and/or conservation of any wild plants that could possibly contribute useful genes, such as disease resistance, to our major crops[61] (see Chapter 16).

Since its success with the grassy stunt virus, IRRI has gone on to use wide crosses to transfer several other disease-resistance genes into cultivated rice. New cultivars have been produced that are resistant to a particularly debilitating bacterial blight, as well as to blast and tungro diseases. Wide crossing with another wild species, *Oryza officinalis*, has also produced four new rice varieties, each of which carries resistance to the brown planthopper, which is a particularly serious pest in its own right (as well as being a viral vector) in Vietnam. Here, the new rice varieties were found to have several additional benefits. First, they saved on the use of expensive and potentially environmentally harmful pesticides, and second the wide cross also coincidentally produced resistance to the grassy stunt virus. This means that, thanks to wide crossing, Vietnamese farmers are now getting two important benefits from this new non-transgenic variety of rice that now contains 'foreign' genes from a different *Oryza* species.

Wide crosses in brassicas

Much use has been made of wide crosses in the improvement of the polyploid brassica crops, and especially oilseed rape. Rape is a tetraploid species formed a few thousand years ago by the spontaneous hybridisation of a variety of cabbage (*Brassica oleracea*) with a type of turnip (*Brassica rapa*). Unfortunately, the varieties of cabbage and turnip that combined to form the original version of oilseed rape had a rather restricted repertoire of disease resistance genes. In particular, there is relatively little useful variation in resistance to some of the most common pathogens in the regions where rape crops are mostly grown today. Two of the most serious of these diseases are blackleg and stem rot. Both are caused by fungal pathogens that have become endemic in many areas of rape cultivation. A bad outbreak of either fungus can cause yield losses of as much as 20–30%, particularly after the sort of heavy infestation that can occur after cool, damp weather. Blackleg, or stem canker, is caused by *Leptosphaeria maculans* and stem rot by *Sclerotinia sclerotiorum.*

The good news for the crop breeder is that there are hundreds of wild populations of *Brassica oleracea* and *Brassica rapa*, many of them growing around the Mediterranean Basin, that are extremely genetically variable. Even better, many of these wild brassicas are also resistant to some of the most virulent fungal pathogens that plague oilseed rape itself. In order to gain access to these useful genes, the breeder must first recreate, a new 'artificial' version of oilseed rape. This is done by crossing wild cabbage and turnip plants that happen to contain the desired disease resistance genes. Because such a cross is interspecific, the initial hybrid progeny may be sterile and therefore techniques such as embryo rescue and chromosome doubling may be called for. Once the fertility problems have been resolved, the breeder is left with a fertile cabbage/turnip hybrid, referred to as 're-synthesised' oilseed rape. This hybrid plant will carry the useful disease resistance genes, but also other unsuitable genes from its wild parents.

In this respect, the hybrid is similar to the initial cross between the wild Indian rice and cultivated rice that was described above. In both cases, the initial hybrid plant must be repeatedly crossed with an elite cultivar of the crop in question, i.e. back-crossed, in order to produce a new crop variety that consists mainly of the elite cultivar, but which also contains the desired genes from the wild relative. As with the example from rice that we have just considered, the laborious process of producing a fertile interspecific brassica hybrid, and then backcrossing it for six or more generations to generate a new elite cultivar, is both expensive and time-consuming, taking as long as 7–10 years. Because they·are so lengthy and require so much skill, wide crossing programmes tend to be used only when there is a serious problem in a cultivated crop, e.g. a particularly virulent disease, and there is no available genetic diversity within the gene pool of the crop species itself. The recently developed technique of marker-assisted selection (see the next chapter) will speed up wide crossing programmes in the future, although wide crosses are only possible if the breeder has access to wild relatives or other genotypes that carry useful genes that can be transferred to the crop.

In this chapter, we have seen how plant breeders of the twentieth century were able to create new variation in a rational manner. This approach was given a tremendous boost by the new insights offered by Mendelian genetics. In many cases genetic variation could now be accessed from wild relatives of the crop, as well as variants within the same species. The use of wide crosses extended still further the reach of the crop geneticist, who could now access genes from plants in completely different genera and transfer them into the crop. This was done via the relatively convoluted, lengthy, but ultimately effective process of hybridisation and backcrossing that enabled breeders to create a host of improvements in all the major commercial crops. In the next chapter, we will see how these techniques were improved still further by the advent of a series of new technologies, largely based on tissue culture, that were most effectively deployed by breeders after the 1940s. In the 1980s, these horizons were enlarged again when tissue culture technologies were married with the new recombinant DNA technologies to produce the first transgenic plants. With this development, the breeder now has (at least in theory) access to any gene from any organism for possible use in a crop plant. In a way, transgenesis therefore represents the logical conclusion of the search for variation by plant breeders. Equally important for the breeder was the revolution in screening and selection technologies since the 1950s. We will now look a little closer at this brave new world of modern high-tech breeding.

3

Modern high-tech breeding

> What a joy life is when you have made a close working partnership with Nature, helping her to produce for the benefit of mankind new ... fruits in form, size, color, and flavor never before seen on this globe; and grains of enormously increased productiveness, whose fat kernels are filled with more and better nourishment, a veritable store-house of perfect food – new food for all the world's untold millions for all time to come.
>
> Luther Burbank (1925)
> *Lecture in San Francisco*

Introduction

The ability of the plant breeder to create new genetic variation was enormously increased in the mid twentieth century by the invention of tissue culture and the use of growth regulators.[62] Attempts at wide crossing, as discussed in the previous chapter, were often frustrated by the incompatibility of genomes from relatively distant species. Embryo rescue could sometimes help, but one of the most crucial advances came with the development of chemically induced chromosome doubling, which has been the key to the success of many crop breeding programmes. As well as making possible much wider genetic crosses, chromosome doubling has enabled the use of powerful methods such as somatic hybridisation and haploid breeding, which have been especially useful in developing countries. In the past few decades, the technique of mass propagation has also been of considerable benefit in breeding programmes for tree crops, most of which are too long lived to be accessible to the sorts of approaches developed for the much shorter lived annual crops. The development of methods to prevent seed propagation is another important target for many commercial breeding programmes. Over the past century, new techniques have been devised either to induce fruit or seed sterility, or to prevent seed saving by using hybrid varieties.

In this chapter, we will also see that tissue culture was the key to enabling the transfer of exogenous genes (transgenesis), which is the basis of modern genetic engineering. Indeed, even today, more than two decades after the first transgenic plants were produced, the efficiency of gene transfer in many crop species is still limited by the capacity of the plants to be cultured and regenerated in vitro, rather than by the ability to transfer exogenous genes per se. Tissue culture has been used in breeding programmes for over 50 years and is now used widely for the improvement of many of our most important crops, including all of the major cereals as well as potatoes, brassicas and even some trees.[63] Meanwhile, we should remember that the creation of variation is only one of the twin foundations of plant breeding. Breeders also need effective and efficient methods for the identification and selection of variants likely to be agronomically useful. Many traits, such as disease resistance or flour-milling quality, are invisible to the naked eye. There has been striking progress in this frequently overlooked aspect of breeding. Therefore, we will finish the chapter by looking at some of the important developments in screening and selection of the huge numbers of new genetic variants created by all of the above technologies. From gas chromatography and mass spectroscopy to automated sequencing and polymerase chain reaction (PCR), a host of new analytical and screening technologies has enabled breeders to progress from crudely processing a few dozen samples a day to automated, round-the-clock analyses of many thousands of plants in exquisite molecular detail.

Tissue culture technologies

Chromosome doubling

One of the most important technologies that has made possible the creation of fertile varieties of interspecific hybrids, such as triticale, brassicas, and many other wide crosses, is the chromosome doubling method. Wide-hybrid plants are often sterile, which means that their seeds cannot be used for further propagation. This sterility is due to the great differences between the sets of chromosomes that these plants inherit from their two divergent parental species. Typically, their chromosomes are so different that they are unable to form stable pairs during meiosis (the special form of cell division that precedes formation of haploid pollen and egg cells). However, if the number of chromosomes can be artificially doubled, the hybrid can produce functional pollen and eggs and should therefore be fertile. But how could a breeder possibly persuade a cell to double its chromosome number without dividing into two daughter cells? The answer came in 1937 from a rather surprising direction, namely an attractive European meadow flower, *Colchicum autumnale* (naked lady or autumn crocus), that produced a toxic alkaloid called colchicine.

Colchicine was initially used as a pesticide because of its particular toxicity to insects. Further research soon revealed that this chemical also had the interesting property of causing the number of chromosomes in a cell to double, without inducing cell division. Plant breeders began to experiment with its use for rescuing the fertility of wide hybrids. As a toxin and a mutagen, colchicine did not always give the desired results, and this is still the case today. For example, there are some important crop species, including maize, where its use is not very effective. However, by and large, breeders have been able to minimise the undesirable side-effects of colchicine, while still achieving the desired chromosome doubling in cultured cells and tissues. And despite its toxicity to humans, colchicine is also used to treat several serious diseases, including chemotherapy for some cancers and for inflammatory conditions like chronic gout. As early as the 1940s, colchicine was being used routinely to double chromosome numbers in plants, and thereby to restore fertility to otherwise sterile wide hybrids. Within a few years, the chromosome doubling technique had been applied to more than fifty plant species, including most of the important annual crops and many fruits. Colchicine treatment can also be used to create seedless fruits. Among many commercial fruits developed with the aid of colchicine is the highly popular seedless watermelon. The technology is also widely used in the production of ornamental plants. Colchicine is also important for production of wide crosses and somatic hybrids, as discussed below. More recently, several additional chromosome doubling agents, all of which act as inhibitors of mitotic cell division, have been identified and used successfully in plant breeding programmes.[64] To date, thanks to colchicine and chromosome doubling technology, dozens of our most important crops have been improved and hundreds of new varieties have been produced around the world.

As with mutagenesis, it is interesting to speculate on the reception that colchicine technology would get if it were discovered today. Because it was originally developed as a tool in the 1940s, colchicine use is nowadays regarded as part of 'conventional' plant breeding. And yet, as breeders have always known, this technology entails the addition of a known toxin and carcinogen to artificially cultured plant cells, in order to disrupt and manipulate their genomic DNA. The resulting chemically induced chromosome doubling does not happen 'naturally' (i.e. in the absence of human intervention). The technology involves the use of toxic chemicals to force the production of artificial hybrids from unrelated species in order to transfer alien genes into crops.[65] One suspects that anti-GM campaigners would vehemently condemn colchicine technology as constituting an 'unnatural intrusion into the integrity of a species', or some such formulation. That would be a pity. Consider the immense public good that we already know has come out of colchicine technology, not to mention many of the other genetically intrusive manipulations of twentieth century plant breeders. Once again, there are some interesting parallels to be drawn with the contemporary debate on GM technology for crop improvement.

Mass propagation

Another application of tissue culture that had been of great utility for breeders is the mass clonal propagation of certain types of mainly non-annual crops, particularly some of the important, long-lived plantation species. Until relatively recently, very little systematic breeding had been done on such perennial or tree crops, and hardly anything was known about their genome organisations. This is despite the fact that perennial or tree crops include nearly all of our most important sources of edible fruits, nuts and several vegetable oils. Some of the most popular fruit trees include the traditional citruses, such as lemons, limes and oranges, as well as many new citrus crops, such as tangelos, mandarins, satsumas, sweet grapefruit and navel oranges. In recent years, breeders have also started to develop a vast range of new vine-like cultivars of former orchard crops like apples and pears. Popular tree nuts include walnut, almond, pecan, hazel, macadamia, Brazil and pistachio.[66] Major oil-bearing tree crops include olive, oil palm and coconut. A particular problem with the breeding of trees is their long lifespans compared to the annual crops. For example, an oil palm tree will not bear a useful crop of fruit until it is about seven years old. After that, the trees have a commercial lifetime averaging about 25 years, although an oil palm tree can still produce fruit for over 50 years.

These factors have made it impractical to set up classical breeding programmes for trees, but there is an alternative. Rather than laboriously trying to breed a tree species as if it were a short-lived annual crop like rice, the breeder can use mass clonal propagation as a much faster and cheaper alternative to multiplying up the best genetic stock. Based on traits such as yield, quality and disease resistance, breeders will typically select a few of the best-performing trees, or sometimes just one especially good tree, for propagation. Tissue cuttings, typically of stems or leaves, are then taken from the chosen tree(s) for cultivation in a mixture of nutrients and growth regulators until tiny plantlets are regenerated. The plantlets are subcultured on a massive scale until thousands, or even millions, of new seedlings have been produced. Finally, the batch of plantlets is taken for replanting in the field. In this way, a single elite tree can give rise to an entire plantation, or even a whole series of plantations, in a very short period.

The obvious downside to this technique is that all of the trees from a particular propagation programme could well end up being genetically identical clones. This may be fine if the clones behave exactly like their elite clonal parent, although in the long term their genetic uniformity might still render them dangerously vulnerable to new diseases to which they may have no resistance. However, there is another risk with the use of mass clonal propagation and that is the creation of abnormalities during the tissue culture process itself. As we saw in the previous chapter, tissue culture of plants can result in somaclonal mutations that are sometimes deleterious, leading to abnormal growth or sterility of the trees after they are planted in the field. Despite the potential drawbacks, mass propagation of clonal lines from a few elite individuals is now commonplace in the breeding of many improved orchard crops, as well as some of the new biomass crops like miscanthus.[67] The need for the rapid multiplication of millions of in vitro produced seedlings of many crops has now led to development of high-tech, automated methods for their clonal propagation.[68]

Clonal propagation has not always been commercially successful, however. In the 1980s, a commercial scheme to mass propagate millions of oil palm plantlets from a superior breeding line foundered when many of the maturing trees were discovered to have a serious abnormality in their floral development that had been induced during tissue culture.[69] This so-called 'mantling' phenotype led to a failure of fruit formation and, since the major products of the crop are fruit oils, the trees were effectively useless.[70] In the case of oil palm, the problem was compounded by the fact that fruits do not normally appear on the plant for about five years. This meant that the abnormalities were not discovered until the trees were already established in mature plantations that had been expensively maintained for several years. Although some of these challenges have now been rectified by further research, commercial confidence in clonal propagation has not fully recovered and relatively little planting of clonal oil palm was done during the succeeding decades. It is only in recent years that relatively small-scale clonal propagation programmes have resumed in some of the more advanced plantations.[71] This episode illustrates some of the problems that can arise from tissue culture. Many of the chemicals used in culturing and regenerating plants can cause developmental abnormalities, and even mutations. Despite these disadvantages and a few expensive setbacks, tissue culture and mass-propagation technology has proved to be an immensely valuable addition to the toolkit of the modern plant breeder. Today, clonal forestry is widely used in the management and improvement of a range of commercial plantation crops, including poplars, eucalypts, acacia and cedar.[72]

Somatic hybridisation

Somatic hybridisation is used to introduce novel genes into a crop genome from a donor species with which the crop will not normally interbreed.[73] In this respect, it is similar in its aims to the forms of conventional assisted hybridisation that we have already considered, although, as we will see, somatic hybridisation involves a more radical technological approach. Somatic hybridisation is yet another way of enhancing variation in crop species by importing genes, or even whole chromosomes, from other species that are not related closely enough for normal sexual crossing.

For example, if the crop species and the donor species are not closely related, it may not be possible to get the pollen of one of the species to fertilise eggs of the other. In such cases, even embryo rescue techniques are of no avail because there is no fertilisation and hence there are no embryos to rescue. An analogy in animals would be attempting to hybridise a species like a mouse with a human – because the sperm and eggs of these two species cannot fuse, such a hybridisation is normally impossible, even in vitro. However, the development of sophisticated microinjection and cell fusion techniques in the 1960s and 1970s allowed researchers to fuse whole cells or parts of cells to create a new composite cell.

In the case of plants, it is possible to take a cell from one species and to fuse it with another cell from a totally unrelated species. When this happens, the nuclei from the two different species will also fuse to create a hybrid nucleus that contains both sets of parental genes. From such hybrid cells, new adult hybrid plants can be regenerated. Scientists had been attempting to fuse plant cells since the beginning of the twentieth century. As early as 1909, German botanist E. Kuster was able to fuse plant cells whose cellulose walls had been removed (i.e. protoplasts), although the products did not survive further culture. The first report of interspecific hybridisation via protoplast fusion in two species of *Nicotiana* was published in 1972.[74] In addition to fusing two complete cells together, it is also possible to perform what is termed an asymmetric cell fusion. This technique involves the use of micro-dissection to transfer part of the nucleus of one cell into another cell. In this way, it is possible to transfer a small number of chromosomes from, say, a wild species that is unrelated to a particular crop into a cell from the crop plant. The resultant asymmetric hybrid cell can be treated with colchicine to induce chromosome doubling, hence stabilising the distinctly odd new genome. The next challenge is to coax the hybrid cells to divide and then to differentiate into a new adult hybrid plant.[75]

Since the 1970s, protoplast fusion has been used to create new types of plant from combinations of unrelated species, and the technique has also been very useful in several areas of basic research.[76] In at least one case, cultured human cells were fused with tobacco protoplasts, which demonstrated the potential for using somatic hybridisation to transfer genetic information between virtually any species of eukaryotic organism.[77] Somatic hybridisation was introduced into crop breeding programmes in the early 1980s and has so far been used to create several new commercial varieties of potato and oilseed rape. The technique has been attempted with many other crops but the main technical hurdle at present is the instability of the new genome combination created by the fusion of chromosomes from two dissimilar species. To a great extent, somatic hybridisation has been replaced over the past decade by transgenesis, with its greater precision, fewer problems with genome instability and higher overall success rate. However, transgenesis is only of use when there is a known gene(s) to be transferred. Many useful traits are controlled by unknown sets of genes and can only be transferred into a crop by adding an entire donor genome, or at least a substantial portion thereof. In such cases, we come back to the various forms of hybridisation as the only recourse for the breeder.

In recent years, breeders have started to return in greater numbers to explore the potential of somatic hybridisation, especially in fruit crops like the citrus group.[78] The reasons for this development are threefold. First, it has become obvious that transgenesis is by no means a quick and easy option for variation enhancement in crops, especially perennial species like trees. Second, tissue culture and molecular marker techniques have improved considerably over the past decade, which has increased the rate of success in the regeneration of genetically stable progeny from somatic hybridisation programmes. Third, somatic hybridisation is not regarded by regulatory authorities around the world as 'genetic manipulation' (as in GM or transgenic). This means that varieties produced by this technology are not subject to the same burden of regulatory approval and testing as transgenic varieties, which has created a business opportunity for breeders in private sector companies, who possess the necessary cell culture expertise to exploit these hybrids commercially. An example of such a company is Green Tec GmbH, which is a spinoff from a noted public research institute in Germany, the Max Planck Institute for Plant Breeding Research. The advantages of somatic hybridisation over transgenic technology are summarised on the Green Tec website as follows:

The somatic hybrids created by our technology are not considered as GMOs (genetically modified organisms) and are not regulated according to the genetic engineering directives of the EU. Therefore, the market approval is not restricted by lengthy and costly investigation imposed for genetically engineered food and feed varieties but only by normal variety registration regulations. Another advantage of this technology is, that it is not necessary to identify and isolate specific genes or promoters. As with genetic engineering technologies, the transfer of genes beyond the restrictions of sexual reproduction and crossing species boundaries has made somatic hybridization an important tool in combining useful characteristics of different organisms. Another limitation of genetic engineering, the transfer of only a restricted number of genes and traits, is avoided by somatic hybridization.[79]

Haploids and doubled haploids

During the 1920s, several species of spontaneously produced haploid plants were discovered. These plants had a single set of chromosomes, from a single parent, rather than two sets of chromosomes from two parents. The following decades witnessed advances in tissue culture that by the 1950s allowed breeders to produce their own haploid versions of crop plants, using methods such as microspore culture. In 1964, a new way of applying hybrid technology to crop breeding was discovered by the Indian breeders, Guha and Maheshwari.[80] Synthetic haploid plants were exposed to the chromosome doubling agent, colchicine, to produce a so-called 'doubled haploid' plant. Nowadays, haploid cells can be induced to undergo chromosome doubling by a range of treatments in addition to exposure to chemical toxins like colchicine. These include thermal shock and incubation with the sugar, mannitol.[81] In the latter case, mannitol acts as an osmoticum and appears to induce the fusion of haploid nuclei from adjacent cells to form a new doubled haploid cell. Doubled haploid plants are very useful to breeders because they are 100% homozygous and any recessive genes are readily apparent. Since the 1970s, doubled haploid methods have been used both for basic research and for crop breeding. The technology has so far been used to create new cultivars of barley, wheat, rice, melon, pepper, tobacco and several *Brassicas*.[82] One of the main centres of commercial haploid breeding is Australia, where several new types of barley and wheat have been produced. In the developing world, the main centre of such breeding work is undoubtedly China, where numerous crops have already been released on a large scale and many more are being developed. By 2003, China was cultivating over two million hectares of doubled haploid crop varieties, the most important of which are rice, wheat, tobacco and peppers.[83]

Sterile plant varieties

As we have seen in the previous two chapters, many manipulations by plant breeders can result in the production of sterile varieties that cannot be propagated. Sometimes this can be a useful trait for a commercial plant variety and so will be deliberately engineered by breeders. Such practices are especially common in commercial horticulture where the breeder may have several reasons to prevent the propagation of plants after their sale to customers. One example is where new ornamental plants, perhaps from another continent, are introduced into a region in which they had not been grown previously. If the newly introduced plants were to escape, they might become an invasive species with a potential to wreak great damage on the native flora. Escaped exotic plants are an increasingly serious problem across the world.[84] In the UK alone, the government estimates that its (frequently ineffective) control methods for a single invasive species, the Japanese knotweed *Fallopia japonica*, cost about $3 billion per year.[85] Many millions are also spent in combating other damaging exotics, such as the giant hogweed, *Heracleum mantegazzianum* and the rhododendron, *Rhododendron ponticum*.[86] Another case where sterility is a useful trait is in fruit crops, such as oranges and other citruses, where consumers demand seedless fruits.

Finally, seed sterility is analogous to F_2 hybrids or other non-propagable varieties in its utility to commercial seed companies. This is because the farmer or gardener is prevented from saving seed from their crop for replanting the next season and is obliged to return to the company to purchase new seed every year.

One of the most rapid and cost-effective approaches for inducing sterility in a plant is the creation of polyploids, especially triploids. Triploids have an additional reproductive barrier in that their three sets of chromosomes cannot be divided evenly during meiosis. This leads to unequal segregation of the chromosomes, resulting in a non-viable cell. Even in the very unusual case when a triploid plant is able to produce a seed (as can sometimes happen, for example, in apples), the seedlings rarely survive. In most cases, these triploid plants will grow and develop normally, except for their inability to set seed and therefore to reproduce. Development of triploids in some plant species can be complicated owing to a specific mechanism that prevents the normal development of a triploid embryo. However, embryo culture can be used to overcome this problem and produce sterile triploid plants. An alternative approach for the creation of triploid plants is regeneration of plants from the endosperm tissue that is present in the developing seeds of all flowering plants. The endosperm is a nutritive tissue, which supports the growth of the developing embryo in the seed and is unique in consisting exclusively of triploid cells. In some plant species, it is possible to dissect out a portion of the endosperm tissue from a maturing seed and to culture it in vitro. Following a series of hormone and other chemical treatments, the endosperm cells can be induced to develop into triploid embryos, from which triploid adult plants can be regenerated. This technique has been used successfully to create a range of triploid cultivars of many key fruit crops including most of the citrus fruits, acacias, kiwifruit (*Actinidia chinensis*), loquat (*Eriobotrya japonica*), passionflower (*Passiflora incarnata*) and pawpaw (*Asimina triloba*).[87]

Transgenesis

Transgenesis is the addition of small segments of exogenous (externally derived) DNA sequences and their incorporation into the genome of a recipient organism, such as a crop plant or an animal. In the case of plants, DNA is normally added to cells by one of two techniques. First, the DNA can be added directly by propelling small gold particles coated with exogenous DNA into plant tissues. This technique, called biolistics, can be used for any plant, crop or otherwise, but is a rather hit-and-miss affair that does not always result in incorporation of the DNA into the plant genome.[88] Alternatively, the DNA can be added in a more controlled fashion by means of a bacterial vector, such as *Agrobacterium tumefaciens*, that is able to insert a specific region of DNA into the genome of the plant.[89] Despite their various limitations, each of these methods of DNA transfer, or transgenesis, can frequently be much more efficient in delivering desired genes into crops than alternative methods of crop genetic manipulation discussed above, such as induced mutation or wide crosses.

Comparison with other technologies for variation enhancement

In comparison with the relative simplicity of transgenesis, the creation of a new variety via radiation, chemical or somaclonal mutagenesis normally involves the repeated exposure of tens of thousands of tissue explants or seeds to drastically damaging or even toxic external agents. These extremely disruptive treatments result in widespread DNA damage throughout the genome and the creation of hundreds of mutated genes. The vast majority of these mutations will have undesirable, and sometimes lethal, consequences to the crop. It takes many more years of back-crossing and selection to obtain a plant that only carries a mutation in the desired gene(s), but not in other essential genes. Even then, it is still possible that there may be undetected, cryptic, mutations that only manifest themselves in later generations, as the crop is tested in the field or grown in commercial cultivation. Another serious limitation of mutagenesis is that the breeder can only manipulate genes that already exist in the crop genome. No new genes can ever be added by this method. Fur-thermore, nearly all mutations result in a loss of gene function. This means that mutagenesis is more usually concerned with reducing the effect of unwanted genes, rather than increasing the expression of desirable genes.

Wide crossing is a more flexible method than mutagenesis for creating new genetic variation, because it can be used to transfer desirable genes from completely different species or genera into a crop plant. Thanks to elaborate technologies like embryo rescue and chemically induced chromosome doubling, wide crossing has opened up a much greater range of genetic variation for crop improvement. However, as we have seen, wide crossing is relatively laborious, time consuming, expensive, and the out-come is often uncertain. As with induced mutagenesis, wide crossing may result in the transfer of unwanted genes, as well as the desirable genes, into the crop cultivar. Even after seven or eight rounds of backcrossing, the breeder cannot always be sure that some undesirable cryptic genes from the wild donor species are not still lurking in the genome of the new crop variety. For example, it took strawberry (*Fragaria ananassa*) breeders no less than 18 years of laborious wide crossing with the cin-quefoil *Potentilla palustris* before they finally succeeded in producing an ornamental pink-flowered strawberry that is now widely available from seed suppliers.[90]

The drawbacks of the earlier technologies of variation enhancement made the prospect of the new, more direct, method of transgenesis appealing, both to basic plant researchers and to scientists from commercial companies. The first reports of transgenic plant cells were published, by European and US groups, in 1983.[91] By 1987, the commercial utility of the technology had been demonstrated when it was shown that copies of a bacterial gene could be transferred to plants where it conferred resistance to certain insect pests. In 1992, the first transgenic crop, the FlavrSavr® tomato, was released by Calgene in the USA. This tomato variety was not commercially successful, however, mainly owing to a lack of appreciation of the importance of plant breeding by senior staff in this molecular biology dominated company.[92] The next group of transgenic crops, which were released on a steadily increasing scale after 1996, has proved to be a more lucrative and enduring phenomenon. Four major crops, soybean, maize, oilseed rape and cotton, have been bred to express two groups of simple genetic traits, namely herbicide tolerance and insect resistance. For the first five years or so, commercial cultivation of these transgenically bred crops was largely limited to North America but the technology has now been adopted more widely, especially in South America and China.

The major drawback of transgenesis is that the breeder must know, or at least have a good idea, about which gene should be transferred to the crop. Also, transgenesis remains an expensive and relatively time consuming technology, albeit not as much as some of the alternative approaches, such as wide crossing. Therefore a breeder on a tight budget might not wish to devote limited resources to a transgenic approach unless there is a high prospect that the transfer of a particular gene will produce the desired outcome. This obviously limits the utility of the technology to more highly resourced breeding programmes and ones where the genetic traits in question are both well characterised and are regulated by a very small number of genes. So far, only a very limited portfolio of extremely simple traits has been amenable to the transgenic route, which means that while transgenesis has had some modest successes over the past decade, its scope for effecting a wide range of crop improvements remains somewhat restricted at the present time.

In this section, we have surveyed the main strategies that are used in modern plant breeding for the enhancement of variation. However, variation is only one of two prerequisites for successful breeding. Charles Darwin described variation as the 'raw material' of evolution, whether the latter process occurs via natural selection or is due to human intervention. In the former case, the totality of the abiotic and biotic environment acts as the selective 'agent', whereas in the latter case, the agent is the plant breeder. However, in order to take advantage of genetic variation, whether preexisting or manmade, the breeder must have an effective method of identifying and selecting the 'best' variants. This brings us to the second component of any breeding strategy, namely the selection of those desired variants and the elimination of undesired variants. We will now look at some key technical developments that have revolutionised the processes of screening and selection in plant breeding over recent decades.

Screening and selection

Selecting non-visible traits

Farmers and crop breeders can readily select suitable plants and varieties on the basis of easily visible characteristics like height, branching, seed size, tuber shape etc. However, many of the most important attributes of a crop, such as the quality traits that determine taste and nutritional content, are often invisible and can only be determined after harvest. The earliest potato farmers in the South American Andes would not have known if they had a suitable low-alkaloid variety until their tubers were cooked and ready for eating. The most important criterion at this point would have been taste. If the potatoes tasted bitter the alkaloid content was probably dangerously high and the entire harvest might have to be rejected. In the case of wheat, an important criterion was the bread-making ability of the flour. This character depends on the presence of a particular ratio of gliadin and glutenin storage proteins in the seed. Again, the presence or absence of this kind of quality trait would not become apparent until well after harvest. One can imagine the difficulty of attempting on an empirical basis to select for any useful variation in such traits. Not only are such traits invisible in the growing crop, they are also frequently regulated by numerous unlinked genes. This made for exceedingly slow progress in the selection of many useful quality traits before the advent of scientific methods of screening and analysis.

Over the last century, the use of chemical methods of analysis has enabled breeders to select new crop varieties that are largely free of the many toxins and other anti-nutritional agents that our ancestors were forced, through either ignorance or a lack of suitable breeding technology, to endure. For example, all brassica crops contain various forms of a large family of compounds called glucosinolates. When digested by animals, including people, glucosinolates are broken down to form isothiocyanates, some of which can cause human diseases such as goitre. Other glucosinolate derivatives can be beneficial and may even have anti-carcinogenic effects in people.[93] Glucosinolates are especially prevalent in the seeds of oilseed rape, and the toxicity of many of them prevents the use of this otherwise nutritious crop as a feed for monogastric animals such as chickens. As of the late 1980s, most industrialised countries enacted legislation setting maximum levels of glucosinolates in rape seeds. However, this was only possible thanks to the development of accurate and reliable technologies for their chemical analysis. The ability to identify glucosinolates allowed us to recognise the nature of the toxicity problem in the first place; it then enabled us to breed low-glucosinolate varieties; and, finally, it gave us access to a rapid and rigorous method of mass-screening harvested seeds on a commercial scale.[94]

   Chemical analysis has also made it possible to develop entirely new crops, more or less from scratch. One of the most impressive recent examples of this was the development in the public research laboratories of Agriculture Canada of a new form of oilseed rape in the 1960s. Prior to this time, oilseed rape, and indeed all the other brassica species, produced a type of seed oil that consisted mainly of a very long-chain fatty acid, called erucic acid. This oil was normally used for non-edible purposes and oilseed rape was very much a minor crop with a limited and not very profitable market. In the 1960s, the Canadians were looking for new crops to grow on their huge prairie farms and one possibility was to breed an edible form of oilseed rape. At this time, plant breeder Keith Downey led a small team of geneticists and biochemists in the Agriculture Canada research centre in Saskatoon. One of their goals was to find a way of reducing the amount of the unwanted erucic acid in rape seeds, and instead to increase the amount of a much more useful fatty acid called oleic acid. Oleic acid is the main ingredient of olive oil and is one of the premium monounsaturated fatty acids most highly recommended by nutritionalists.[95] The Canadian group used two innovative methods in order to select mutant seeds with low erucic acid contents. First, the seeds were screened using a non-destructive technique that involved carefully removing a tiny tissue fragment for analysis, while keeping the rest of the seed to produce a new plant. Second, the fatty acid composition of each of the thousands of tiny dissected seed fragments was analysed in detail by the recently developed technique of gas–liquid chromatography.

   Prior to the development of gas–liquid chromatography, it required about 200 000 whole seeds (1 kilogram) and about two weeks of labwork to perform a single fatty acid analysis. Now, it was possible to analyse a tiny fragment of tissue from a single seed, weighing less than 2 milligrams, in just 15 minutes. Thanks to this 650 million-fold improvement in analytical efficiency, breeders could accurately screen many thousands of seeds in the search for that rarest of events: a spontaneous mutation in just one or two of the genes that controlled erucic acid content in rape seeds.[96] By 1964 the team had developed the first zero-erucic acid variety of oilseed rape, which they christened 'canola'. For the past thirty years, canola has been one of the mainstays of Canadian prairie agriculture and is now a major export earner for the country. Canola-standard oilseed rape has also been adopted as an edible oil crop around the world, with an annual value in excess of $6 billion.[97] Largely thanks to this team of Canadian breeders, oilseed rape is now a globally important crop used to make salad oil, cooking oil and margarine, as well as being a key ingredient in all manner of food products from biscuits and cakes to curries and pies. It is interesting to reflect that this single rather modest new crop, developed over about a decade by a very small team of public sector breeders, has already earned far in excess of all the profits of the entire agbiotech industry over the past two decades.[98]

More recently, researchers have developed a host of ever more accurate and rapid techniques for the simultaneous screening and selection of thousands of different compounds in plants. These technologies, often referred to as metabolomics, take advantage of microprocessor-controlled robotic systems that are able to automate the processes of sample collection and processing for analysis. Hence, it is possible to run automated, round-the-clock screening programmes, e.g. for the presence or absence of a particular compound or group of compounds. Chemical analysis has also been revolutionised in the past twenty years by the development of techniques that are more accurate, faster, cheaper and require much smaller equipment than previously. Examples are the various forms of spectroscopy and chromatography that are now routinely used in both biomedical research and in hospital practice. Important spectroscopic techniques include mass spectroscopy (MS), plus the various forms of infrared (IR) and nuclear magnetic resonance (NMR) spectroscopy. A related method, NMR imaging, is now used extensively to complement X-ray analysis in hospital diagnoses.[99] Two of the most powerful methods of metabolite screening are gas–liquid chromatography (GLC) and high-performance liquid chromatography (HPLC).

DNA marker-assisted selection (MAS)

Another relatively new biotech method that is proving to be extremely beneficial to plant breeders is selection by means of DNA-derived molecular markers. The basic technology used in DNA marker-assisted selection in plant research and crop breeding is the same as is used in DNA fingerprinting methods that are used to such great effect in forensic analysis in criminology, and genetic profiling in biomedicine.[100] Molecular markers can save breeders both time and money in their crop improvement programmes as follows. In order to select most characters of interest to the plant breeder, it is normally necessary to grow up and analyse each new generation of the crop before it is possible to measure the character, or phenotype, and select the appropriate plants. Obviously, the potential performance of disease resistance or salt tolerance traits in a new variety cannot be measured until the plant has been grown, often to full maturity, and then tested in the field. The advent of marker-assisted selection has changed this as breeders can now select those plants that are likely to express the required characters without having to wait until they have completed their development. Molecular markers have now been developed for most of the major commercial crops, including several tree species. These markers can be used to track the presence of valuable characters as part of a crop breeding programme.[101]

For example, if a useful trait like disease resistance or high oil yield can be linked with a specific marker, many hundreds or even thousands of tiny plantlets can be screened for the likely presence of the trait without the necessity of growing them to maturity or doing costly and time consuming physiological and biochemical assays. Only plantlets carrying the marker are selected and all the unsuitable candidates can be eliminated.[102] By using molecular markers, therefore, a breeder can screen many more plants at a very early stage and thereby save several years of laborious work in the development of a new crop variety, as well as significantly reducing costs. At present, large-scale use of marker-assisted selection in crop breeding is largely restricted to a small number of economically important temperate crops.[103] However, the potential utility of this technology for other crops has encouraged public sector initiatives and public–private partnerships to develop cheaper, easier systems of marker-assisted breeding.[104] Earlier hybridisation-based methods were useful for researchers but, owing to their high cost and complexity, had more limited applications in practical breeding. But in the past decade, molecular marker technology has benefited from development of more efficient methods including DNA/DNA hybridisation, the polymerase chain reaction (PCR), and DNA sequencing. Today, the most practical molecular marker technologies are based on PCR methods, such as sequence-tagged microsatellites and single nucleotide polymorphisms (SNPs).

Although largely limited to the major temperate crops at present, molecular marker technology can be applied to the breeding of any crop and even to domesticating entirely new crops (see Chapter 19). A good example of the potential for marker-assisted selection can be seen with tree crops, many of which are important export earners for developing countries. Examples include oil palm, coconut, coffee, tea, cocoa and the many tropical fruit trees like bananas and mangoes. Owing to their relatively long life cycles and large size compared to annual crops, research into the biology of tree crops has always been a difficult and expensive undertaking. For example, one cannot fit many adult palm trees into the sort of conventional research glasshouse that could accommodate many thousands of wheat or rice plants. Hence the application of modern breeding methods that have been used for many decades with non-tree crops, is only just beginning for tree crops. By using DNA markers in conjunction with other new breeding technologies like clonal propagation, it should be possible to make rapid strides in the creation and cultivation of greatly improved varieties of many of these important tropical crops. In the medium-term future, straightforward marker-assisted selection could well evolve into what has been termed 'genomics-assisted breeding'.[105] Here, the whole panoply of bioinformatics-supported genomic resources can be deployed in support of a crop improvement programme, providing of course that such resources and infrastructure are available for the crop in question: a moot point, perhaps, in the case of subsistence crops like manioc or millet.

One of the main challenges in developing programmes of marker-assisted breeding is the relatively high cost and technical sophistication of the initial investment, even

after the various improvements of the past decade. For each crop, mapping populations need to be created, genomic markers assembled, and genetic maps created.
A study from the CIMMYT in Mexico has analysed the cost/benefit considerations
involved in using marker-assisted selection in resource-limited public breeding programmes.[106] The conclusion of this and related studies is that the justification for
developing a marker-assisted breeding programme depends critically on the nature of
the crop, including its genomic organisation, the availability of the requisite technical
infrastructure, and the availability of external capital to meet the set-up costs.[107] Such
considerations are of particular importance when deciding whether to invest in these
new technologies in developing countries, where potential profits from improved
subsistence crops may not even pay off the additional investment in the technology.
Nevertheless, there may be considerable societal benefits from such improvements.
Similar considerations may also apply to the acquisition and deployment of transgenesis technology. Both transgenesis and molecular marker technologies are
becoming progressively cheaper, but are still relatively expensive compared to conventional alternatives for many major crops, especially in developing countries. As
stated in a recent review: 'the breeding paradigm that has served the industry well for
many decades will be touched, but not overturned, by genomics-driven MAS. Wheat
breeding will continue to be driven primarily by selection in breeders' plots, rather
than by detection in microtitre plates.'[108]

New technologies for high-tech breeding

There are concerns that the current fascination with some of the more glamorous
transgenic approaches to crop improvement will distort our perspective and lead to
the neglect of alternatives such as mutation breeding or the various tissue culture
based technologies that may sometimes give better results in the long run. However,
plant researchers and breeders are continually developing advanced tools for crop
improvement and there are whole journals devoted to new fields such as 'molecular
breeding'.[109] One of the most interesting new technologies is rather engagingly
known as TILLING.[110] Despite its rather homely agrarian connotations, TILLING
is an extremely sophisticated method based on molecular genetics – the acronym
stands for 'Targeting Induced Local Lesions IN Genomes'.

We can regard TILLING as an updated and much improved version of mutation
breeding, a technique that has been practised for decades by plant breeders. The big
difference with TILLING lies in the method by which useful mutations are detected.
First of all, mutagenic agents, such as alkylating agents or irradiation, are used in the
normal manner to create a large, genetically diverse population consisting of thousands of mutagenised plants. Next, the second (or M2) generation of these mutants is

screened by a semi-automated, high throughput, DNA-based method in order to detect mutations in genes of interest. The third and final step is to evaluate the phenotype of those mutant plants that were selected following the screen in step two. The screening stage involves the use of PCR to amplify gene fragments of interest, and the rapid identification of any mutation-induced lesions by looking for mismatches in duplexes with non-mutagenised DNA sequences. The advantages of TILLING are that mutations can be detected without needing to grow up the plant and screen for an observable phenotype, such as plant height or disease resistance. TILLING is also amenable to automation and the use of high-throughput robotic screening systems, making it especially suitable for some of the large and complex polyploid genomes of major crops like wheat. Large pools of variation can be produced in a short time for introduction into breeding programmes.

The potential of TILLING for creating useful new variation might eventually rival that of alternative germplasm resources available to plant breeders – including landraces and undomesticated relatives. TILLING was initially developed in 2000 as a research tool, especially in the genetically simple model plant, *Arabidopsis thaliana*. Within a relatively short time, however, a few breeders were able to apply the method to one of the most challenging crop genetic systems, namely tetraploid and hexaploid wheats. In 2005, these efforts were rewarded with success when TILLING was used to identify variants in a gene, known as *Waxy*, which plays an important role in determining flour and bread quality.[111] This paper demonstrated that TILLING can work in crops, i.e. it provided proof of concept for utility of the technology. The company that first reported the use of TILLING in a major crop plant, Anawah Inc. of Seattle, Washington, was subsequently acquired by Arcadia Biosciences of Davis, California and the technology is now reportedly being used to produce commercial crop varieties that might be available as soon as 2008.[112] USDA researchers have also applied TILLING for the improvement of oil quality in soybeans and predict that commercial varieties will be available by 2009.[113]

It is likely that we will hear a lot more about TILLING and other advanced breeding systems in the next few years.[114] As with other technologies, TILLING will eventually get cheaper, will be applied to a wider range of crops, and will become more routine and accessible to the relatively less trained technician. However, the wider applications of this and other new technologies, including transgenesis, depend critically on how and where they have been developed. For example, the use of chemical and radiation mutagenesis was pioneered in the public sector and these technologies have subsequently been disseminated around the world. In contrast, other technologies, like maize inbred-hybrids and transgenesis, were commercialised by the private sector and have been diffused less widely. In the case of TILLING, it will be important to maintain a balance between protection of the legitimate commercial interests and research investment of the exploiting companies, while enabling the technology to be disseminated widely for non-commercial, basic research and public-good applications.

In Part I, we have explored the scientific basis of modern plant breeding, emphasising the importance of genetic variation and screening/selection methods that enable breeders to identify and capture such variation. In the remainder of this book, we will examine the wider societal basis of plant breeding and plant science research over the past two centuries, and their prospects to contribute to future crop improvements in the coming years of the twenty-first century.

Part II
The societal context of plant breeding

The fruitful pursuit of scientific truth and its application, once discovered, is not just a matter of talented individuals ... These are very important, but the cultivation of science is a collective undertaking and success in it depends on an appropriate social structure. This social structure is the scientific community and its specialised institutions

Stevan Dedijer (1911–2004)
Minerva[115]

4

Rise of the public sector and the US pioneers

I know of no pursuit in which more real and important services can be rendered to any country than by improving its agriculture, its breed of useful animals, and other branches of a husbandman's cares.

George Washington (1732–1799)
Letter to John Sinclair

Introduction

In this and subsequent chapters of Parts II and III, we will follow the evolution of plant breeding research over the past two centuries. Many changes in the conduct of plant breeding have occurred in response to the increasing input of scientific knowledge. But it has also been deeply influenced by wider changes in the organisation of the societies and cultures in which such research activities are embedded. In the modern world, it seems positively invidious to make a distinction between those who discover new knowledge and those who apply it to create new and useful products, such as improved crops. Both are part of a single process of the acquisition and exploitation (for whatever purpose) of knowledge. They form a single continuum, the extreme points of which may appear quite distinct, but are in reality linked with each other by a series of tightly linked intermediate stages. The processes involved in the discovery and subsequent application of scientific knowledge in a subject such as plant biology have always been inextricably linked. It seems that these lessons may have been forgotten by some scientists, as their disciplines were professionalised and in some cases removed from an immediate context of societal relevance as the twentieth century progressed. The rather precious notion of the pure knowledge seeker, who operates within a completely unbiased frame of reference that has no connection with the education, socialisation, and institutional context, would have been manifestly untrue to scientists such as Darwin, Einstein, and the Curies – and even more so to earlier pioneers such as Newton or Galileo.[116]

59

In early Western societies, the development of new technologies that arose from new scientific discoveries was historically carried out by the private sector.[117] For the vast majority of such technologies, the public sector did not become significantly involved in their development and exploitation until well into the twentieth century. Agriculture seems to have been, at least partially, an exception to this rule.[118] Agricultural improvement in sixteenth and early seventeenth century England certainly started as an exclusively private enterprise and many of the later well-known and much copied innovators, such as Townshend and Coke, began as strictly private entrepreneurs.[119] Improvement was seen as a path to wealth and prosperity, as stated succinctly in 1560 by English poet, Thomas Tusser:

> Ill husbandry braggeth. To go with the best:
> Good husbandry baggeth. Up gold in his chest.[120]

Even at the level of the individual landowner, we can observe a transformation from the medieval husbandman closely following traditional practices, to the more dynamic and experimental farmer-entrepreneur between the sixteenth and eighteenth centuries.[121] Such people would be more alert and receptive to any opportunity for betterment of their crops, whether from a public or a private source. These folk were the original driving force of post-Renaissance agricultural improvement.[122] However, the state was also beginning to assert a role in the exploitation of what was increasingly seen as a potential source of both wealth and power. During the seventeenth and eighteenth centuries, several major European powers developed increasingly effective state institutions for the exploitation of a vast range of plant resources across the globe. Several of these public institutions became important centres for botanical research that was both fundamental and applied. One of the best examples is the Royal Botanical Gardens at Kew, London, which was established in 1759.

New technologies, old problems

During the nineteenth century, it became apparent that the efficient harnessing of the emerging knowledge of plant science for the benefit of agriculture was beyond the capacity of the private sector alone. One of the mechanisms for the promotion of the more systematic application of plant science in agriculture was the establishment of various clubs and learned societies. Societies for agricultural improvement had been established in Britain as far back as 1723, but they became especially popular and widespread in early nineteenth century America. It was members of some of the agricultural societies in New England, such as Turner and Morrill (see below), who played a particularly prominent role in the establishment of the unique network of public sector crop research and extension centres throughout the USA during the latter part of the nineteenth century. Another useful mechanism to harness science for the public good was for a government to issue a call for research to achieve a particular national objective, together with a suitable incentive, such as a personal prize or even cash funding for the best proposals.[123] Throughout the nineteenth century, European Academies regularly awarded prizes for the exploitation of new discoveries in plant reproduction to assist crop breeding (see Chapter 2). Such calls were made to the general research community, which at that time largely consisted of individual scientists, often working alone, who were financed either by wealthy benefactors or from their own private means. Sometimes, such appeals were very successful in addressing a specific challenge, especially if the scientific issues were clearly defined and amenable to solution by the efforts of a single individual or very small team.

An interesting example of a public appeal that indirectly stimulated a new area of agriculture comes from the seemingly unlikely case of margarine in 1860s France. Following an appeal from Emperor Napoleon III for a substitute for butter, a chemist called Hippolyte Mège Mouriès produced the first version of a new synthetic edible fat in 1869.[124] This marvellous new fatty substance formed lustrous, white, spherical drops reminiscent of pearls, which are known as margarites in Greek.[125] Hence, the fat was called margaric acid from which we get the word margarine.[126] Demand for margarine eventually stimulated the development of a new group of oil-bearing crops, including soybean, oil palm, oilseed rape and sunflower. These crops are some of today's most valuable agricultural commodities, with an annual global turnover of $50 billion.[127] Margarine soon spread around the world and, just ten years later, the American author, Mark Twain, overheard the following conversation between two businessmen aboard a Cincinnati riverboat: 'Why, we are turning out oleomargarine now, by the thousands of tons. ... Butter don't stand any shows – there ain't any chance for competition.'[128] Similarly to many other agricultural innovations, including hybrid maize in the 1920s and transgenesis in the 1990s, margarine did not succeed overnight and was soon under attack from people with vested interests, who castigated it for its novelty and seeming artificiality.

The backlash against margarine was most marked in the USA, which had a large and prosperous dairy industry. Across the country, the dairy industry mounted a counterattack against a product that they believed would be their ruin. Thanks to the political clout of the dairy lobby, especially in strong milk-producing States like Wisconsin, margarine was classified as a 'harmful drug' and was subject to restricted sales. Margarine was also heavily taxed and stores required a licence before they were allowed to sell it. Finally, as a way of making it even less attractive, some States did not allow yellow margarine to be sold, meaning that shoppers had to purchase unsightly, off-white slabs of fat.[129] But, as with alcohol and tobacco, margarine was promptly bootlegged by enterprising middlemen seeking to meet a public demand that persisted despite the prolonged, and sometimes hysterical, propaganda war waged against this inoffensive fatty spread. Ironically, in view of its origins as a food for the French army, the US government stubbornly refused to purchase margarine for its armed forces. Amazingly, the punitive federal taxes on margarine were not abolished until 1950, and yellow margarine could not be sold in Wisconsin until 1967. Some of this prejudice and hysteria about margarine may seem comical today, but there are clear echoes in some of the current debates about certain aspects of the use of agbiotech for crop improvement.[130] The ironic twist is that much of today's new agbiotech and its most fervent boosters are from the USA, while some of its most trenchant critics are associated with vested interests in Europe, and especially in France.

Agricultural research as a public good

During the second half of the nineteenth century, it became increasingly apparent that a more systematic and professional approach to crop improvement was warranted. Even before the rediscovery of Mendel's work on heredity in 1901, there had been immense, and often overlooked, advances in the understanding of plant reproduction and the potential of hybrids for the creation of new varieties (see Chapter 1). What was lacking at this time was a cadre of specialists who could effectively comprehend and then utilise these new discoveries in plant science to create better crops for the benefit of a particular country or region. Such specialists could not themselves act as entrepreneurs because of the long-term nature of their projects and the uncertainty of efficiently harnessing the fruits for profit. It became apparent that specialist plant scientists would need to be accommodated within a new form of career structure within institutions that would be established specifically for research and breeding. In the absence of a coherent mechanism to apply a private sector led, margarine-like solution to the problem, these innovations would have to be financed either by the public purse or by wealthy private benefactors.

Somewhat surprisingly, the public sector solution to crop improvement was adopted most enthusiastically in the supposedly free market orientated, and private sector dominated, nation that was the USA. In contrast, the UK initially opted for a very hands-off and rather ad hoc mechanism financed overwhelmingly by private sector philanthropists. As well as a need for more research, the US pioneers of public sector investment in agriculture recognised the need for more and better education of farmers, in order to apply the increasingly complex technical innovations in a more efficient manner. As noted below by George Washington, well-educated entrepreneur-farmers had emerged in England in the early eighteenth century but, decades later, such a culture of knowledge-based improvement had yet to spread to the USA. It was the perceived need to disseminate best practice in commercial agriculture that inspired the activities of a succession of public spirited and influential (but also pragmatic) American citizens. This movement eventually led, in the latter half of the nineteenth century, to what became the most significant, far reaching, and successful innovation in public sector scientific research and education that had yet been witnessed anywhere in the world.

Emergence of public sector research in the USA

Before the mid nineteenth century, agricultural improvement in the recently independent colonies of the United States was a very localised private venture, which was much influenced by the work of English entrepreneurs, such as 'Turnip' Townshend. At this time, England led the world in the application of scientific and technological principles to crop improvement. The first president of the nascent American republic, George Washington, was himself a wealthy Virginian land-owner with an immense interest in the new techniques of crop cultivation. Washington, like many other innovative estate owners, purchased large quantities of the best available seed from British suppliers. He also imported tropical plants from the British West Indies, including breadfruit, which the sugar cane growers were then testing as a food for the slaves who worked their plantations.[131] Among the pamphlets that Washington acquired from his English contacts was one that was entitled *A new system of Agriculture, or a speedy way to grow rich*.[132] The title of this pamphlet is an important indicator of the attitudes of these late eighteenth century improvers. They may have been knowledgeable and enlightened men, who sought to use science to promote greater agricultural efficiency, but the whole point of their enterprise was to get rich (or richer). They were unabashedly open about this ambition in a way that today seems rather naïve, if perhaps refreshingly so. Washington himself was highly critical of his fellow American farmers for their lackadaisical attitude to crop improvement, and their reluctance to exploit the latest knowledge of plant science. In a letter to Arthur Young in 1791, Washington deplored their conservatism as follows:

The aim of the farmers in this country (if they can be called farmers) is, not to make the most they can from the land, which is or has been cheap, but the most of the labour, which is dear; the consequence of which has been, much ground has been scratched over and none cultivated or improved as it ought to have been: whereas a farmer in England, where land is dear, and labour cheap, finds it his interest to improve and cultivate highly, that he may reap large crops from a small quantity of ground.[133]

Washington and his fellow enthusiasts for agricultural innovation, including his English mentors[134] and his colleague Thomas Jefferson,[135] still regarded their work as a private undertaking. However, this mindset began to change during the first half of the nineteenth century. As we shall see, the role of the US public sector gradually became more pronounced in all matters pertaining to crop improvement, until by the end of the century it had become by far the dominant player in this enterprise. At first sight, it may seem odd that the most systematic, and by far the most extensive and consequential, public investment into agricultural improvement anywhere in the world occurred in the supposed bastion of aggressive, *laissez-faire*, 'Yankee capitalism' that was mid-late nineteenth century America.[136] So, why did the USA, which was best known at this time for its notorious oligarchy of 'robber baron' private entrepreneurs,[137] indulge in such (apparently uncharacteristic) public largesse for the benefit of its farmers? Part of the answer to this question is that the USA has, of course, never really been the home of unbridled free and private enterprise of popular myth.[138] Another factor was the immense importance of agriculture in the US economy.

In 1860, fully half of the US population still worked on farms; and agricultural products, such as wheat and cotton, were the nation's major export earners. At the turn of the nineteenth century, in the heyday of the earlier generation of agricultural innovators like George Washington and Thomas Jefferson, the USA was an even more agrarian nation, with over 75% of Americans working on the land. At the beginning of his career and throughout the revolutionary war and beyond, Jefferson was convinced that the future of the new republic lay in agriculture rather than manufacturing: 'the United States ... will be more virtuous, more free and more happy employed in agriculture than as carriers or manufacturers. It is a truth, and a precious one for them, if they could be persuaded of it.'[139] However, as the years passed, Jefferson changed his mind to support a more mixed economy: 'I am quoted by those who wish to continue our dependence on England for manufactures. There was a time when I might have been so quoted with more candor, but within the thirty years which have since elapsed, how are circumstances changed!'[140]

While the nation might be gradually reducing its dependence on agriculture in the mid nineteenth century, the politicians of the time were also cognisant of the potential power that could be wielded by a major food supplying country over less fortunate recipient nations, as the British had already demonstrated over half a century before.[141] Thanks to the foresight of these American agricultural pioneers, and the consequent improvements in crop production, the full power of food as a political tool for the US government would emerge within a few decades.[142] By the mid nineteenth century, the more perceptive American observers could also see that the increasingly successful application of scientific knowledge in Europe was beginning to act as a major spur to agricultural improvement in France, Scandinavia, and the newly unifying German States. However, the Europeans were still mostly relying on small networks of gentlemen scientists and breeders, many of whom were virtual amateurs or had their fingers in many other pies at the same time. A notable example is Charles Darwin, who privately researched plant reproduction and selection, as well as writing extensively on crop breeding and domestication. Darwin corresponded with a large group of like-minded enthusiasts, some of whom published work that led to valuable insights into agriculturally important areas, such as the mechanism of hybridisation and the selection of domestication traits in cereals.[143]

Although these enthusiasts made some of the most important theoretical contributions to plant science of the time, they were not themselves directly involved in agricultural enterprise. Such coteries of learned dilettantes were no substitute for a focused professional body of scientifically trained experts.[144] But where were such professionals to be trained and where might they practise their discipline? At this time, there was an almost complete dearth of research and education in plant and agricultural sciences at universities around the world.[145] This was a considerable handicap to the rigorous application of science to crop improvement and to the training of a well-educated generation of scientifically literate farmers and breeders. Such a deficiency was bad enough in heavily industrialised Britain, but it was potentially more serious for the much more agrarian US economy of the time. In 1850, just over 20% of the British population worked on the land, but productivity was high and this relatively small country, with a population of 21 million, was still able to grow 80% of its own food. In contrast, the census of 1850 showed that over 50% of the US population of 23 million worked on the land and productivity per unit area and per worker was much lower than in Britain.

Morrill and the land grant institutions

During the early nineteenth century, the campaign to extend higher education to the artisan classes, in order to spur on agricultural and industrial innovation, was championed for example by Jonathan Baldwin Turner, a Yale educated farmer, newspaper editor, and college professor. Turner's cause was taken up by Justin Smith Morrill, a representative, and later a senator, from Vermont. It was Morrill who went on to sponsor the land-grant legislation that bears his name today. The ideas of these men were regarded by many as radical at the time, especially their desire to extend higher education to a class of manual workers such as farmers. However, there was also a deeply pragmatic vein among many nineteenth century American thinkers that sought to harness and utilise all available resources, including science and education, in order to underpin economic advancement. Such economic progress would doubtless be for the ultimate good of all, but in the meantime it would also result in the private enrichment of many a savvy innovator. In other words, it was expected that Morrill's radical innovations would lead to a happy coincidence of public weal and private profit.

With such a promising prospect for the nation's future, Turner and Morrill were soon joined by a broad spectrum of influential and articulate people, including some of the larger farmers, bankers and industrialists. This impressive coalition recognised the new possibilities presented by the application of scientific knowledge to agriculture. They also knew all too well about the threat from European competitors who might beat them in the race to apply such knowledge.[146] Moreover, these enlightened but realistic Americans were also cognisant of a unique opportunity to establish a coherent network of agricultural research and education that extended from the scientist in the laboratory to farmer in the field, whether that field was in the relatively arid Great Plains or in lush subtropical Florida. It was this last innovation of linking scientists to farmers that really set the American initiative apart from anything being attempted by the Europeans of the time. Perhaps even more impressive is the fact that, notwithstanding the many vicissitudes of the past century and a half, the legacy of Morrill and his colleagues has largely endured in the USA to the present day.

Like most quality products, the Morrill Act did not come cheap, but the sponsors wisely avoided the profligate (and deeply unpopular, then as now) expenditure of tax dollars to fund their scheme. Instead, they proposed to use the donation and sale of public land in the expanding Republic to finance the building and maintenance of the new centres of research and education. Accordingly, Morrill *et al.* established:

An Act donating Public Lands to the several States and Territories which may provide Colleges for the Benefit of Agriculture and the Mechanic Arts. Be it enacted ... that there be granted to the several States, for the purposes hereinafter mentioned, an amount of public land, to be apportioned to each State a quantity equal to 30,000 acres for each senator and representative in Congress. ... And be it further enacted that all moneys derived from the sale of the lands ... shall be invested in stocks of the United States, or of the States, or some other safe stocks, yielding not less than 5% ... and that the moneys so invested shall constitute a perpetual fund, the capital of which shall remain forever undiminished ... to the endowment, support, and maintenance of at least one college where the leading object shall be, without excluding other scientific and classical studies, and including military tactics, to teach such branches of learning as are related to agriculture and mechanic arts, in such manner as the legislatures of the State may respectively prescribe, in order to promote the liberal and practical education of the industrial classes in the several pursuits and professions in life ...[147]

Morrill and colleagues did not envisage their network as providing for applied research alone, or for a mere technical education curriculum. They also insisted that these new public universities should undertake the full spectrum of scholarly research, and give their students a broadly based liberal education.[148] The Morrill Act of 1862 led to the establishment of dozens of Land Grant Universities over the next fifty years. These universities include some of the most famous and successful academic institutions in the USA, including Purdue, Rutgers, Cornell, Texas A&M, Iowa State, Michigan State, and the University of California campuses at Davis, Berkeley and Riverside. However, while these universities generated new scientific knowledge and provided a new educated class to drive the development of agriculture, there was still an unfilled gap between the scientific research and its practical application.[149] This gap was filled in 1887 with the establishment of the State Agricultural Experiment Stations (SAES) which enabled new crop varieties that might emerge from research at the universities to be tested under realistic field conditions and eventually multiplied up for dissemination to farmers.

Scientists at Land Grant Universities were encouraged to support agricultural development, expansion and intensification through applied, practically orientated research as well as through their teaching. The results of the research from such universities were to be made public and were to be actively disseminated, not just via learned journals, but also through more accessible public bulletins, circulars, as well as through direct speaking engagements to interested groups, and more formal classroom education. In other words, this initiative should not allow researchers and scholars simply to produce yet more esoteric knowledge for the benefit of their peers. Land-grant scientists also had an explicit mission to communicate with and educate the common man and to share the fruits of their newly won knowledge. But the task facing the new land grant institutions of the 1860s was formidable. In the words of J. M. Gregory, Regent of Illinois Industrial University in 1869: 'Looking at the crude and disjointed facts which agricultural writers give us, we come to the conclusion that we have no science of agriculture. It is simply a mass of empiricism.'[150]

During the twentieth century, the various Land Grant Universities developed in many different ways, with some of them diverging more significantly from their original missions than others. For example, some of the Midwestern institutions such as Iowa State and Purdue, which are located in important agricultural regions, are still very much focused on crop development. In contrast, others like UC Berkeley and Cornell have moved away from their agricultural roots, and are now rather more academic and broadly based in their research and teaching orientations. The farming client-base of the State Agricultural Experiment Stations served by the universities is also very variable across the USA, and its profile has changed greatly over the past few decades.[151] And finally, the private sector has become much more prominent in many aspects of crop improvement. However, despite their continual evolution and the need to adapt to the ever changing times, the land grant institutions as a whole remain vigorous centres of plant science research and applied agricultural innovation to this day.[152]

The USDA and its botanisers

The United States Department of Agriculture (USDA), which was established in 1862, became an increasingly effective public agency for crop improvement in the latter half of the nineteenth century. In an echo of the activities of the British and Dutch imperial botanisers of the seventeenth and eighteenth centuries, the USDA dispatched 'special agents' overseas to search for new crop germplasm.[153] The earliest of these 'bio-prospectors' were Niels Hansen and Mark Carlton who, from 1898, were sent on crop-hunting expeditions to Russia.[154] At that time, agricultural experts in the USA were predicting food shortages and famine within about thirty years, because of the increasing population that would supposedly eventually outstrip the ability of the nation to produce enough food. This may sound like a rather familiar refrain. Indeed, it seems that public concerns about overpopulation and imminent famine have been a recurring theme amongst pundits of every hue for at least three centuries. In the next chapter, we will discuss a similar, and equally ill-founded, series of Malthusian prognostications about the ability of agriculture to cope with increasing populations at about the time of the Green Revolution in the 1960s and 1970s. We will explore this theme yet again in Chapter 14, this time in its contemporary guise of agonising about 'feeding the world' and the putative role of transgenic crops as unique saviours of agriculture.

Meanwhile, back in 1898, USDA special agent Carlton eventually returned from his first plant-hunting expedition in Russia, having been successful beyond the wildest dreams of his bosses in Washington DC. Among the many hundreds of cereal samples that he brought back were several new varieties of durum wheat and hard red wheat. These wheat varieties were particularly resistant to the increasingly common droughts that were then plaguing many Midwestern and Great Plains farmers. When European settlers first arrived in the Great Plains in the 1870s and early 1880s, there was an atypical run of relatively moist summers and cereal farming flourished. This climatic anomaly ceased after a few years and the sudden advent of the more normal dry summers resulted in the collapse of crop yields and abandonment of many farms.[155] The situation was particularly acute in the Prairie States, where in some regions there was a wholesale retreat from farming after the drought years of 1884 and 1894. The effect of the new Russian wheat varieties on Prairie agriculture was dramatic. Within five years of their introduction into the US cropping system, the annual wheat production had soared from a mere 1600 tonnes to 550 000 tonnes. Not only did the drought tolerance of these new varieties open up vast tracts of the Great Plains and the Northwest for large-scale wheat cultivation, the Russian durum wheat also tasted much better in pasta, and the hard red wheat made much improved and tastier bread.

The USDA official who sent Carlton on his collecting expedition in 1898 commented later as follows: 'We have forgotten how poor our bread was at the time of Carlton's trip to Russia. In truth, we were eating an almost tasteless product, ignorant of the fact that most of Europe had a better flavored bread with far higher nutritive qualities than ours.'[156] Of course, Carlton's 'Russian' wheat was not originally from Russia itself. Both durum wheat and bread wheat originated in Southwest and Central Asia, several thousand kilometres from where Carlton had collected his samples. Carlton and other USDA collectors subsequently returned repeatedly to Russia and Central Asia in the early years of the twentieth century, collecting thousands of samples of cereals, fruit crops and legumes such as alfalfa. They were particularly interested in cold-hardy varieties that would be suitable for cultivation in the immense expanse of the newly opened prairies of West-Central USA. Interestingly, the Russian government later returned the compliment to the US government, in a manner of speaking, thanks to a series of official plant-collecting expeditions under the renowned botanist Nikolai Vavilov, who toured the USA during 1930 and 1932–1933. Like Carlton, Vavilov also discovered thousands of potentially useful new plants and varieties, which he took back to Russia for further appraisal.[157] Just as Carlton *et al.* successfully introduced Russian crops into North America, so Vavilov and colleagues collected some moderately useful New World crops and successfully adapted them to serve as major crops in the Soviet Union.

Some idea of the benefits of such unhindered reciprocal exchanges of germplasm can be seen from the story of the sunflower, *Helianthus annuus*. Nowadays sunflowers are one of the most successful European crops and are grown across the continent, from the Atlantic to the Urals. But sunflower is native to North America and was unknown in Europe until relatively recent times. Sunflowers were grown by indigenous peoples in the Americas, but wider commercial cultivation of the crop was restricted by its susceptibility to late frosts. Russian collectors first obtained sunflower samples from the USA in the nineteenth century. Russian and Ukrainian breeders then produced a series of greatly improved chilling tolerant varieties that were especially suitable for cultivation in the rich farmland of the Ukraine. These breeders also succeeded in almost doubling the seed oil content, a most impressive feat that greatly added to the value of the crop as a source of very high quality edible oil. More than half a century later, after the end of World War II, American farmers belatedly realised the potential of sunflowers as an edible oil crop. The circular odyssey of the sunflowers was duly completed when the chilling resistant, high oil Russian varieties were taken to North America and commercial cultivation began there in the 1970s. Thanks to those Russian breeders, sunflowers are now a common sight in the Northern Prairie States of the USA and across the border in parts of Southern Canada.

Probably the doyen of the early USDA crop collectors was Frank Meyer, who explored throughout Eurasia from 1905 to 1918, eventually bringing back thousands of new specimens of useful plants, ranging from alfalfa and Chinese cabbage to new sorts of apples and pears. Many of his varieties are still widely grown today. Any potential new crop variety that Meyer and his successors imported via the USDA germplasm collection programme was first sent for field testing and, if successful, for multiplication, at the appropriate State Agricultural Experiment Stations. Dozens of new crop species and thousands of new varieties have been introduced into US agriculture in this manner over the past century. It was Meyer who introduced soybeans as a major crop in the USA. Before his expedition to China, only a few varieties of soybean were grown in the USA, and it was generally used as a rather poorly regarded forage crop. Meyer introduced no fewer than 42 new soybean varieties into US agriculture and these have in turn spawned thousands of different cultivars, including all of the present-day commercial varieties.

Largely thanks to Meyer, soybeans are now the major oilseed crop in the world.[158] Soybeans are also the second most profitable US commodity crop, after maize; they even managed to outstrip that former giant of temperate agriculture, namely wheat, after World War II.[159] In at least one respect, however, Meyer proved to be rather ahead of his time. He observed how the Chinese used the soymilk to make a cheese-like food that they called 'tofu'. The word 'tofu' is often presumed to be of Japanese origin, but the word is more likely derived from the Cantonese word, dóufu or 豆 腐. Tofu is made by coagulating soymilk to produce a white solid that has the consistency of a soft cheese but little, if any, taste. It is, however, a versatile food product often used as a nutritious 'filler' or alternatively it can be flavoured and eaten alone. As well as being rich in protein, tofu also contains calcium, iron and vitamins B1, B2 and B3. Meyer tried, but failed, to introduce tofu into the US cuisine in the early twentieth century. However, his instincts about the value of tofu eventually proved to be sound. It might have taken almost another century for him to be proved correct, but the US consumer has now embraced tofu with a vengeance and this nutritious foodstuff is widely enjoyed across the country.

Extension services

Agricultural extension services are the key point of delivery of new information and advice, normally originating from research and breeding centres, which must eventually be communicated to the working farmer. Modern farmers require specialised, up to date, and accurate information about new crop varieties, fertilisers, control agents, cropping methods and so on. In the USA, at around the turn of the twentieth century, an additional series of new crop-based institutes was established in a systematic manner to complement existing state and private ventures and to connect more explicitly with the now much better educated farmer. For example, the USDA set up a series of regional research laboratories, now known collectively as the Agricultural Research Service (ARS), to work alongside the existing Land Grant Universities and State Agricultural Experiment Stations (SAES). The USDA also set up a network of extension services and a Cooperative State Research Service to work directly with farmers and local State Experimental Stations in order to facilitate the more efficient transfer of knowledge from plant research to the improvement of US crops.[160] This public sector initiative replaced a whole host of largely uncoordinated and localised private sector ventures that had hitherto attempted to fulfil the extension requirements of growers.[161]

Perhaps surprisingly, prior to these public initiatives in the USA, a great deal of agricultural extension work was organised and often generously resourced by the private sector. It is even more of a surprise to learn that some of the most active private sector agricultural extension ventures of the early twentieth century were carried out by that most pragmatic and hard-headed group of moneymaking organisations, the US railroad companies. Most railroad companies had their own in-house agricultural departments that worked with scientists to facilitate the uptake by farmers of the latest agronomic improvements. For example, between 1904 and 1911, the Burlington and Rock Island Railroad ran 62 'seed corn specials', carrying Iowa State University agronomists to 740 lectures that were attended by almost 940 000 people. Lest we doubt the capitalist credentials of the railroad companies, however, we can be reassured by the sobering words of R. B. White of the Baltimore and Ohio Railroad, who was clearly concerned lest his company be accused of the heinous crime of philanthropy: 'The Railroad's interest in problems of the farm is not prompted by any philanthropic motive, but purely because we believe it good business to take an active interest in what the territory we serve produces. The wisdom of such a policy is indicated in the greatly increased traffic of farm supplies and farm products.'[162]

The establishment of nationwide government-run extension networks in the USA soon made such company schemes redundant and they were quickly wound up. The extension system was the final link in a chain of knowledge and technology transfer from the plant research scientist, via breeders and agronomists, to the individual farmer. Only a century after George Washington and Thomas Jefferson had called upon the American farmer to become more of a botanical entrepreneur, the days of the improving farmer were virtually done. New developments in breeding and agronomy had rendered that function too specialised and time consuming for the individual farmer. But farmers still had to make complex and scientifically informed decisions about which crop variety to plant and how to manage inputs for their crops. Extension specialists were the key conduits for such information, as were State organised agricultural shows and exhibitions. This public sector network from laboratory to field soon proved its worth in underpinning the dramatic gains made by US agriculture in the succeeding decades.[163] The model was widely copied, albeit with local adjustments, in many other industrialised countries.

5

Public sector breeding in the UK

> You know, this applied science is just as interesting as pure science
> and what's more it's a damned sight more difficult.
>
> William Bate Hardy[164] (1864–1934)
>
> *Letter to Henry Tizard*

Introduction

Following the lead of the Americans in the late nineteenth century, dozens of specialised public sector agricultural research institutes were set up around the world during the early years of the twentieth century.[165] In countries like Germany and France, state-run agricultural experimental units had been established as early as the 1850s, but these were largely desultory, uncoordinated affairs in comparison with the US Land Grant and State Agricultural Experiment Stations networks. Most of the new European agricultural research centres were relatively small and tended to specialise in local crops, often under the control of a regional administration. This meant that breeders in such centres were often less aware of scientific developments in the wider world, many of which might have had useful application to their own crops. Unlike the USA, where much of the post nineteenth century agriculture was starting with a clean slate on largely virgin land with new crops, most of the farming in Europe occurred in the context of centuries of local and regional traditions that complicated broader strategic management by the nation state.

One solution that was applied in the Netherlands was to establish a single national centre of crop innovation. This Dutch initiative occurred at about the same time the US system was finally being completed in the early twentieth century. Following the US lead, albeit on a smaller scale, the government of the Netherlands decided to establish a systematic network of research and extension operations. The difference was that such a small country could easily base its entire network in a single centre. The centre was duly built at Wageningen, near Arnhem, which is in the largely agricultural province of Gelderland. This happened in 1912 when the Institute for Breeding of Field Crops was established as a unit of the State College for Agriculture, Horticulture and Silviculture. Shortly thereafter the name was changed to the Institute for Plant Breeding (IvP), an institution in which all aspects of breeding were concentrated, including education, research, variety development, registration and testing. Unlike the still very much extant US system, however, the Dutch network was largely dismantled during the privatisation/rationalisation manias of the 1980s and 1990s. Today its functions have been dispersed to a multitude of successor organisations which, in their contorted complexity, rather resemble the often bewildering collage of plant research centres that existed in the UK during the early 1980s.[166] We will now look in some detail at these centres and the rather different ways that UK public sector plant research evolved, particularly compared with the successful US model examined in the previous chapter.

The UK – a *laissez-faire* approach

In countries that lacked the tradition of systematic top-down governance that we have seen manifest in the USA and the Netherlands, public sector research institutes tended to evolve in a less coherent and more ad hoc manner. This was particularly true in the UK where a variegated patchwork of institutes and research centres gradually arose over the course of the twentieth century. To get some idea of the very different evolutionary pathway of crop research in the UK, as compared with the USA, we will now consider the development of a few of the better-known British agricultural research institutes.[167] This development took place within the context of a relatively 'hands-off', *laissez-faire* attitude to state intervention in the UK that contrasts strongly with the more interventionist policies found in the USA and much of the remainder of Western Europe. For example, the Royal Agricultural Societies in the UK received patronage but no government support until well into the twentieth century.[168] And it was not until 1910 that the administration of Lloyd George introduced the first systematic funding programme for agricultural research and education, almost fifty years later than in the USA.[169] In the meantime, it was up to the private individual, the successor of 'Turnip' Townshend, Thomas Coke, and their ilk, to step into the breach.

For this reason, we find that nearly all of the early agricultural research centres in the UK were originally set up as privately owned and operated establishments. This fact is reflected in their often convoluted ownership and governance, a situation that has persisted in many cases to this day. These research centres were often established following an initiative by a wealthy individual, rather than as any considered aspect of government policy. This led to the growth of a hotchpotch of institutes around the country that developed in their own sometimes rather eccentric ways, largely in the absence of any sort of coherent overall national plan. The early institutes were independent of one another and, at least initially, of the state. Despite being brought under a measure of public control later in the twentieth century, the UK institutes always retained a degree of autonomy and distance from government that was quite different from the more closely scrutinised and heavily managed US and European agricultural research centres.

There were advantages and disadvantages to these different arrangements. In the UK, a lack of central direction could have affected overall efficiency: for example, a given institute might duplicate some activities of research centres elsewhere in the country. On the other hand, the greater freedom of UK researchers enabled them to pursue innovative avenues of investigation, largely on their own initiative.[170] UK institutes were also more effectively shielded than their US or European counterparts from political or other pressures, whether at local or central levels. On the other hand, separation of most UK research institutes from universities, in contrast to the close relationship between SAES and Land Grant Universities in the USA, was a distinct disadvantage. Unlike the more welcoming attitude to technology related studies at most US universities, UK academia tended to be rather hostile. The attitudes of influential thinkers like John Henry Newman and John Stuart Mill are typical of the late nineteenth and early twentieth centuries: For Newman, a university existed to produce: 'a cultivated intellect, a delicate taste, a candid, equitable, dispassionate mind, [and] a noble and courteous bearing' in the student.[171] Universities were not concerned with applied science, least of all anything linked to industries such as agriculture. Surprisingly for such a vigorous proponent of utilitarianism, Mill's views on the role of the university were equally unequivocal: 'It is not a place of professional education. ... Their object is not to make skilful lawyers and physicians or engineers, but capable and cultivated human beings.'[172] Note that both of these learned gentlemen stress the role of a university in producing 'cultivated' persons – but certainly not cultivated plants.

The contrast between these sentiments and those of their transatlantic contemporaries, such as Morrill, could not have been starker. Thus we find that most of the early agricultural institutes in the UK tend to be private ventures, with few enduring ties with the unwelcoming universities. Surprisingly for a country whose wealth was largely based on industrial innovation, this myopic academic hostility was not restricted to agriculture; it often applied to manufacturing industry in general. In a few cases the unreceptive attitude of UK universities towards industry led to the establishment of private technology colleges by various company consortia. An interesting example is the mining industry, which underpinned the entire UK economy for much of the nineteenth and twentieth centuries. Mine owners were the oil-rich Arabian Sheikhs of their day, commanding a similar measure of disposable wealth. Following the failure of negotiations with Cardiff University, the mine owners of South Wales set up a private college at Treforest in 1912 with the enormous bequest for the time of £250 000. This sum was far more than the entire budget of the much larger public university in the principal Welsh city of Cardiff.[173] The Treforest School of Mining went on to be a premier centre of mining technology innovation, to the detriment of the universities, until it joined the public sector in the 1950s, and evolved into the University of Glamorgan in the 1990s.[174]

One of the few private initiatives in the UK to emulate the Land Grant Universities of the USA came to grief in the 1890s due to academic obstinacy. This happened when Alfred Palmer, founder of the profitable Huntley and Palmer biscuit company, attempted to establish a college in Reading, near London, to meet the needs of local farmers and horticulturists. He was joined by Leonard Sutton, who was owner of Sutton's Seeds, one of the oldest seed companies that dated back to 1806. Sutton later became a keen supporter of the application in plant breeding of the recently redis-covered Mendelian principles of heredity. These two well-heeled and highly motivated gentlemen joined up with an equally wealthy landowner and stockbreeder of the area, Lord Wantage. The college was duly established and went on to become Reading University, which is still one of the tiny number of UK universities with any kind of agricultural focus. However, once they were in charge, the faculty members at Reading rejected the US model and sought to define their organisation and course structures according to conventional UK academic criteria.[175]

In this generally unwelcoming atmosphere from both academia and the state, it was up to private individuals in the UK to set up their own initiatives to ensure the more systematic application of scientific knowledge to crop improvement. Britain had no Jonathan Turner or Justin Morrill and, even if such a person had existed, they would almost certainly have been ignored or spurned. Fortunately for UK agriculture, however, the late Victorian and Edwardian eras were times of immense prosperity for large numbers of successful industrialists, many of whom went on to become influential scientific philanthropists. It was to such men of independent means that the country owes the origin of almost all of its early ventures in applied plant science. These ventures include such well-known centres as the Rothamsted, John Innes, and Welsh Plant Breeding institutions. We will now look briefly at these British centres of crop research and innovation.

Rothamsted

The earliest UK research centre, and almost certainly the oldest extant agricultural research station in the world, was established at Rothamsted, near London. The Rothamsted laboratory has an interesting provenance. In 1842, James Murray, an

Irish doctor who dabbled in chemistry as a hobby, discovered that acid converts calcium phosphate into a soluble mixture of calcium hydrogen phosphate and calcium dihydrogen phosphate. This mixture is the basis of superphosphate, an extremely effective slow-release fertiliser. Murray then patented a process for the large-scale manufacture of an inexpensive form of the fertiliser. The industrialist and amateur scientist, John Bennett Lawes, purchased the patent from Murray and soon the new fertiliser was in widespread use in England. As a reward, grateful farmers offered Lawes the choice of having a laboratory built for his scientific hobbies, or the equivalent value in silver plate. Like any good scientist, Lawes chose laboratory over lucre, and it was duly erected on his country estate at Rothamsted just north of London.[176] Lawes was one of those classical mid-Victorian scientist-entrepreneurs who went on to use the profits from his business, in this case the manufacture of artificial fertilisers, to fund his personal scientific interests. Lawes appointed the chemist Joseph Henry Gilbert as his chief scientific collaborator. The two of them went on to carry out a singularly impressive amount of pioneering research, in which they established many of the principles of crop nutrition. Therefore the well-respected research institute at Rothamsted owes its origins to a judicious mixture of private enterprise, philanthropy and scientific expertise. A similar mixture of motives and mechanisms was behind many other agricultural and technological advances during this period before the advent of 'big science' and 'big business'.

Until 1900, the Rothamsted research centre was funded and managed by a trust financed from the profits of the Lawes fertiliser business. Gradually, over the twentieth century, Rothamsted began to increase its reliance on public funds, but its staff remained employees of the trust until 1991 and the site and buildings are still privately owned.[177] Rothamsted is now one of the few remaining public institutes in the UK where some crop research is still done, although even this work tends to be mostly strategic, rather than applied, in nature. Strategic research is often relatively fundamental in nature but is regarded as likely to lead to an eventual application. In contrast, fundamental, basic, or blue-skies (these are all synonyms) research is often purely curiosity driven, with no notion of any application. Therefore, an investigation of the genome of wheat would be regarded as strategic research, whereas the study of an obscure Patagonian moss would be basic research. Although Rothamsted currently includes research on crops in its portfolio, this work is of a strategic, rather than applied, nature and so, for example, the institute does not produce new crop varieties for farmers.

John Innes Centre

Another well-known institute that was originally a private crop research venture is the John Innes Centre, now located near the city of Norwich in Norfolk. This institute was founded, in 1910, as the John Innes Horticultural Institution in the London suburb of Merton. The institute was funded by a bequest from John Innes, a wealthy Victorian merchant in the City of London. Innes was a successful property and land dealer, and a founder of the City of London Real Property Company. Following his death in 1904, Innes bequeathed his estate to be used for 'the promotion of horticultural instruction, experimentation and research'. During its early days as a Horticultural Institution, John Innes scientists produced many new varieties of ornamental plants. In the 1930s, the Institution also pioneered the use of sterilised gardening composts for the more controlled cultivation of plants, especially indoors. To this day, John Innes composts are a mainstay of gardeners and indoor growers throughout Britain.[178] In 1945, the institute moved to Bayfordbury, Hertfordshire, and then, in 1966–1967 to its present site near Norwich. Like Rothamsted, the John Innes Centre (as it is now known) was only brought fully into the public domain in the 1990s; and even today the John Innes Foundation (a private entity) still owns the land and buildings where the Centre is based. So, although the John Innes Centre is now regarded as a publicly funded research organisation, the Foundation still owns all its estate and almost 30% of its current funding is from private sources.[179] Today the John Innes Centre focuses largely on basic studies of plant and microbial science.

Welsh Plant Breeding Station

Meanwhile, in Wales,[180] the Welsh Plant Breeding Station was established in 1919, largely thanks to a private donation from Lord Milford, who was a successful shipping magnate in South Wales. Lord Milford provided a capital grant of £10 000 plus an annual maintenance grant of £1000 for ten years. Thanks to the lobbying of breeder George Stapleton, who went on to become the first Director of the Station, Milford's bequest was matched from government funds. However, Milford made his funding conditional on the institute being run along commercial lines, with the profits going to fund the scientific research.[181] In the immediate aftermath of World War I, during which Britain had suffered serious food shortages (but was saved by imports of US food), there was much concern that food security be improved by stimulating agricultural production, especially in regions like Wales.[182] Because Wales is a largely non-arable region, the Plant Breeding Station focused on pasture crops such as clover and grasses.

One of its most notable successes was to breed an exceptionally productive variety of ryegrass (called S23) for use in fattening pasture for livestock. This ryegrass variety was on the UK National Seed List for an unprecedented sixty years, from 1933–1993. Despite the immediate focus of the Station on the improvement of grazing pasture for the local community of small dairy and sheep farmers, Stapleton was also an enthusiast for the fullest and most efficient possible exploitation of all available land. In practice, this meant the conversion of pasture to more productive arable use, and the opening up of new lands by bringing the plough back to the highlands of Scotland, England and Wales. In the latter venture Stapleton was supported and funded by government in the guise of the Development Commission, and by industry in the form of ICI, the largest fertiliser manufacturer in the British Empire.[183] Although most people today would react with horror to the notion of ploughing up the highlands of Britain, much of which has now been converted to National Parks, many of these lands were intensively farmed in pre-Roman times.[184] If wheat could be grown in the Scottish Highlands and the Yorkshire Dales several thousand years ago, there seemed to be no reason not to re-establish arable farming, especially in the context of the need to improve national self-sufficiency in food production. In the end, Stapleton's crusade foundered, largely owing to the inability (due to lack of cash) of highland farmers to invest in intensive farming methods, and the idea was quietly dropped. We will come back to the sometimes controversial issue of winning new arable land, whether from virgin areas or from pasture, in the context of the current debate on 'feeding the world' in Chapter 14.

During the 1940s, the Station moved to the University of Wales at Aberystwyth and passed into public ownership.[185] In 1953, the Station finally moved to Plas Gogerddan, north of Aberystwyth, and the Plant Genetics Department was created with the aim of studying inheritance of physiological characteristics affecting crop production. This initiative marked a shift in research from its previous focus on immediately applied targets to a more strategic focus that would not necessarily result in any practical applications in the short or medium term. Such a shift in research ethos gradually became more and more pronounced in agricultural institutes in industrial countries over subsequent decades, as we will explore in subsequent chapters. From 1964, more immediately practical work on establishing hill pastures was carried out at upland field centres in mid-Wales, rather than at the Station itself. In 1990, this collection of research centres changed its name to the Institute of Grassland and Environmental Research (IGER). Today, most of the research at IGER relates to basic and strategic, rather than applied, plant science.

Cambridge Plant Breeding Institute

One of the earliest UK research centres to be established de novo with the support of government funding was the Plant Breeding Institute (PBI) in Cambridge. Originally established by the UK Board of Agriculture in 1912, the Institute was first located within the Cambridge University School of Agriculture. The genesis of PBI harks back as far as the late 1880s when the newly established rural county councils of England sought to improve agricultural education by persuading Cambridge University to train teachers and researchers in the subject (doubtless mindful of the impact of the Land Grant Universities then being established in the USA). However, the University opposed this development by arguing (in agreement with Newman and Mill, as described above) that: 'education in technical and commercial subjects was incompatible with university education'.[186] It took several decades of further struggle before this attitude was overturned and PBI was ushered into being.

The first PBI Director, Sir Rowland Biffen, was one of the earliest breeders to apply Mendel's laws to manipulation of economically important traits in crops. He insisted that breeding for traits such as improved grain quality should be based on sound research into crop genetics and physiology. In this respect, Biffen set himself against a considerable number of breeders who thought that most of the key traits in crops like wheat, including yield and milling quality, were far too complex to be amenable to the simplicities of the Mendelian approach, i.e. what was now called 'genetics'. Biffen was influenced by William Bateson, the Cambridge botanist who was one of the researchers who had rediscovered Mendel's work in 1900 and who had then invented the term 'genetics' to describe what he called the new 'science of heredity'. Bateson subsequently went on to be the first Director of the John Innes Institute in 1910. Following Bateson's advice, Biffen proceeded to gainsay his critics by developing new disease resistant wheat varieties using a Mendelian approach.[187] His most notable achievement was his observation that resistance to the wheat yellow rust (a common fungal disease of wheat) was inherited in a simple manner that was similar to the round and wrinkled traits of peas so famously described by Mendel, i.e. this particular resistance trait was monogenic (controlled by a single genetic locus). Using this knowledge, Biffen was able to identify plants that carried the fungal resistance trait and to cross it into an elite high-yielding line of wheat to create the variety 'Little Joss', which was both high yielding and resistant to rust. 'Little Joss' was launched in 1910 and was an immediate success, most particularly in the rich wheatlands of East Anglia that were immediately adjacent to the breeders in Cambridge.

It was this demonstration of the power of Mendelian genetics that finally persuaded both industry and government to support the establishment of PBI at Cambridge in 1912. The new institute was an almost immediate success. Biffen's Mendelian approach to disease resistance paved the way to the identification and development of a whole series of new wheat varieties with resistance to many of the endemic microbial pathogens that had hitherto plagued UK farmers and often drastically reduced their yields. From its earliest days, PBI was the key centre of crop breeding in the UK. One of its most successful varieties was a wheat cultivar called

'Yeoman' that soon set the standard for yield and grain quality. Yeoman wheat was a breakthrough for British farmers and millers because it was a so-called 'strong wheat' variety. In this context, strength refers to the high protein content of the grain, which gives the sort of good dough elasticity that is ideal for breadmaking. In contrast, weak wheats have extensible, inelastic dough more suited for biscuit manufacture. Hitherto, British wheats had a relatively poor breadmaking quality, forcing bakers to import strong wheat flour from North America. As a new domestic source of strong wheat, Yeoman transformed prospects of farmers across the country, and greatly endeared Biffen and PBI to them. Following its release in 1916, Yeoman remained on the UK National Varieties List for an impressive 41 years.

As well as wheat, PBI developed many of the best-used varieties of other major UK crops such as barley, peas, sugar beet, brassicas and potatoes. In 1955, PBI moved to a new and larger site at Trumpington, on the outskirts of Cambridge, and became independent of the University.[188] Like most other research centres in the UK, PBI was largely autonomous in its operations and was funded from a variety of sources that often changed from year to year. After the 1950s, PBI received significant funding for both basic and strategic research from three separate government departments, as well as some large contracts from private companies. During its 75 years of crop research, PBI produced over 130 new varieties of wheat, barley, oats, triticale, potatoes, field beans, maize, oilseed rape, clover, sugar beet and grasses. As we will see in Chapter 9, PBI was eventually dismantled, and many of its most precious assets were sold off in the late 1980s, following the first major privatisation of an agricultural research centre in the UK.

Order versus chaos or control versus initiative?

From this brief overview, we can see the enormous differences in how the major centres of agricultural research developed in the UK, compared with their counterparts in the USA. Most of the UK plant breeding centres started out as privately funded, philanthropic institutions, only gradually becoming overwhelmingly publicly funded by the 1960s. As late as the 1920s, the four main UK plant breeding research centres received only 39% of funding from the public sector. By the 1930s, this increased to 54%, and by the 1950s to 77%. The peak of public sector funding occurred in the 1960s, when it reached 97% of the total. But, by 2005, the proportion of public funding had fallen back to nearer the levels of the 1950s.[189] The organisation, or rather the lack thereof, of agricultural research centres in the UK appears both cumbersome and inefficient. Unlike US researchers, whether in Land Grant Universities and SAES or in the various USDA regional laboratories, UK crop scientists worked in organisations with complex, sometimes positively Byzantine, histories. UK centres have also been regularly restructured, and frequently relocated throughout the past century. The amazing thing about this intricate and constantly changing system was that, not only did it work, but it produced some of the most innovative crop-related research anywhere in the world, especially during the half century from the 1930s to the late 1980s.

In the industrialised, and relatively wealthy, countries of Northwestern continental Europe, the organisation of agricultural research in the twentieth century has tended to lie somewhere between the two paradigms of the relatively top-down, government dominated USA model on the one hand, and the rather more anarchic and organic UK model on the other. For much of the first half of the twentieth century, European plant research tended to be somewhat loosely organised, often on a regional rather than a national basis. However, the need to rebuild research structures that had been so greatly disrupted during World War II led to the establishment of a series of more closely coordinated, state directed networks in most European countries after 1945. Examples in France include INRA (Institut National de la Recherche Agronomique, set up in 1946), and CNRS (Centre National de la Recherche Scientifique, set up in 1939, but which did not carry out scientific research until 1945). In West Germany a network of Forschungs Bundesanstalten, or federal research institutes, was established after 1945.

As with much of its industrial infrastructure, the scientific research networks in the UK had survived World War II without experiencing anything like the scale of physical disruption suffered by most countries in continental Europe. This meant that the UK missed a unique opportunity to reform much of its dispersed and somewhat disorganised agricultural research base. As we shall soon see, this omission came back to haunt British policymakers and scientists in later decades. It certainly contributed to vulnerability of the research institutes to the whirlwind of restructuring and privatisation that enveloped them in the 1980s and beyond.[190] In the meantime, the period after World War II witnessed the extension of the public sector, public-good paradigm of plant science research from the USA and Europe to the rest of the world. As we will see in the next chapter, scientific breeding was about to go truly global, with momentous consequences for food production and the welfare of the rapidly growing populations in many developing countries.

6

Breeding goes global: the Green Revolution and beyond

It is science alone that can solve the problems of hunger and poverty ... of a rich country inhabited by starving people. The future belongs to science and to those who make friends with science

Jawaharlal Nehru (1961)

PNIS India[191]

Introduction

By the mid-1940s, agricultural research and breeding centres had already been operating successfully for more than half a century in most industrialised countries. The benefits of this public sector led approach were universally obvious. Yields of each of the major commercial crops increased to the extent that food surpluses were generated in many of the main producer countries. Even tiny Britain, with its relative dearth of useful arable land, and a large, rapidly growing, urban/industrial population, was able to produce over three quarters of its food requirements. This achievement was largely thanks to a combination of assiduous attention to quality breeding and the introduction of intensive farming practices. By this time, the yield benefits arising from farming mechanisation (e.g. tractors and harvesters), that had been so prominent in the 1920s and 1930s, were being outstripped by gains conferred by biological improvements, i.e. from plant breeding.[192] By harnessing plant genetics, breeders could also design crops that were specifically suited to the new high-input, fertiliser/pesticide regimes, and were also adapted to mechanised cultivation and harvesting. Improved breeding practices also extended further the prospects for even greater gains in yield and productivity. Following World War II, the promise of more productive crops was urgently required, not so much in the well-fed industrialised countries, but in the developing world, where huge population increases were putting ever more serious burdens on existing production networks.

Developing countries faced many challenges that hampered their aspirations to take advantage of the benefits of modern scientific breeding techniques. Public sector plant breeding centres had been set up in a few non-industrialised countries as early as 1905 but, prior to the 1950s, these centres were often either ineffective or simply acted to service non-indigenous cash crops for the benefit of a colonial power.[193] Many developing countries were relatively small, all were economically poor, and many had only recently won their independence and therefore lacked experienced civil administrators and technical experts. They all suffered from a dearth of both physical (roads, ports, power etc.) and scientific (universities, research centres etc.) infrastructure; and many were also hobbled by endemic instabilities, both political and climatic. At this time, there were no coherent mechanisms for the translation of scientific advances made in the industrialised nations into practical application in such developing countries. However, unlike the commercially applied mechanical technologies, much of the new breeding work had been carried out in, and still resided in, the public domain. This meant that most of the scientific knowledge and germplasm resources were freely available for immediate use in developing countries. In contrast, agricultural hardware, such as motorised harvesters and crop processors, were relatively expensive and were produced via privately developed/owned technologies.

Hence, while it would be prohibitively expensive to supply hundreds of thousands of tractors and combine harvesters to a country like India, breeding expertise was, in contrast, relatively cheap and the raw materials, the seeds or cuttings, could be obtained freely from public centres. As we saw above, the biological potential residing in such seeds was becoming much greater than the potential of mere mechanical improvements to agriculture. This provided a golden opportunity to use a relatively inexpensive resource, i.e. seeds and breeding knowledge, to effect a totally disproportionate series of improvements in food production in order to secure the futures of millions of people. At this point, a very small investment had the catalytic potential to yield massive results. The only question was, who would be willing to make such an apparently altruistic investment? As we will now see, the surprising answer was a comparatively small but highly influential and persistent group of American philanthropists.

US philanthropy exported

It was largely thanks to the initiatives of a few private individuals that the postwar and post-colonial period of the 1950s and 1960s witnessed a major turning point in the fortunes of crop breeding in developing countries. Following World War II, there was a gradual expansion of public-good crop research across the world, especially into regions that did not traditionally benefit from strong scientific research infrastructures. To begin with, most of these new initiatives originated in the private (but non-commercial) sector, and were funded by charitable foundations or other philanthropic bodies. This process began in the 1940s with an alliance between a US-based charity, the Rockefeller Foundation, and the Mexican government. As we will now see, the motives of these philanthropists were not solely altruistic, and they did not operate without the knowledge and consent of the US government. Whatever the motives, however, the achievements of these US philanthropists were momentous.

As I alluded to above, philanthropy is rarely a totally disinterested activity, and there were mixed reasons for its operation in the context of plant breeding. In the case of the Rockefeller Foundation, the initial interest was, if not exactly inspired by, certainly congruent with the commercial interests of several major US seed companies that saw potential new markets being created as a consequence of the updating of crop production in Latin America. In addition to the Rockefeller Mexican Agricultural Program that began in 1943, over the next eight years the USDA and several Land Grant Universities organised hybrid maize breeding ventures in eleven other Latin American countries, from Cuba to Argentina. Following surveys of market potential for maize hybrids in Brazil by Rockefeller in 1946, a US company linked to the Foundation invested heavily in hybrid seed production there. The giant grain merchant, Cargill, then followed suit by moving into Argentina in 1947.[194] The pattern here is clear: acts of doubtless genuine philanthropy by a private foundation are closely coupled with public sector and private sector initiatives with the aim of simultaneously improving crop yields and business opportunities in developing countries, to the eventual benefit of all of the stakeholders involved.

These relatively small-scale, privately funded initiatives catalysed a series of impressive advances in crop production. In particular, they enabled some talented and highly motivated breeders from industrialised countries to use their expertise in the service of farmers in the developing world. This in turn stimulated the development of a number of some very effective nationally based public sector research centres e.g. in Brazil, India and Pakistan. Without the stimulus from the original philanthropic initiatives, it is unlikely that the Green Revolution of the late 1960s and 1970s would have happened in time to stave off a crisis in food production. Once the potential of the scientific approach to improve international agriculture became apparent, the initial relatively limited sponsorship from the US Foundations was considerably augmented in the 1970s, by funding from governments and other public organisations, e.g. the United Nations (UN). This enabled the establishment of the global network for crop improvement known as CGIAR. We will now chart the evolution of this process that started in a dusty field in Mexico and culminated in the most effective programme of crop improvement that the world has ever seen.

CIMMYT and wheat in Mexico

The initiative that led to the creation of CIMMYT was largely the brainchild of US Vice President, Henry Wallace, who had tried in the 1940s to win Congressional funding for a joint crop-breeding venture with Mexico. When the US Congress denied him funds, Wallace successfully approached the Rockefeller Foundation instead, and the programme was then launched with their backing. As we saw in Chapter 2, Wallace himself was a successful plant breeder who had helped to produce the first commercially successful sterile hybrid crop (maize) in the 1920s. He also founded one of the world's largest seed companies, Pioneer Hi-Bred. This company began by marketing hybrid maize in Iowa and soon flourished as it came to dominate the increasingly lucrative maize seed business across the Midwestern croplands during the 1930s. At the time of Wallace's 1940s initiative, Mexican agriculture was in an especially parlous state. This once bountiful land, the home of maize domestication and former centre of several mighty agriculturally based empires including the Toltecs and Aztecs, was now bedevilled by inefficient production and low crop yields.[195] To make matters worse, the decline in agricultural productivity coincided with greatly increasing demand for food from the country's rapidly growing, but relatively impoverished, population. The spectacular successes of crop breeders in the USA, despite the near catastrophe of the Great Depression and dustbowl years of the 1930s,[196] encouraged the view that the application of similar crop improvement techniques might also help the Mexicans to regain their self-sufficiency in food production.[197]

Initial results from a small pilot programme established in 1943 were so encouraging that the Rockefeller Foundation quickly moved to establish a full-scale institute; this was the body that is now known as CIMMYT. The aim of the new institute was to breed new crop varieties for immediate use in improving the agricultural output of Mexican farmers. The CIMMYT breeders were aware from the start of the great social importance of their research for the Mexican people, a fact that doubtless helped them to retain their relentlessly practical focus on crop improvement. From the outset, CIMMYT was a genuinely international undertaking that was largely independent of its host government in Mexico. A succession of experienced cereal breeders from the USA, with access to many of the latest techniques and germplasm, worked alongside scientists from Mexico and many other countries from around the world. Following the achievements of CIMMYT in the 1950s, breeders from as far away as India and Pakistan travelled there to learn about its innovative techniques of cereal breeding and cultivation. As we will see below, the crop breeders at CIMMYT were largely responsible for starting the 'Green Revolution' in world agriculture during the 1960s. The outstanding success of the

CIMMYT team was largely due to the vision and determination of Norman Borlaug, its founder and Director until his retirement in 1979.

Borlaug himself was especially focused on practically orientated research on crop improvement. He restricted his staff to projects that were 'relevant to increasing wheat production.' Borlaug later recalled that: 'Researches in pursuit of irrelevant academic butterflies were discouraged, both because of the acute shortage of scientific manpower and because of the need to have data and materials available as soon as possible for use in the production program.'[198] This intensely practical ethos also underlined much of the research in public sector institutes across the world in the immediate postwar period. These researchers recognised that there was an urgent job to be done, and they duly got on with it. However, as in the industrialised countries, the strong focus on practical breeding in many of the international institutes was gradually eroded, especially in the immediate post Green Revolution period after the 1980s. Partly this might have been a case of researchers switching priorities to less practical concerns, once their most pressing tasks had, to some extent, been achieved. But, as I will be arguing later, the change in priorities was doubtless also influenced by the events in industrialised countries. Here, the discipline of applied crop research started to become devalued relative to more esoteric, but more highly esteemed, basic research on plants. Such highly attractive, but less immediately useful, research projects were the *'academic butterflies'* so frowned upon by Borlaug and his ilk. The seemingly relentless global advance of basic research into plant science, at the expense of its practical applications, is discussed further in Chapter 10.

IRRI and rice in the Philippines

The second major postwar initiative in the field of international agricultural research and crop breeding was the establishment of the International Rice Research Institute (IRRI) in 1960. Like CIMMYT in Mexico, IRRI started out as a joint venture between a developing country government and American philanthropic organisations. In this case, the Ford and Rockefeller Foundations joined with the government of the Philippines to build and run a research and breeding centre at Los Banõs, near Manila.[199] A major difference between IRRI and CIMMYT is that work at IRRI is exclusively focused on rice, which is a staple for well over one billion mostly poor people across the world. Because rice growers and consumers are a relatively impoverished constituency, the crop had hitherto been almost completely ignored by the private sector. This in turn meant that some sort of public-good initiative was the only mechanism for unlocking the potential of scientific breeding to improve rice agriculture. In contrast to rice, which is very much a 'poor man's crop', wheat is grown throughout the industrialised world and there was always considerable private sector interest in its improvement. Hence, there was much greater scope for commercial companies to invest in wheat research, and the richer governments of such countries would also be more likely to donate public funds to improve this archetypical crop staple. Wheat breeders like Borlaug could therefore call upon large resources of existing germplasm and expertise from many public sector programmes in the industrialised world.

But nothing comparable existed for rice. The nearest equivalents to the dozens of global wheat breeding programmes for rice were a few isolated and poorly resourced breeding centres in Japan. However, while the rice breeders faced formidable challenges, the urgency of their mission was also plain. It was in the mainly rice growing regions of Asia that the population was growing most rapidly and there was the most acute need for increased food production. Research at IRRI began in 1962 and is now estimated to have affected the lives of almost half the world's population. In addition to its research, IRRI is involved in education and in community outreach projects to improve the livelihoods of rice farmers. Today, about half of all the rice varieties cultivated in the major growing regions of South and Southeast Asia originate from breeding programmes at IRRI. Moreover, the successes of the Institute in increasing rice yields, and hence overall production, mean that prices in world markets have fallen by no less than 80% over the past 20 years. This has resulted in national food security for major rice producing countries like India, China and Indonesia. It has also significantly underpinned the recent economic progress made by these countries. Following the success of institutes like CIMMYT and IRRI, donor groups set up further international centres in South America and Africa. For example, research centres for tropical agriculture were established in Colombia (Centro Internacional de Agricultura Tropical, CIAT) and Nigeria (International Institute of Tropical Agriculture, IITA) in 1967. A full list of these international centres is available in the Glossary.

National research organisations

At the same time as CIMMYT and IRRI were being set up, many governments in developing countries started to establish their own networks of plant research and crop improvement centres. These were largely modelled on the successful US and European networks that had been set up many decades previously. Two of the largest of the new government-run networks were organised in Brazil and India. In Brazil, the Brazilian Agricultural Research Corporation (Empresa Brasileira de Pesquisa Agropecuária, or EMBRAPA) was set up in 1973 as a nationwide organisation of 37 research centres employing a highly qualified staff of more than

8500.[200] Unusually for a government institution, EMBRAPA coordinates the work of most private and public entities that are involved in agricultural research in the country. This is done via the National Agricultural Research System of Brazil. As we will see in Chapters 14 and 19, EMBRAPA scientists have recently set the stage for massive increases in agricultural productivity in Brazil by showing that millions of hectares of the uncultivated lands of the *cerrado* (savannah) can be converted to arable use, as well as developing indigenous transgenic crop varieties.

In India, the Indian Council for Agricultural Research had been established by the British in 1929, but did not carry out any significant research during the colonial period. The Council was reorganised in 1965 and 1973, and now coordinates an impressively large network of research centres with a combined scientific and support staff of about 30 000. The Council also supports teaching, research and extension education activities in thirty Agricultural Universities employing a further 26 000 scientists. The postwar initiative in India was coordinated by the Minister of Agriculture, C. Subramaniam, who has become known as the father of modern Indian agriculture. Meanwhile, the wheat breeding programme was led by the doyen of the Green Revolution on the subcontinent, M. S. Swaminathan (see next section). Minister Subramaniam ensured that the Indians were able to grow improved wheat varieties imported from CIMMYT in Mexico to stave off the immediate food crisis of the late 1960s, but he also reorganised the national agricultural research system and ensured that the breeding of locally adapted cereal varieties remained a major priority. While CIMMYT provided the raw material for the Indian Green Revolution, it was breeders like Swaminathan who translated this foreign germplasm into finished varieties that were adapted for cultivation by Indian farmers. Thus the Indian breeders developed long-term solutions to their food needs that enabled the country to move from incipient famine to reliable food surpluses in just over a decade. The Indian Council for Agricultural Research has also developed many new varieties of high-yield seeds, mainly wheat and rice but also millet and maize.

Lack of space precludes detailed mention of the many other examples of effective national research centres in developing countries, but many of these institutions have worked in synergy with the international crop improvement network, CGIAR, that was established in the 1970s.[201] Before discussing CGIAR, however, it will be useful to look briefly at what is probably the finest achievement of practical plant science to date, the Green Revolution.

The Green Revolution

Arguably the most dramatic and far reaching manifestation of the benefits of public-good science during the twentieth century, at least in human terms, was the so-called

'Green Revolution' that had its main impact during the 1960s and 1970s. The term 'Green Revolution' was coined by William Gaud, Director of the US Agency for International Development. At a meeting in Washington DC in 1968, Gaud said: 'For the last five years, we've had more people starving and hungry. But something has happened. Pakistan is self-sufficient in wheat and rice, and India is moving towards it. It wasn't a red, bloody revolution as predicted. It was a green revolution.' There were two separate strands to the Green Revolution: first the development of improved wheats that were mainly used in Southern Asia, and second the breeding of the 'miracle rice' for the benefit of Eastern and Southeastern Asia. The international institutes, CIMMYT and IRRI, played a leading role in beginning the Green Revolution, but public sector breeders in other national agricultural research institutes across the world also made decisive contributions to its local application in many countries.

Wheat

The Green Revolution probably saved many tens, and possibly hundreds, of millions of people from severe hunger or starvation during the explosive population growth in developing countries following World War II. The key research that led to the Green Revolution was carried out at CIMMYT by a team of wheat breeders led by American Midwesterner, Norman Borlaug. When Borlaug first arrived at the deserted site that was to become CIMMYT in 1944, he found that the local wheat crops had been devastated for several years by a series of rust epidemics. The major pathogen of the Mexican wheat confronting Borlaug in the 1940s was stem rust, a fungus known as *Puccinia graminis*, which had an established reputation for virulence. Almost two millennia earlier, the Roman author Pliny the Elder (23–79 CE) had called stem rust 'the greatest pest of crops'. Wheat stem rust is also a serious disease in the USA, where there have been at least eight major epidemics since 1916. Due to the ravages of stem rust and other agronomic shortcomings, the average national yield in Mexico was a paltry 0.75 tonnes per hectare, even though much of the land was irrigated and well suited to wheat production. By this time, Mexican farmers had become so disenchanted with their ineffectual local cereal breeders that the latter were regarded as 'little better than parasites'.[202]

Borlaug gradually recruited and trained a team of breeders and set about collecting as many different wheat varieties as possible, in the search for the one plant in a million that carried genes for resistance to the fungal rust pathogen. Once candidate plants had been found, they had to be hybridised with good agronomic varieties to produce a plant that grew well as a crop, but also carried the traits that conferred resistance to the rust fungus. It is far from easy to breed rust resistant varieties in a self-pollinating species like wheat and it took Borlaug's team more than nine years and over 6000 crosses before they produced a series of resistant hybrids that could be tested in large-scale cultivation. As soon as plants were ready, Borlaug's team rushed them to farmers. This was very much a 'quick and dirty' approach to crop breeding. In Borlaug's own words: 'We never waited for perfection in varieties or methods, but used the best available each year and modified them as further improvement came to hand.'[203] This methodology may have lacked refinement but it was spectacularly successful. Moreover, despite the focus on immediate results, the work was also scientifically rigorous and Borlaug regularly published the results of his breeding research in well-regarded scientific journals.[204] But it was always practical results that were the primary interest and motivation for the work of the CIMMYT team.

Initial field trials with the help of Mexican wheat farmers soon showed that the breeders had achieved something special. Indeed, they were so successful that Mexico progressed from importing half of its annual wheat requirement in the early 1950s to complete self-sufficiency in the crop by 1956. By 1964, Mexico was even able to export an annual 500 000 tonnes of wheat. Nowadays, the average wheat yield in Mexico is an impressive 6.5 tonnes per hectare. This is almost nine-fold higher than the average yields back in those dark days of the mid-1940s, when Norman Borlaug first arrived to set up his new-fangled breeding programme. Contemporary Mexican wheat yields are more than double the meagre US average of 2.5 tonnes per hectare and are significantly higher than the heavily subsidised European Union average of 5.0 tonnes per hectare.[205] The CIMMYT team in Mexico made several additional innovations in their wheat breeding programme that were to have a profound effect on food production around the world, especially in some of the poorest nations in Southern Asia.

First, they developed the group of high-yielding, semi-dwarf varieties of wheat that are still the mainstays of present-day cereal agriculture. The semi-dwarf wheat was based on a Japanese variety called Norin 10. Shorter wheat varieties tend to be much higher yielding than taller varieties because the shorter plants put much less energy into their vertical growth and are therefore able to devote more resources to grain production. Originally developed in Japan, Norin 10 wheat was subsequently crossed by USDA breeder Orville Vogel with Washington State winter wheats, to create the especially high-yielding Gaines variety. Vogel shared this wheat with Norman Borlaug in Mexico, who crossed it with local Mexican wheats, as well as with varieties from other countries. This series of genetic crosses finally gave rise to the high-yielding, semi-dwarf varieties that transformed wheat farming worldwide and are the universally cultivated form of wheat today. The CIMMYT group managed to combine this particularly valuable semi-dwarf trait with disease resistance and other useful characters to produce a variety that was the 'killer-app' of wheat cultivation, resulting in the doubling or tripling of crop yields.[206]

In addition, the breeders at CIMMYT worked with agronomists to establish a high-input system of fertilisers, irrigation and mechanisation to ensure that the yield of the crop was able to approach its maximum biological potential. Finally, the CIMMYT researchers worked out a system called 'shuttle breeding', which allowed them to grow two field crops a year, hence halving the time needed to produce a new crop variety.[207] In a coda to this story of global wheat improvement, Borlaug's 'Green Revolution' semi-dwarf spring wheats eventually found their way from Mexico back to the USA, where they have been widely used in breeding programmes and have also allowed substantial gains to be made in US wheat production. Word soon spread about the remarkable achievements at CIMMYT and, in the mid-1960s, Borlaug's team was approached by the Indian government for assistance with their own wheat breeding programme. India experienced a succession of poor harvests in the early 1960s, which was also a time of rapid population increase. There was much concern in India about the inability of the country to feed itself from its own produce. As well as the very real threat of famine, the necessity to import food from overseas imposed a crippling financial burden that India, and most other impoverished developing nations, could ill afford.

It is interesting to recall that, in the late 1960s, many western 'experts' and other less qualified pundits were talking earnestly about the imminence of a series of global famines in which untold millions of people would surely perish, most notably in India. One of the most famous and influential books of the period was *The Population Bomb* by the biologist, Paul Ehrlich.[208] As a child of the 1960s, I recall feeling rather depressed, in the year of promise that was 1968,[209] to read Ehrlich's confident assertion that: 'The battle to feed all of humanity is over ... In the 1970s and 1980s hundreds of millions of people will starve to death in spite of any crash programs embarked upon now.' Apparently unaware of Borlaug's work, or of the as-yet barely tapped potential of scientific crop breeding in the developing world, Ehrlich went on to say: 'I have yet to meet anyone familiar with the situation who thinks India, will be self-sufficient in food by 1971 ... India couldn't possibly feed two hundred million more people by 1980.'[210] Luckily for humanity, and thanks to the work of the CIMMYT breeders and their Indian colleagues, Ehrlich's original thesis was proved spectacularly wrong within a few short years.[211]

The process of improving the Indian wheat crop had started in 1966 when the government imported 18 000 tonnes of the new disease resistant, drought tolerant seed from CIMMYT in Mexico.[212] This was something of a stopgap measure until the Indians could breed their own locally adapted wheat varieties. Even so, the results were both rapid and successful beyond the most optimistic predictions. In the single year from 1967 to 1968, the Indian wheat harvest increased from 11.3 to 16.5 million tonnes, and production rose steadily thereafter.[213] Pakistan soon followed suit by importing wheat seed from CIMMYT and, between 1966 and 1971, both countries had doubled their wheat production. The predicted famine in the Indian subcontinent had been averted and the whole region was well on its way to regaining self-sufficiency in food production. Today, despite a population of over one billion, India is self-sufficient in all the major food grains and earns additional valuable foreign exchange by exporting some of its more productive crops, including wheat. Its wheat production has tripled since 1968 and this has certainly contributed to the nine-fold growth in the Indian economy over the same period.

Rice

During the 1960s and 1970s, Henry Beachell and colleagues at IRRI in the Philippines pursued a similar breeding strategy to the wheat breeders at CIMMYT, with the aim of improving rice yields in eastern Asia.[214] After screening thousands of plants and crossing several promising varieties for several years, the IRRI researchers eventually came up with a new variety that became known as 'miracle rice'. The new rice was a short, thick-stemmed variety, called IR8, that had much higher yields than conventional tall rice. The IR8 variety was itself derived from a cross between a popular rice variety called Peta and a new Chinese variety called *Dee-geo-woo-gen* (*Dee-geo* means 'short leg' in Chinese). This Chinese variety carried a spontaneous mutation in the sd_1 gene, causing a dwarfing phenotype.[215] But there was a remarkable added bonus to this semi-dwarf rice, which Borlaug had also observed when he produced his new Mexican wheat varieties. In both cases, the growth and development of the new crop varieties were unaffected by day length. In practical terms, this meant that the growing season of IR8 rice was a full 60 days shorter than that of normal rice.

For rice growers in the tropics, this had immense consequences. Thanks to this dramatic reduction in the growing season, many farmers would now be able to grow and harvest two crops a year, instead of just one. That fact alone meant a potential doubling of rice yields. However, the potential yields were increased further by the additional grain production of the shorter, semi-dwarf plants, and the much greater responsiveness of these varieties to added fertilisers. The shorter, semi-dwarf rice and wheat plants were much sturdier than their tall cousins and therefore less likely to fall owing to wind or rain – a phenomenon called lodging that can cause severe crop losses in late-season storms. Addition of fertiliser tends to make most crops grow tall and spindly and increases their susceptibility to lodging. However, semi-dwarf varieties respond to fertiliser by making more grain rather than a wasteful increase in height. This made it worthwhile to apply higher doses of fertiliser and hence to get a correspondingly high grain yield. Following the release of the new 'miracle rice', yields across Asia more than doubled from the mid-1960s to 1990. By 2004, over 90% of the semi-dwarf rice varieties grown in tropical Asia were derived from the *Dee-geo-woo-gen* mutation.[216] Between 1966 and 2000, worldwide rice production more than doubled, from 257 to 600 million tonnes. Over 90% of this increase occurred in Asia and was directly attributable to improved breeding and related achievements of the Green Revolution.[217]

Global impact

By as early as 1968, it was already becoming clear that CIMMYT and IRRI crop breeders were in the process of achieving something momentous for world agriculture, and the term 'Green Revolution' was duly coined. The pioneering role of Norman Borlaug in carrying out the research, and then reducing it to practice in such an effective manner, was recognised just two years later. In 1970, this intensely practical but unassuming Midwesterner (now also a Mexican by adoption) accepted the Nobel Peace Prize amongst the glitterati of Stockholm.[218] In a typical comment, he later said (only half jokingly) that the Nobel Prize had been a 'disaster' for him.[219] The recognition certainly proved to be a distinctly mixed blessing as the resultant worldwide fame and constant invitations distracted him from his first love, namely the breeding of better crops for poorer countries. The impressive successes of the Green Revolution were not only due to improved crop varieties; they also involved the more intensive use of fertiliser and the use of other new management methods by farmers. These measures have sometimes been controversial, especially given evidence that many poorer farmers could not afford them and hence did not always share in the direct benefits of the Green Revolution.[220] However, the higher crop yields also resulted in lower food prices and greater economic growth and job creation. Many poorer farm workers benefited indirectly from cheaper food and better job prospects, often moving to the rapidly growing urban areas that were springing up across the developing world.

   During the 1960s and 1970s, the same package of measures, involving newly bred crop varieties and more intensive management, was adapted for use in dozens of other countries across the world with often spectacular success.[221] Despite localised famines, nearly all of which are essentially manmade, the overall per capita production of food has more than kept pace with increasing populations.[222] Hence, despite a doubling of the human population since the late 1960s, global food production has more than trebled. Across the world, in 2003, farmers produced over 40% more food per capita of total population than they did in 1950.[223] The achievements of the Green Revolution were nearly all realised because of the free and open application of public sector research in an international context. The research was primarily driven by talented and motivated scientists who were supported by a few far-sighted philanthropists and policymakers. One may cavil about the deeper motivations of some of the US philanthropists, who were undoubtedly in cahoots with interested parties in companies and government, but nobody can doubt the eventual wider efficacy of their sponsorship of centres like CIMMYT and IRRI.

Meanwhile, breeders like Borlaug obtained seeds freely from other public sector breeders in the USA; the seed was improved by the multinational team at CIMMYT; and improved seed was then passed on to Indian and Pakistani breeders. It was this process of open and free exchange of both germplasm and expertise that really made possible the rapid development of the Green Revolution cultivars of wheat and rice. There was some contribution by the private sector, but this was very minor indeed. For example, in some countries, private companies were involved in developing commercial seed lines that were specifically adapted to local conditions. However, such firms were few in number and they all relied on a supply of improved germplasm that had been initially developed in the public sector and then disseminated as a public good.[224] Following the initial success with the major cereals, wheat and rice, it was realised that it would be necessary to widen the portfolio of improved varieties to other staples, many of which were grown by poorer farmers in less favoured agricultural areas. This was especially true in Africa where the Green Revolution had far less impact than elsewhere.[225] In order to achieve such a task, a considerable expansion of the international agricultural research effort would be necessary, and this eventually led to the establishment of the organisation known as CGIAR.

The emergence of CGIAR in the 1970s

By the end of the 1960s, the two main funders of international agricultural research, the Ford and Rockefeller Foundations, realised that the very success of their venture was endangering its future prospects. As their research increasingly bore fruit and became recognised around the world, CIMMYT and IRRI had to expand their operations. A steady stream of requests for assistance came in from countries in Asia, Africa and South America. More staff and new field stations were needed to test crop varieties destined for different climatic regions. It was also necessary to expand the list of crops in the breeding programmes. In particular, it would be necessary to include hitherto 'orphan' species, such as sorghum, millet, barley, yams, taro, cassava and the various pulses. These were known as orphan crops because almost no modern breeding had been carried out on them up to that point. Therefore such crops suffered from chronically low yields that lagged far behind their true potential, as was evident from comparisons with the major cereals like wheat and maize.

The two Foundations had realised that the magnitude of the task now confronting them was far beyond the resources of a couple of charitable bodies. For this reason, they recruited two of the most powerful international bodies that could potentially contribute both funding and organisational expertise, namely the UN Food and Agriculture Organisation and the World Bank. Between 1969 and 1971, these four organisations and other interested parties held a series of high-level meetings and developed a strategy to ensure the long-term support of international agricultural research.[226] The process culminated in May 1971 with the establishment of the Consultative Group on International Agricultural Research (CGIAR) as an umbrella organisation to coordinate agricultural research in developing countries worldwide. As part of the establishment of CGIAR, the World Bank agreed to support the overall organisation, while the two Foundations continued their direct support of the two centres at CIMMYT and IRRI. Other donors, such as the United Nations and governments of several industrial countries, also assisted in this effort and soon CGIAR was on a firm financial basis and could begin its scientific work.[227]

One of the first tasks for CGIAR was to establish priorities for a new series of research programmes aimed at the improvement of agriculture in developing countries and the coordination of such research. An organisation like CGIAR was needed for this work because the vast majority of developing countries lacked the resources to carry out their own research in an effective and efficient manner. As we have seen, some of the larger countries, including Brazil and India, had already established effective national research networks, but many other countries were too small and too poor to afford such facilities. Furthermore, many of the problems faced by individual countries were not unique to them alone but affected whole regions of the world. The same could be said of many of the potential solutions to these problems. Neither crops nor their enemies recognise national boundaries. By pooling physical and intellectual resources, and by coordinating the activities of its many research centres, the relatively modest sums that CGIAR has had at its disposal over the past three to four decades have been used very effectively indeed.

The principal objective of CGIAR at its inception was to 'increase the pile of rice', i.e. the food supply, in those tropical countries that faced the most serious scarcity.[228] At first, the highest priority was given to research on cereals. Soon, however, the research portfolio was broadened from rice, wheat, maize, cassava and pasture species to include other crops, such as chickpea, sorghum, potato and the millets, extending eventually to a list of almost thirty commodities. The emphasis on boosting food production brought great benefits to developing countries but, as time went by, other aspects of agricultural development, such as economic and social factors, were also taken into account in the CGIAR mission. This more holistic mission of CGIAR has greatly enhanced its effectiveness in improving the overall lot of the general population of many developing countries. From its rather modest beginnings in 1971, CGIAR gradually expanded to include the coordination of the activities of 15 international agricultural research centres (see Glossary). Each of these centres is an independent institution, with its own charter, international board of trustees, director general and staff, but they act in concert under the aegis of CGIAR to coordinate their research. By the early twenty-first century CGIAR expanded to support 8500 scientists working in more than one hundred countries. One of the great benefits of CGIAR is that the central group and its component research centres are largely independent of the governments of the countries in which they operate. This freedom allows flexibility in research policy and largely insulates its scientists from political and other pressures. In contrast, scientists working in government funded institutes, many of whom are classified as civil servants, can sometimes find it more difficult to focus on research priorities that are not shared by their state paymasters.

In Part II, we have looked at the development of several public sector based models of plant research, crop breeding, and agricultural improvement that were established in the nineteenth and twentieth centuries. These societal developments provided the institutional frameworks within which long-term research and development could flourish. In particular, they allowed plant scientists to translate emerging knowledge from genetics, reproduction and physiology into a practically useful context within crop breeding programmes. As a result, food yields increased steadily, more than keeping pace with population pressures in the industrialised world. In the mid twentieth century, a small group of US philanthropists catalysed the transfer of these crop breeding achievements to the developing world, which was then facing an unprecedented population explosion. This led to the spectacular achievements of the Green Revolution and the establishment in the 1970s of CGIAR and a series of national research centres across the world. In Part III, we will investigate the surprising story of how, while apparently at the peak of its success, this public sector paradigm of plant breeding began to unravel during the 1980s and beyond.

Part III

Turmoil and transition: the legacy of the 1980s

If there were nothing else to trouble us, the fate of the flowers would make us sad.

John Lancaster Spalding (1840–1916)
Aphorisms and Reflections

7

Resurgence of the private sector

Government has laid its hand on health, housing, farming, industry, commerce, education ... But the truth is that outside of its legitimate function, government does nothing as well or as economically as the private sector of the economy. What better example do we have of this than government's involvement in the farm economy over the last thirty years. One-fourth of farming has seen a steady decline in the per capita consumption of everything it produces. That one-fourth is regulated and subsidized by government. In contrast, the three-fourths of farming unregulated and unsubsidized has seen a 21% increase in the per capita consumption of all its produce.

Ronald Reagan (1911–2004)
Campaign speech, 27 October 1964[229]

Introduction

By the 1970s, the public sector was the overwhelmingly dominant force in plant science research and crop breeding throughout the world. In countries like the USA, there had been over a century of such a tradition, and most other industrialised countries had also witnessed many decades of public sector led agricultural improvement. The public sector model was also being embraced with enthusiasm by most developing countries, and the newly established CGIAR network was steadily extending the public-good improvement paradigm across the world. So why is it that, especially since the mid-1980s, we have witnessed such a strong resurgence of private sector involvement in crop improvement? Why is it that the private sector agenda, and especially the agbiotech paradigm, seems to dominate so much of today's plant breeding research and development (R&D)? Where have all the new agribusiness ventures come from, and from where do they derive their expertise in plant science? What has changed about the business model of crop breeding that now makes it so much more attractive, and therefore apparently so profitable (at least potentially) for the private sector? And finally, how is it that individual private companies can possibly compete with the well-established public sector breeding effort, with its complex international infrastructure, its many tens of thousands of scientists, and its decades of technical expertise?

These are questions that have puzzled many of those who work both in basic plant science and in more applied areas of crop improvement. Part of the reason for the emergence of large agribusiness companies can be found in the new political climate that was beginning to emerge in the 1960s, as exemplified in the above quotation from Ronald Reagan. The political aspirations that were alluded to in Reagan's highly influential speech of 1964 eventually came to fruition in the 1980s with the implementation of monetarist economic policies in many major industrialised countries. At this time, there was an emerging groundswell of hostility to much of the public sector as a matter of principle. Doubtless, many companies derived comfort and support from the gradual movement towards deregulation and privatisation of those parts of the economy (such as agriculture) where the public sector still played a large role up to the 1980s. As we shall see, the increasingly benign regulatory environment regarding the private ownership of plant genetic resources after the 1960s also served as a considerable stimulus to private sector growth and the eventual evolution of what we will refer to as 'agbiotech', i.e. commercial use of DNA-based methods, such as transgenesis, as major tools for crop modification. My thesis here is that these socio-economic trends have acted in concert with new scientific developments to create the particular form of agbiotech phenomenon that we see today. Indeed, to a great extent, it is not so much the science of agbiotech that has engendered such hostility as its broader societal context. This is why it is important to explore agbiotech, not merely as a scientific and commercial phenomenon, but also in regard to its broader socio-economic contexts.

The increasingly overt role of the private sector as a major player in plant breeding had a considerable impact on the work of the public sector and on the very nature of plant research. To add to the complexity of this picture, the structure of the private sector itself has also changed dramatically over the past few decades. While there was an initial explosion of small agbiotech startup companies in the 1980s and 1990s, this was followed by a series of mergers and takeovers that resulted in the demise of most small companies and the emergence of an oligopoly of large multinational agbiotech/agrochemical corporations by the turn of the millennium. Ironically, the benign patenting regime that initially encouraged so many small companies into this marketplace has now been complemented by some very restrictive national regulatory regimes governing the testing and cultivation of transgenic crops.

These regulations have made it almost impossibly time consuming, expensive and risky for most small agbiotech companies to commercialise the fruits of their research. The adverse effects of this situation have been compounded by the rejection of transgenic crops by special interest groups in regions such as Europe. While the larger Europe-based agbiotech multinationals, such as Bayer and BASF, can afford to sit out the consumer rejection of GM crops, this is a luxury not afforded to the smaller and more vulnerable companies with their reliance on rapid uptake of their R&D and an uninterrupted cash flow. Moreover, although several small companies were awarded key agbiotech patents in the early days, almost all of these patents had been subsequently acquired by a few large companies via acquisitions and mergers later in the 1990s. In several areas, the existence of these broad patents acted as a further disincentive to the entry of new companies into the marketplace. The end result has been the reinforcement of the big-company oligopoly with the attendant risk that innovation in this area will be stifled in the longer term.

In the meantime, there is also a perception that some of the ways in which private sector breeding has evolved over the past few decades have been regressive rather than progressive, particularly in terms of the wider and more long-term project of worldwide crop improvement. For example, during the 1990s, the sometimes inordinate focus on transgenic technologies impacted negatively on the perception and use of alternative methods of variation enhancement in breeding programmes. Private sector breeding has led to a multiplicity of individual crop varieties, but an overall impoverishment of the genepool of many of the major crops. There has also been a greatly increased focus on the training and resourcing of plant molecular biologists. This in itself is not necessarily a bad thing, but it can have adverse consequences when it comes at the expense of expertise in a more holistic and widely based approach to plant breeding, especially in highly resource-limited developing countries. In the remainder of this chapter, we will look at these issues in an attempt to shed some light on the events of the last couple of decades that have left the world of plant science and crop breeding in such a topsy-turvy state today.

A phoenix reborn

In a sense, private sector involvement in plant research and breeding never quite died, but it did become mightily eclipsed by an overwhelmingly dominant public sector in most industrialised countries early in the twentieth century. There were a few niche areas in which the private sector maintained a prominent presence, such as with hybrid maize in the USA and in the retail seed market in most industrial countries. But virtually all of the research and innovation occurred in the public sector. As we saw in Part II, the public sector steadily expanded its involvement in agricultural R&D throughout the twentieth century, and accelerated this process after World War II. Many of these initiatives were supported, often with funding, by seed companies or other private entrepreneurs who recognised the efficiency gains that could be realised from larger national research efforts. The private sector also recognised that this was a useful mechanism to get state funding for research that they might otherwise be obliged to finance themselves. But there was always a tension in the public sector between basic curiosity-driven, pre-competitive research, and its more practical application for wealth-creation activities, such as developing new commercial varieties.[230] As we have seen, international plant research went truly global with the establishment of the CGIAR network in the 1970s. However, some of these initiatives, such as the establishment of CIMMYT in the 1940s, occurred in parallel with private sector activities in the same region. Hence, the private sector of the early-mid twentieth century was a relatively junior, but nonetheless active, partner in crop improvement that always retained the potential to seize any new opportunities that might arise to re-enter the market in a big way.

We can trace the most recent revival of the private sector as a major player in applied plant research back to the 1960s, and particularly to the introduction of breeders' rights. Prior to this period, the private sector never really saw itself as taking over or substituting for the wide range of activities that were being undertaken by the public sector. Rather than attempting to duplicate this vast portfolio of work, private companies have always sought to focus their efforts on a relatively limited set of new technologies where specific business opportunities have been identified. In the 1920s and 1930s, such opportunities were presented by the hybrid maize varieties, where farmers were obliged to re-purchase seed from the companies each year. Several decades later, in the 1960s, new opportunities arose for the private sector with the enactment of legislation establishing stronger forms of legal protection for new seed varieties.

In the 1980s and 1990s, yet more opportunities came from genetic engineering technologies, whereby transgenic varieties could be granted utility patents, just like mechanical devices. The ability to patent new plant varieties meant that the private inventor of a transgenic variety had a form of legal protection which was much stronger than the 1960s version of plant breeders' rights. In turn, this gave inventors an enhanced means of extracting profit from new plant varieties. The congruence of this new 'high-tech' approach to crop improvement, with the ability to patent the resulting transgenic seed varieties, stimulated much of the private sector renaissance in the agribusiness sector. Between the mid-1980s and the late 1990s, the private sector duly emerged as the dominant force in many aspects of crop research and breeding across the industrialised world. The dominance of the private sector has been especially marked in those crops that are traded as major commodities on world markets. Examples include maize, wheat, soybean, oilseed rape and cotton.[231] For some of these crops, public sector breeding work declined dramatically as the companies expanded their market share.

In the USA, the private sector increased its funding for R&D related to plant breeding almost five-fold between the 1960s and the mid-1990s, while the public sector share declined by three quarters over the same period.[232] Similar shifts in funding were happening across the industrialised world. It should be stressed, however, that the private sector did not suddenly emerge into the crop business after the 1980s, owing to the invention of genetic engineering technology for plants. Rather, there was a gradual rise in private sector interest in crops throughout the twentieth century, but more especially in the period after the 1960s, as legislation was enacted that gave increased protection to breeders of new crop varieties.[233] We will now trace the co-evolution of private sector plant breeding with that of increasingly benign regulatory regimes governing the ownership of plant varieties.

Favourable regulatory environments

Patents and breeders' rights

Ever since the days of Adam Smith, it has been widely believed that the self-interested actions of individuals and companies in a perfectly competitive marketplace can provide optimal levels of consumption and investment. Smith posited that it was the 'invisible hand' of self-interested private individuals or groups that worked in truly competitive marketplaces to ensure optimal production and consumption of goods, plus investment and savings for the future.[234] However, in an ideal market with a completely free flow of information, innovation would eventually become pointless, because there would be nothing to stop competitors from freely copying new products that may have been developed at great cost to the original innovator. In 1954, the economist Joseph Schumpeter noted that perfectly free markets, with no restriction on information flow, are inimical to innovation.[235]

In fact, this had been recognised centuries previously when the first patenting system was established in Europe by late-medieval rulers in order to protect new inventions. The first recorded instance of a formal patent being issued for a new invention was in 1449, when John of Utynam was awarded a 20-year monopoly for a glass-making process previously unknown in England. In return for his guaranteed but temporary monopoly, John of Utynam was required to teach his process to native Englishmen. That same function of passing on information to the public domain is nowadays fulfilled by the publication of a patent specification. The long-term interests of the market are therefore generally regarded as being best served by imposing a limitation on the free flow of information in markets, and by state intervention to protect the rights of innovators to create and exploit new products. In order to get around the problem of monopolies by patent owners, patent protection is nowadays generally limited to a period of 10–15 years (depending on the country). This allows the innovator to extract a worthwhile profit, while also ensuring that the innovation is eventually available for all to copy and use ad libitum.

As we will see, one of the key problems confronting today's newly reconfigured marketplace for crop innovation is that the nature and extent of such legislative interventions are in serious danger of unbalancing and destabilising the entire sector, with adverse impacts on both its public and private components. Private sector breeders have been attempting to gain some sort of patent protection system for their newly developed varieties since as early as 1885.[236] In 1930, the US Congress passed the first Plant Patent Act, but this was restricted to asexually propagating species and excluded most commercial crops.[237] Hence, the Act applied to potatoes, which are one of the few major crops that are normally propagated asexually, but did not apply to wheat, maize, barley, oilseeds, rice, cotton and most of the other important commercial crops that are sexually propagating species.

After World War II, many seed companies started to increase their investment in all aspects of breeding research. They were mindful of the huge potential of the market, as demonstrated by the example of hybrid maize in the USA. Across the industrialised world, agriculture was undergoing enormous structural changes that were to transform it from a labour-intensive, small family operation into a large-scale, capital-intensive corporate enterprise. Seed companies joined with chemical manufacturers of inputs such as fertilisers and pesticides. Part of the reason for this was an attempt by chemical companies to diversify away from petrochemical products in the wake of the oil shocks of the mid-1970s. This gave the newly merged seed/agrochemical companies, such as Monsanto and Dupont, a lot more political clout than the original relatively small-sized seed companies. Because they had already been engaged in laboratory-based R&D on agrochemicals, the new agribusiness combines also had access to the kinds of resources, both financial and scientific, that were necessary to establish their own up-to-date plant science research facilities. Such companies now had an especially useful combination of expertise in biochemical and molecular disciplines, as well as in plant sciences.

However, before serious money could be made from plant breeding, there were two significant barriers to the extraction of significant value from any new varieties that were developed.[238] The first barrier was biological, i.e. the prevention of saving and repropagation of improved seed by farmers. The second was regulatory, i.e. ensuring the maximum level of legal ownership of the seed and its propagules.[239] Surmounting the biological barriers to profitability has proved to be rather formidable. The quest to produce sterile hybrids or inbred-hybrids that effectively preclude farmer saving has involved a long and arduous struggle for private sector breeders. One problem, as we have seen previously, is that these biological attributes can only be manipulated on a crop-by-crop basis. Hence the production of commercial hybrids in maize proved to be relatively easy but has been much more challenging for wheat. Sterility technologies have also been difficult to develop, despite the invention over the past four decades of many ingenious methods, ranging from chemically induced male sterility to transgenically produced seed sterility.[240] In the end, companies marketing transgenic crops have resorted to written contracts that commit farmers to repurchase seed each year for a specified period. Predictably, some growers have ignored their contracts and have saved seed anyway. The pursuit and prosecution of such transgressors by some agbiotech companies has resulted in a great deal of adverse publicity for the industry as a whole.[241] So the biological challenge remains to be satisfactorily resolved.

The second barrier to the extraction of value from transgenic crops, i.e. the regulatory environment, was much easier to overcome than the biological challenges.[242] In essence, it was simply a matter of persuading the right people to enact the right legislation (i.e. lobbying), which is normally a lot simpler than manipulating a complex biological system like plant fertility. The first steps in the extension of legal ownership rights to plant germplasm came in 1961 when six European nations established the Union for the Protection of New Varieties (UPOV).[243] It took the best part of another decade before the US Congress followed suit and approved the Plant Variety Protection Act (PVPA) in 1970.[244] Other industrialised countries passed similar legislation establishing plant breeders' rights in the 1960s and 1970s. This encouraged the entry into the market of several new seed companies, some of which were later to play a part in the development of transgenic crops. However, the wider impact of the PVPA on commercial crop improvement remains controversial to this day. For example, on the one hand, Janis and Kesan argue that 'results indicate that the Plant Variety Protection Act (PVPA) rights are burdensome to acquire, and yet the expected post-issuance licensing and enforcement activities ... are virtually non-existent,'[245] while in contrast Naseem et al. have averred that 'at least for cotton, the PVP Act has served to encourage a greater flow of innovation and the development of more productive cotton varieties'.[246]

The early entry of seed companies into a market that had hitherto been dominated by public research institutes was most marked in the Netherlands, which introduced Plant Variety Rights soon after the end of World War II. This led to the formation of a plethora of successful seed companies, some of which still dominate much of private sector breeding for vegetables and grasses. Examples include Rijk Zwaan and Royal Sluis (vegetables), Mommersteeg and Van der Have (grasses) and Joordens (grass and forage crops). The Dutch were then followed by the UK, which established some of the strongest forms of plant breeders' rights legislation during the 1960s.[247] There is good evidence that the introduction of plant breeders' rights in Europe has increased the incentives for private companies to develop new varieties, although these studies did not look at the effects of these new varieties on actual crop yields.[248]

By the 1970s, US companies were also beginning to ramp up their investment in plant breeding. From the late 1960s to the mid-1980s, the annual private sector expenditure on plant breeding in the USA increased from about $20 million to over $200 million.[249] Even at this early stage of the growth of the private sector, there was a wave of acquisitions and mergers in the seed industry, which was commented on at the time by critics of the new legislation on breeders' rights.[250] But these consolidations were relatively small affairs when compared with what was to come in the 1990s. During the 1960s and 1970s, private sector involvement in plant breeding was still relatively small compared to the large public commitment, although the former was growing steadily. For example, in 1978, the public sector in the USA was still producing the overwhelming majority of commercial varieties of important commodity crops such as soybeans (89%), wheat (86%), barley (95%), oats (86%) and rice (92%).[251]

Prior to the 1970s and 1980s, most of the innovation in crop improvement occurred outside commercial marketplaces and relatively little value was extracted from the direct products, i.e. the new varieties of seeds or other propagules. Since the 1980s, there has been an increasing tendency for governments in industrial countries to encourage and facilitate greater private sector involvement in crop improvement. This soon went well beyond plant breeders' rights and resulted in the extension of the far more valuable utility patent protection to plants in the USA in 1985. Europe eventually followed suit in 1999 when the European Patenting Office decided to grant patents on transgenic crops.[252] Ownership of this kind of patent protection gave companies a much more restrictive set of rights, and hence more scope to commercialise any resulting plants, than did the earlier legislation on plant breeders' rights.

In other initiatives to stimulate the private sector, governments have also provided agbiotech companies with free or low-cost access to public research and infrastructure.[253] In some cases, companies seeking to do research into new crop-related products have even been directly funded by governments and/or have been provided with compliant public sector partners to stimulate such research[254] (see Chapter 19). Such government initiatives, totalling many hundreds of millions of dollars, have tended to favour what are perceived as 'high-tech' approaches to crop improvement, e.g. transgenesis, which are seen as intrinsically innovative. In contrast, other crop improvement strategies are sometimes perceived as less 'innovative' and therefore less worthy of state support in sponsored commercial projects, regardless of their real-world prospects for crop improvement.

The effect of these tendencies has been to skew both scientific and commercial attention towards the favoured 'high-tech' methods, without any serious analysis of their real impact or value for taxpayer dollars in the longer term. These structural factors in state regulation and funding support have had the combined effect of favouring the success of the earlier and larger private sector entrants into the crop improvement sector. In particular, earlier entrants were often able to establish dominant ownership portfolios over key technologies, especially in relation to the development of transgenic crops, which then inhibited the entry of newer participants, both public and private, into the marketplace. Most of the smaller companies soon found that the costs of moving from discovery to production were beyond their means, both financially and in terms of business expertise, while the risk of failure has become ever higher. For example, the cost of producing a new transgenic crop variety is estimated at over $1.5 million, while the probability of getting to market is relatively low. Regulatory costs add a further $1–5 million and the process takes a minimum of three years. These costs are beyond the reach of most small companies.[255] Even if a small agbiotech company gets a product to market, it can still fail owing to a lack of wider forms of scientific or business expertise. A classic case of this was the FlavrSavr® disaster that destroyed Calgene as an independent enterprise in the 1990s. Calgene had some good expertise in laboratory-based biotechnology, but was let down by deficiencies in its conventional breeding and business acumen.[256]

It is ironic that some of the more convoluted regulatory hurdles that now hobble the smaller agbiotech companies came about in the early 1990s following discussions between US government agencies and private companies like Calgene and Monsanto. The government had sought to provide a framework for the commercialisation of the first transgenic crops that were about to be released by these companies. At that time, there was an inclination among many in government to treat transgenic varieties in the same way as those produced by other modern genetic modification technologies, such as radiation mutagenesis or wide hybridisation. In the end, it was largely at the insistence of Calgene scientists, who wanted their new transgenic crop to be as distinctive as possible, that this class of crops came to be treated so differently. As a result, a series of increasingly confusing, expensive, and lengthy approval processes for transgenic varieties have become established across the world. In the USA, many of these regulations evolved in an ad hoc manner during the early 1990s and have been criticised ever since for their lack of coherence, their inconsistency, and the spread of responsibility for their implementation across no fewer than three separate federal agencies.[257]

According to a recent report from the respected Pew Initiative, the current regulatory environment acts not only to inhibit the commercialisation activities of smaller companies, it also impedes more basic field research by university scientists.[258] Similar conclusions have also been reached by numerous agricultural economists.[259] In this context of burdensome regulation, it is also worth considering that the most serious cases of transgression of the regulations, namely the StarLink scandal and the more recent Bt10 affair (see Chapter 13), were due to errors and poor management in the commercial arena rather than due to any failures in the regulatory system itself. In other words, the stringency, or lack of it, in a regulatory regime is no guarantee of its efficacy, unless companies acting under such a regime take the trouble to implement adequate management controls over their products and supply chains. Meanwhile, the burden of the legislation that uniquely falls on this particular crop improvement technology only serves to reinforce the current quasi-monopoly and discourage diversity and innovation in the marketplace.[260] There is also evidence that the comparatively limited opportunities for researchers to work with patented plant varieties, especially in the USA, acts to restrict public sector access to germplasm for future breeding.[261]

The realisation that small companies would face significant hurdles in commercialising their wares undoubtedly contributed to the rash of acquisitions and mergers in the mid-late 1990s. This 'feeding frenzy' (as it was described at the time[262]) dramatically reduced the number of companies and also greatly increased the market dominance of the few remaining players. This led to an increasing perception in the industry of significant barriers to the entry of new companies into the marketplace. Another disincentive to innovation has come from the tendency of patenting bodies to approve patents that have inappropriately broad, and sometimes overlapping, claims. As we will see, this practice has tended to favour early entrants into the market, and/or the tiny number of large companies who could afford to buy up the original owners of such excessively broad patents. Many of the latter were small research-based companies, such as Calgene (USA) and Plant Genetic Systems (Belgium).

The problem of broad claims

It may have been excusable, although still regrettable, that patent granting agencies were relatively lax in their approval of biotech related claims during the early years of the biotech era. After all, patent reviewers of that era rarely had a comprehensive training in biology, and often had little or no expertise in the technicalities of the new molecular genetics. Added to this lack of expertise, there was a febrile atmosphere abroad during the late 1980s and early 1990s that was replete with hyperbolic expectations for the prospects of agbiotech. This meant that the relatively naïve patent clerks and attorneys, and even many scientific assessors, probably believed many of the inflated claims that were often then duly approved. However, it is troubling to note that more than twenty years after the first commercial agbiotech patents were granted, and the more recent dawning of a new hard-headed reality about the real potential of biotechnology in general, the problem of the inappropriate award of claims has certainly not gone away. This was demonstrated by data from a review of 75 gene-based patents from the USA that was published in 2005.[263] In this study, the authors found that more than one third of these biotech patents did not satisfy the legal requirements of novelty, utility and non-obviousness. Another recent study of human genome patenting has revealed that almost 20% of the human genes have been explicitly subject to intellectual property rights (IPR) claims in the USA.[264] Moreover, some genes are the subject of as many as 20 separate patents asserting various rights from different organisations and companies. This is a situation that may lead to 'high costs on future innovators and underuse of genomic resources'.[265]

The 75 gene-based patents examined in the above-mentioned US study all related to biomedicine rather than agbiotech. Biomedicine is a field in which patenting has been extremely active since the early 1990s, and therefore large numbers of case studies and precedents have accumulated. There have also been many more commercial successes and a broader diffusion of biotech products in the case of biomedicine than in agbiotech. However, in the case of agbiotech, the size of the patenting community is much smaller than in biomedicine. This means that there has been less demand for the training of suitably qualified patenting experts with expertise in plant molecular biology. It is therefore likely that the problems recently highlighted with biomedical patents are even more serious in the case of patents related to agbiotech. For example, as stated in a report by the Nuffield Council on Bioethics: 'We recommend that national patent offices, the EPO (European Patent Office) and the WIPO (World Intellectual Property Organization) draw up new guidelines for patent offices to discourage the over-generous granting of patents with broad claims that have become a feature of both plant and other areas of biotechnology.'[266]

As highlighted in several recent publications, the situation regarding over-broad patent claims and lack of access to IPR seems to have improved relatively little since the Nuffield report was published in 1999.[267] In addition, the concentration within a few large companies of broadly defined intellectual property rights relating to key areas of transgene technology is not only a disincentive to private sector innovation and exploitation; it also serves to exclude the public sector from a much greater involvement in this facet of crop improvement. The difficulty in realising such commercial opportunities is one of the factors that has driven many plant scientists away from applied research and towards more basic, academically orientated studies, as discussed in Chapter 10. However, it is not all bad news on the patent front. In 2005, the US Court of Appeals made a landmark ruling that specific plant DNA sequences of as-yet unknown function (called expressed sequence tags or ESTs) could not be patented by Monsanto.[268] This ruling will affect over one hundred other agbiotech and biomedical patent applications that are also based on ESTs. In addition to these regulatory developments, as we will see in Chapter 17, there have been several recent scientific advances that may help to alleviate the current anti-competitive situation of much of agbiotech. These new initiatives include the development of open access technologies for production of transgenic plants, and the establishment of public consortia that patent and exploit their own proprietary crop improvement technologies.

Startups and multinationals

During the period from 1980 to 1995, the combination of potentially powerful transgenic technologies and an especially benign regulatory climate, encouraged several types of private sector entrants into this newly attractive marketplace. Broadly speaking, two distinctive types of company were initially interested in the commercial application of plant breeding. The first was the large, well-established firm that was typically already in the agricultural business as a chemical manu-facturer or a seed supplier.[269] The second type was a new breed of small startup company that was largely research led, rather than product orientated. Such small companies were often spinoffs from university laboratories and tended to focus on developing transgenic varieties of mainstream crops. The smaller biotech companies were often technologically speaking very innovative but less skilled in plant breeding and near-market aspects of product development.[270]

Many of those entrepreneurial scientists who founded many of the dozens of first-generation small agbiotech startup companies in the 1980s were fuelled by the beguiling prospect of the relatively facile isolation of dozens of key genes that could supposedly be used to manipulate key agronomic traits in crops. Having isolated such genes, the new ability to patent their use in crops promised to allow a company either to develop the new crop varieties itself or, as was much more likely, to sell or license their use to one of the large multinational seed companies. Although this vision seemed straightforward enough at the time, and despite the heavily hyped prospects of a quantum leap forward in crop improvement, agbiotech companies have found it difficult to live up to these aspirations. During the 1990s, both small and large agbiotech companies struggled to find a viable business model for the application of their technology in the marketplace.

The first commercial transgenic crop to be approved, the FlavrSavr® tomato in 1992, was developed by the small agbiotech company Calgene but this was a com-mercial (and plant breeding) failure.[271] The subsequent demise of Calgene signalled the end of the brief entry into the crop improvement business of small, independent companies. As we will see in Chapter 17, there are still many small companies in the agbiotech marketplace, but these tend to be niche focused service providers to larger companies, rather than directly involved in crop development itself. Since the mid-1990s, a new breed of large multinational agrochemical/seed company has dominated the rather narrowly based market for transgenic crops. The secret of the success (such as it is) of these multinationals in commercialising the first generation of mainstream transgenic crops has been the development of a simple but effective business model. This model is based on the use of genetically straightforward traits that normally involve single genes, plus the added value of coupling these traits with agrochemical inputs that can be tied together into a single package, for sale with the transgenic seed.

This strategy was possible because, thanks to the favourable IPR legislation in the mid-1980s, US companies could continue to own the rights to their transgenic seed, even after it had been sold to farmers. Hence, companies could enforce exclusive sales contracts for the purchase of a package of [GM-seed + chemicals] that prevented farmers from saving and replanting seed. These developments opened up the alluring prospect of significant new value creation in the seeds business, that would be combined with its guaranteed continued recapture from customers over an extended period. This surely represents the ultimate in the encouragement of brand loyalty among customers. Farmers are now legally obliged to remain loyal to the brand for several years and some of the companies have not been chary about prosecuting malefactors among their customers.[272]

The large multinationals have also been favoured by the fact that herbicide tolerance was one of the earliest commercially developed transgenic crop traits. Although the choice of herbicide tolerance was largely dictated by technical considerations (the traits were controlled by single genes and these genes were readily available to company scientists), it was also fortuitous for business reasons. This was because farmers growing such crops were also obliged to purchase a proprietary herbicide (e.g. Roundup) that was manufactured by the same company (e.g. Monsanto) that sold them the transgenic seed.[273] Hence, the company was able to sell a bio/chemical package of [seed + herbicide] under contractual terms that required the farmer to continue purchasing this particular package for a specified period of years.[274] Another consideration that influenced the decision of Monsanto to base its initial agbiotech strategy so heavily on Roundup resistant crop varieties was that this hitherto profitable herbicide was due to come off patent in 2001. Not only would the agbiotech earnings from seeds help the company to diversify away from the chemicals side of its business, but also the compulsory package of [seed + herbicide] to be sold to farmers would guarantee many more years of sales of its otherwise non-proprietary (and therefore generic and cheaper) Roundup herbicide.

Of course, the farmers who bought into this technology package also benefited. Otherwise they would not have opted in their many tens of thousands to tie themselves to a single seed supplier for several years at a time. Farmers using transgenic seeds gained by yield increases and simpler management practices that translated into additional profits. However, the agrochemical/seed companies who developed these [transgenic seed + herbicide] packages have been by far the greatest beneficiaries in terms of additional profitability. According to a study of transgenic herbicide tolerant canola in Canada, the additional profitability from using the [seed + herbicide] packages is distributed as follows: 57% to the supplier company; 29% to the farmer; and a mere 17% to the rest of the supply chain. The latter figure includes food processors, wholesalers, retailers and eventually the consumer (e.g. the supermarket shopper).[275] Because the consumer is at the very end of a lengthy and convoluted supply chain s/he shares little or none of the profitability of such transgenic crops.

In retrospect, one of the most crucial misjudgements made by agbiotech companies when they implemented the sale of such transgenic seeds was to focus on farmers as the sole customers for their seeds. This ignored the fact that the farmers were not end-users of the seeds but merely one more link in a complex supply chain that was potentially vulnerable to disruption by third parties. The fact that two huge and influential groups of stakeholders in this supply chain, namely supermarkets and consumers, accrued virtually none of the benefits of transgene technology may explain why so many of them have been rather wary about embracing it, especially when they were faced with alleged potential risks, e.g. the Pusztai affair of 1998–1999 in the UK, however farfetched these health risks have since proved to be (see Chapter 17).[276] In the next chapter, we will examine the consequences of the private sector renaissance, and agbiotech, for plant breeding in general.

8

Emergence of a new crop improvement paradigm

Crop improvements like these can help provide an abundant, healthful food supply and protect our environment for future generations.

Monsanto website, 2005

Introduction

We have seen that the major driving force behind the massive private sector expansion into crop development of the 1980s and 1990s was the development of transgenic crops. Unlike other types of crop, transgenic varieties could be protected via the utility patent route, which gave a much more powerful form of ownership than plant breeders' rights.[277] Companies who wished to develop transgenic crops were further assisted by a relatively lax patenting regime, especially before 1995. During this period, many patents were granted that, even at the time, were recognised as being of inordinate breadth in the scope of their claims.[278] Therefore, the emergence of the private sector as the dominant player in crop breeding was stimulated by the conjunction of new legislation and new technologies, the combination of which allowed companies to develop potentially lucrative business models in a hitherto rather unprofitable area of agricultural commerce.[279] The much trumpeted entry of the private sector en masse into the marketplace for crop improvement came at a time when many governments in industrialised countries were seeking to shed much of their public sector enterprises via the mechanism of privatisation. We will examine this topic in more detail, especially in regard to UK plant breeding research, in the next chapter.

Meanwhile, it is germane to look at how the privatisation of public sector functions coincided with apparent emergence of a viable alternative paradigm for crop improvement in the shape of the agbiotech dominated private sector. To many government policymakers, it seemed that these exciting new private sector ventures were delivering advanced and innovative mechanisms for crop improvement, which were based on the latest in high-tech molecular biology. In contrast, public sector breeding seemed like a relatively outdated, cumbersome, expensive and unnecessary function that would be better sold off to these dynamic, modern private enterprises. Government organisations involved in implementing the privatisations of the 1980s and 1990s did not appreciate that private sector firms had neither the capacity nor the desire to assume all of the functions of the institutions that they were purchasing. Rather, companies sought to acquire access to high-quality breeding lines from the public laboratories, into which they could insert their own proprietary genes of interest. The key expertise was deemed to be the initial gene manipulation steps, which could be done in-house by relatively small teams of molecular biologists working in secure company laboratories. The downstream breeding could then be subcontracted to specialist seed companies, many of which were bought up by the major agbiotech giants in the early 1990s.

This led to the entirely new perception that a single form of creating novel genetic variation, i.e. transgenesis, was more or less the be-all and end-all of crop improvement. As we saw in Chapter 3, this viewpoint is very far from an accurate version of the reality of practical breeding. Nevertheless, genetic engineering soon acquired a unique cachet that seduced scientists, entrepreneurs, investors, governments and even those who opposed it, with its promise of virtually unlimited power to transform agriculture. During the 1990s, many people believed that the newly ascendant private sector phoenix was fully capable of both carrying out basic research, and applying innovative techniques to improve crops, in ways that were even more effective than those used previously by public sector institutions. Indeed, over the decade from the mid-1980s to the mid-1990s, the US private sector increased its spending on plant breeding, from an annual $200 million to over $500 million.[280]

Many younger biologists were duly attracted into the glamorous new arena of private sector research. They were often lured by tempting packages of high salaries, high-status science, getting their names on valuable patents, and the prospect of share options or other lucrative rewards in the case of commercial success. Although one or two shareholder-scientists went on to make millions of dollars from agbiotech, such people were extreme rarities. Moreover, their financial fortune was almost never as a result of successful commercialisation of any genetically engineered crops they had developed. Rather it was due to their companies being bought up by larger multinationals, often at wildly inflated prices, in the unreal acquisitions mania of the late 1990s. At times it seemed as if the business plans of some small startup companies consisted mainly of building up a strong IPR portfolio based on patented genes in as short a time as possible. Well-publicised descriptions of these newly isolated genes and their potential commercial uses could then be used to leverage more venture capital or other investor funding.[281] Eventually the value of the company might have been talked up to such a degree that multinationals became interested in buying them out, much to the profit of the original shareholders.

An alternative way for the original investors to bow out with a decent profit was for the company to go public via an IPO, or initial public offering, of shares. A few scientists became instant millionaires in this manner, but it is interesting that hardly any of their 'wonder' genes seem to have seen the commercial light of day. Unfortunately, many agbiotech startups went bust, and even in those that survived there was a huge attrition rate among the scientific staff. A single downturn in the share price of a small company might lead to redundancy of many of its research staff. On more than one occasion, former colleagues of mine found themselves being escorted to the door of their company laboratory, with only a few minutes notice. This grim jungle of ultra short-term and narrowly focused research, within an insecure and ever changing work environment, was the reality of the brave new world that confronted many molecular biology graduates as the 1990s unfolded.

Given the hype that surrounded genetic engineering and agbiotech in the late 1980s, and well into the 1990s, it was quite natural that many company researchers tended to focus on modern molecular-based technologies for crop improvement. This was very much at the expense of work on the relatively unglamorous and unprofitable (because they could not be so readily patented) traditional breeding techniques. During the 1990s, transgenic crop technology was hyped up by everybody, from university scientists anxious for research funding to company PR staff in search of venture capital. Even today, many of the agbiotech companies still tell us that their products will usher in a new Green Revolution. Some companies go even further and say that the only way of averting global famine over the next fifty years is to embrace their technology and, by extension, their business model and their products. Because of their importance for the future of plant breeding, the evidence for these particular claims will be examined in some detail, and then largely discounted, in Chapters 14 and 15.

Meanwhile, in the late 1990s, transgenic technology was successfully applied (after a fashion) to just four major North American commodity crops, i.e. maize, soybean, cotton and canola. In some cases, the commercial application of agbiotech involved the displacement of public sector breeders who had previously developed most or all of the new varieties of a particular crop. To see the effects of these changes on a specific crop species, we can consider the case of oilseed rape in Canada.[282] As we saw in Chapter 3, a highly improved edible form of oilseed rape, called canola, had originally been bred in the public sector during the 1960s. For the next few decades, public research and breeding centres, such as Agriculture Canada, produced new varieties of canola at the rate of about one every two years. The new cultivars were produced using publicly available, non-proprietary technologies. Over this period, there was no private sector involvement in canola breeding. Things began to change in canola breeding in the early 1980s. In particular, between 1982 and 1997, private companies moved aggressively into the canola business, using proprietary technologies like transgenesis.[283] By 1996, private companies were producing three quarters of all new canola varieties and these were now introduced at the rate of more than eight new varieties per year.[284]

One might question whether Canadian farmers have really benefited from their access to 16-fold more varieties than previously, especially given the great similarity between some of these newer varieties. One example would be the number of herbicide resistant varieties, both transgenic and non-transgenic. Then there is the question of what has been alleged to be a process of 'planned obsolescence' in new crop varieties (see below). Most of the Canadian canola crop breeding effort is now private sector owned, and is focused heavily on proprietary transgenic varieties. Despite this apparent dominance of the private sector/transgenesis paradigm for canola improvement, it is noteworthy that the much reduced, but surprisingly effective, activity of public sector breeders in Canada is still responsible for a great deal of innovation in some key agronomic traits of longer term interest, such as salt tolerance and fungal resistance. Indeed, it is questionable whether canola could survive for very long as a competitive, high-yielding crop in the absence of this kind of relatively unglamorous, Cinderella-like public sector breeding programme.

Obsolescence and impoverishment

One consequence of increased private sector dominance in plant breeding programmes has been the emergence of what has been controversially termed 'planned obsolescence' in the production of many new seed varieties.[285] This phenomenon has been described in several major crops in which the private sector assumed a prominent position after the 1960s. Probably the best-researched case is that of wheat breeding in the UK.[286] Private seed companies moved into the wheat market to a much greater extent in the 1960s, following the enactment of favourable new legislation (i.e. plant breeders' rights, as discussed above) that gave them much greater protection for new seed varieties. As the companies increased their market share, the number of newly released wheat varieties rose steadily, from ten per year in the late 1960s to about 34 per year by the late 1990s.[287] Over the same period, the average lifespan of an individual wheat variety in the UK declined sharply, from 13 years to about 5.5 years. Whether such obsolescence was really planned, or simply happened as a by-product of breeding practices has been disputed, although I would favour the latter conclusion.[288]

Something very similar happened with Canadian canola, where the average life-span of a typical cultivar was about ten years during the early 1980s, when canola breeding was dominated by the public sector.[289] By 1995, with private companies developing three quarters of all new varieties, the average lifespan of a canola cultivar had fallen to a mere three years. There is also evidence from the USA that, despite their supposedly greater efficiency, private sector breeders have only been able to achieve the same levels of yield improvement as the public breeding pro-grammes. Evidence of the relatively poor performance of private sector breeders, in the study by Alston and Venner,[290] makes depressing reading for those who champion a 'lean and mean' private sector in contrast to a 'bloated and bureaucratic' public sector. Of course, both views are caricatures, but the study does show that plant breeding is a complex process that is not necessarily amenable to the same sort of economic analysis as other industries such as steel manufacture or coal mining.

In addition to more rapid turnover of new seed varieties, there is evidence that the breeding of commercial varieties of wheat has been accompanied by a narrowing of the genetic resistance to fungal disease.[291] This would not necessarily have been a concern to commercial breeders because they knew that farmers could rely on a large suite of newly developed chemical fungicides. Indeed, in some cases where the breeder had a connection with the agrochemical company supplying the fungicides, it might well be in their mutual interest to de-emphasise genetic resistance to fungi as agronomic desirable traits in a breeding programme.[292] While certain fungal diseases of crops can be controlled adequately by application of chemical fungicides, it is generally accepted by breeders that the development of genetic resistance to such diseases is a superior strategy, especially in the long term.[293] It was the development of genetic resistance to fungal disease in wheat by Biffen in 1910 (see Chapter 5) that was one of the first major successes of modern scientific breeding in the UK. The average susceptibility of UK wheat to fungal disease increased sharply in the late 1970s and early 1980s as more commercial varieties with reduced genetic resistance came to dominate the market.[294]

In contrast, public sector breeders, e.g. at PBI in the UK, had always consciously selected crop varieties with relatively broad ranges of genetically based disease resistance. This meant that the public varieties were more durable in the long term and required less protection by fungicides. Two of the consequences of the new commercial regime were to encourage the increasing use of chemical inputs, and to discourage seed saving as varieties became obsolete more rapidly. These two factors worked to the advantage of the private sector, and were especially useful to com-panies that supplied both agrochemicals and seed to farmers. It is probably no coincidence, therefore, that it was the large agrochemical/seed combines, such as Monsanto, DuPont and Syngenta that dominated much of the seed industry by the early twenty-first century. The very same companies have also ended up as the major developers of transgenic crop varieties, and as the principal proponents of this new paradigm for crop improvement.

Genetic obsolescence normally only affects a few high-profile traits, such as disease resistance, but there is also growing international concern amongst many plant scientists about more generalised varietal erosion as a consequence of virtually all forms of modern plant breeding.[295] This is because we can also observe a very distinct narrowing of overall genetic diversity in the major commercial lines of some of our most important food staples, such as maize, which has been directly linked with serious outbreaks of disease and consequent loss of crop production. Thanks to its well-developed inbred-hybrid system, maize is one of the most highly commercialised and profitable of the major commodity crops. But commercial maize is also one of the least genetically diverse of our major crops. For example, in the 1940s, there were over 100 000 inbred lines of maize available from public and private sector breeders in the USA, but only 40 were used commercially.

By the 1970s, only a dozen lines accounted for the entire national area of hybrid maize. The adverse consequences of a narrow genetic base were demonstrated in the early 1970s, when much of the US maize crop was stricken by the Southern corn leaf blight, *Helminthosporium maydis*. The highly inbred maize varieties had little or no genetic resistance disease to this virulent fungus. Maize crops across the country were devastated, and there was even an outbreak of panic on the stock exchange, as investors tried to buy up remaining crop stocks and futures.[296] The unusually severe nature of this disease episode had been exacerbated by the fact that 85% of the US maize crop contained Texas male-sterile cytoplasm, which made the plants uniquely susceptible to the leaf blight fungus.[297] A subsequent report by the National Academy of Sciences concluded that most US crops were 'impressively uniform genetically and impressively vulnerable'.[298]

Genetic erosion in the major commercial crops has been ongoing ever since highly inbred varieties began to be adopted in preference to local landraces in the early twentieth century. This process is therefore common to both public and private sector breeding programmes. Just to give a flavour of the huge losses in crop germplasm resources over the past century or so, we can consider the percentage loss of varieties of some common fruit and vegetable crops held in 1983 at the US National Seed Storage Laboratory, as compared with the numbers present in 1903.[299] In the 80 years between 1903 and 1983, we lost 97.8% of asparagus varieties, 94.5% of broad beans, 94.2% of radish, 94.1% of beet, 94.1% of onions, 93.6% of spinach, 92.7% of carrots, 92.6% of lettuce, and so on. Losses have continued steadily in the years since 1983, so it is likely that, for many of our most useful crops, we have lost well over 95% of varieties over the past century.[300] The situation is not always as serious as these figures might imply, and such varietal losses are not necessarily the looming catastrophe that some commentators claim. Indeed, breeders can and do cope with some startling extremes of genetic impoverishment in a few crop plants that are vegetatively propagated. In addition, as discussed in Chapter 16, many varieties have been saved by storage in public seed banks.[301]

Probably the most extreme example of a genetically impoverished crop is the commercial banana, *Musa acuminata*. Virtually all the commercially traded bananas in the world consist of a single triploid clonal variety, called Cavendish.[302] This means that all commercial bananas, i.e. the sort that you and I might buy in a store or supermarket, are not just extremely inbred, they are genetically identical to each other. Being triploid, these plants are also sterile and must be propagated vegetatively, which makes it very difficult for breeders to introduce any additional genetic variation into commercial bananas.[303] It would take just one opportunistic pathogen to wipe out the entire global crop, as happened in the 1930s with the previous commercial variety, Gros Michel, after it was attacked by the fungus *Fusarium oxysporum*. Bananas are, admittedly, a special case. No other crop has such a narrow genetic base and, even in the worst-case scenario, the demise of the banana is hardly likely to result in mass starvation. However, as I have repeatedly pointed out in earlier chapters, genetic variation is the cornerstone of crop improvement so its unnecessary loss from any crop is to be regretted.

While private sector breeding does not carry sole responsibility for the dramatic genetic erosion that has occurred over the past century, it has certainly accelerated this trend, both wittingly and unwittingly. To instil a mild note of optimism here, it should be noted that reduction in genetic diversity is neither inevitable nor irreversible in breeding programmes. A recent study of public sector wheat breeding programmes at CIMMYT showed that the diversity of breeding lines declined greatly from 1950 to 1989, but increased again from 1990 to 1999, as novel genetic material was introgressed into the population. Significantly, the best sources of genetic improvement were local landrace cultivars and wild relatives, such as *Triticum tauschii*, collected from centres of origin in the Near East.[304] So, while all is far from lost as regards genetic impoverishment, in many cases, the solution is much more likely to be found, not from the sort of expensive high-tech transgenic varieties currently on offer, but rather from a few scrubby wild relatives eking out a precarious existence in the steppes of Central Asia or in the Andean Altiplano.

Effects on breeding programmes

As we have seen, the private sector may sometimes have different objectives in its plant breeding programmes compared with the public sector. This should not be surprising; after all, private sector breeders have different timescales of operation and different priorities from their public sector colleagues. In reviewing the operation of the private sector over the past six decades or so, we can now come to some conclusions about these differences, and their possible consequences for plant breeding. In particular, we can discern three general tendencies in the way that commercial pressures have affected breeding programmes in the private sector. These tendencies are genetic erosion, adapting crops to commercial agronomy, and a technology focus based on profitability rather than long-term utility and/or sustainability. We will now briefly summarise the nature and effects of these three factors.

Genetic erosion

In brief, we can say that private sector breeding tends to focus on technologies of crop improvement that favour commercial exploitation of the new crop varieties, rather than being necessarily the best way of producing such varieties. This has resulted in a decline in the number of truly differentiated varieties of all the major commercial crops. Such a decline had been occurring to a limited extent anyway throughout the twentieth century, but was greatly accelerated post-1970.[305] Paradoxically, the very success of breeding widely adaptable varieties leads to the problem of genetic erosion once they are widely adopted across whole countries or even across whole continents.[306] In the words of US wheat breeder V. A. Johnson: 'The convergence on a narrow set of improved and well-adapted genetic material is a reflection of commercial breeding programmes being predominantly concerned with short term varietal development rather than long-term goals of germplasm development.'[307]It should be noted here that the decline in genetic diversity in the major commercial crop species significantly predated the advent of agbiotech and transgenesis. The potential impact of the widespread adoption of transgenic crops per se on genetic diversity remains unclear and will be influenced more by policies adopted by the private sector in general than by the transgenic status of individual crop varieties.[308]

Adapting crops to commercial agronomy

This process involves the strong coupling of varietal selection with agronomic factors, such as inputs or mechanisation, rather than focusing on traits like pathogen resistance or crop quality.[309] The results of this second private sector tendency can be seen in the way that crops have been selected specifically to be responsive to fertilisers and other inputs. This sort of selection also applies to traits such as machine harvestability, as was noted in the 1960s: 'Machines are not made to harvest crops; in reality, crops are designed to be harvested by machines.'[310] The process of adapting crops to suit harvesting machinery dates back to the 1920s, when the ubiquitous Henry Wallace proposed to a manufacturer of harvesting equipment that he would develop a 'stiff-stalked, strong-rooted hybrid' variety of maize to suit his particular machinery.[311] But things really took off in the 1940s and 1950s as mechanical harvesters virtually replaced draught animals in industrialised countries. Henceforth, new crop varieties were designed by breeders to cope with the limitations of existing harvesting and processing machinery, rather than the other way around.

This shift in breeding objectives applied to all crops and in some cases led to the breeding of new and dramatically different versions of once familiar crops. For example, the fruit structure of sugar beet has been altered to avoid the requirement for hand thinning and to facilitate automated harvesting. In the 1950s, this entailed the development of monogerm-seeded sugar beet plants that did not need the manual thinning required by conventional mutltigerm varieties, allowing for complete mechanisation of production.[312] In the mid-1960s, thick-walled tomatoes were bred at UC Davis to adapt the fruits for mechanical harvesting and prolonged transportation and storage of this hitherto soft, handpicked crop. The harvesting machinery was also co-developed at UC Davis. As a result, Californian tomatoes are exceptionally thick walled, robust and relatively tasteless.[313] This technology-created problem led to the quest for a new-technology solution in the 1990s when agbiotech company Calgene, also based in Davis, California, developed its transgenic FlavrSavr® tomatoes, designed not only to retain the thick-walled trait for easy harvest, but also to recapture just a hint of how American tomatoes used to taste in bygone years.[314]

A technology focus based on short-term profitability rather than long-term utility

This final tendency from private sector breeding relates to an issue that was raised briefly in Chapter 2 with regard to the success of hybrid maize. Very similar arguments can be applied to the more recent case of transgenic technologies. In both cases, it seems that there has been a strong focus by commercial breeders on a specific form of improvement technology, namely inbred-hybrid development in the 1920s and transgenesis in the 1990s. Both technologies lent themselves to commercial exploitation, which led to their adoption as the breeding methods of choice in private sector programmes. The problem with both technologies is that their apparent success may simply be due to their use, almost to the exclusion of alternative (and possibly better) methods. For example, as with transgenesis today, the earlier success with inbred-hybrid technology in the 1920s did not mean that it was necessarily superior to, or even as good as, selection programmes based on the use of open-pollinated maize varieties. It just so happened that most of the funding and research effort at the time went into the hybrid approach, to the great detriment of investment in alternative approaches.

As we have seen, hybrid maize varieties cannot be saved for replanting by farmers, and it was this attribute, rather than the inherent advantages of the technology as a way of breeding better crops, that determined their choice by the seed industry.[315] This point was well expressed in a classic text by noted UK geneticist, Norman Simmonds: 'the undoubted practical success of hybrid maize is irrelevant; a huge effort has gone into it over a period of nearly sixty years; ... [whereas in outbreeding varieties] ... population improvement is recent in origin and small in scale in comparison, yet it is evidently capable of rates of advance which are at least comparable and may well be achieved more cheaply.'[316] Harvard geneticist Richard Lewontin made the same point even more forcefully in 1982: 'Since the 1930s, immense effort has been put into getting better and better hybrids. Virtually no one has tried to improve the open-pollinated varieties, although scientific evidence shows that if the same effort had been put into such varieties, they would be as good or better than hybrids by now.'[317]

Lewontin, in another (admittedly polemical) text, has also asserted that: 'hybrids open up enormous profits for private enterprises and for this reason all efforts were shifted to the new technique.'[318] However, what seems more likely, and more reflects the opinion of most plant breeders who are close to the subject, is that the development of inbred-hybrid technology in maize happened to come along at a time when the alternative methods were underperforming. In the words of Simmonds, development of hybrid maize was more probably as a result of: 'the historical accident that ideas on hybrids developed at just the time at which progress by other methods seemed poor.'[319] After all, the technology was invented by public sector breeders and was vigorously promoted by them; it also then took well over a decade before the technology became commercially successful. It was during this period that one could say that breeders, especially in the public sector, may have become overly fascinated by the technology to the exclusion of other approaches. But to ascribe the entire process to the private sector, as Lewontin does, is surely to credit the latter with far too much prescience.

One could make a similar point about the cost-benefit equations for transgenic technology as a mechanism for crop improvement over the past twenty years. For those real-world plant breeding objectives that are most relevant to most farmers, such as pest and disease resistance, soil salinisation, yield gaps, and domestication-related traits, it is indeed a moot point whether the transgenic strategy is necessarily the cheapest, quickest or the most effective route to success. It is, however, the easiest to commercialise because of the utility patent protection that is afforded to transgenic varieties, but not to varieties produced by what might be equally innovative breeding technologies. We will come back to explore this topic in more detail in Parts IV and V.

The private sector triumphant?

In the last two chapters, we have seen that there was a gradual increase in the involvement of the private sector in crop research during the latter half of the twentieth century. To begin with, this was largely due to improved business opportunities created by favourable legislation on plant breeders' rights in the 1960s and 1970s. The increasing role of the private sector in crop breeding during this period was reflected by a switch towards varieties that responded well to inputs such as fertilisers and biocides, as well as a much reduced shelf life for the proliferating number of new varieties. However, right through the 1970s, the public sector remained the senior partner in the overall plant breeding enterprise, flourishing especially in the international arena, led by CGIAR and its affiliated research centres. But things changed again during the 1980s, as the private sector received fresh impetus from a range of new possibilities that were suddenly opened up by agbiotech. Unlike previous technological innovations, such as mutation breeding or hybrid production that were essentially public sector enterprises, private companies were at the forefront of the development of agbiotech from the start. These companies benefited from a fairly lax patenting regime and aggressively patented many basic procedures in transgenic crop technology.

By the end of the twentieth century, the private sector was the dominant force in the breeding of a wide spectrum of the most valuable crops, including maize, soybeans, sorghum, cotton, plus various fruits and vegetables, a process that was accompanied by a significant reduction in genetic variation in these species.[320] At almost exactly the same time as the private sector began its most recent growth spurt, public sector plant research in many industrial countries was subject to a massive structural change that largely removed it from many areas of crop improvement per se, for the first time in over a century. Beginning in the 1980s, there was a steady decline in public sector plant breeding, which spread across the globe and, by the early twenty-first century, was even threatening the international work of CGIAR.[321] We will now analyse how and why public sector plant breeding research declined so suddenly, so rapidly, and so soon after its astonishing achievements during the Green Revolution years of the 1960s and 1970s.

9

Decline of the public sector

Habemus publice egestatem, privatim opulentiam
(We have public want amidst private plenty)
Sallust (86–34 BCE)
Conspiracy of Cataline[322]

Introduction

The public sector paradigm of crop improvement has been spectacularly successful in feeding the unprecedented growth of human populations over the past century. The immense achievements of the Green Revolution during the 1960s and 1970s were largely built on the public-good breeding efforts of the previous half century. But just as these achievements were coming to fruition, a new, and superficially attractive, private sector paradigm of crop improvement emerged. The timing of this renaissance of private enterprise could not have come at a worse time for public sector plant science, especially in those industrialised countries that so vigorously privatised many state assets, including centres of scientific research, in the 1980s and 1990s. In no country were the resulting cutbacks in public sector plant breeding more far reaching than in the UK. In a period of a little over a decade, the UK lost virtually all of its public plant breeding infrastructure and much of the related scientific expertise. This process was duly exported to other nations and has even engulfed some developing countries. In December 1989, I was appointed as Head of one of the three departments of the former Plant Breeding Institute (PBI) that stayed in the public sector while the rest of the organisation was privatised. This gave me something of an 'insider's view' of the effects of privatisation and its effects on public sector plant breeding in the UK.

There were both positive and negative consequences of these privatisations but overall most of the effects on UK breeding were negative. By removing the option to pursue applied crop research, the UK has channelled most of its plant scientists towards more academically focused strategic and fundamental research projects. Today, the UK is probably second only to the USA in the output and breadth of its innovative plant science, but the downside is that the UK also has a much reduced capacity for practical crop breeding in both the private and public sectors. In contrast, and despite paying much lip service to the same privatisation agenda as the UK, the US government did not dismantle its plant breeding centres; and in many cases, even university-based scientists are still encouraged to pursue applied, crop-related research. The purpose of the next two chapters is to examine the decline of public sector plant breeding and the different ways in which this has been manifest across the world. We will mainly focus on events in the UK, both to link these more recent developments with our previous accounts of the development of UK plant science, and because some of the most far reaching changes occurred there. In particular, I will discuss two separate, but interconnected, processes that have had an especially profound impact on UK plant science, namely privatisation and academisation. These events in the UK will then be linked with more global developments in the conduct of public sector plant breeding research.

Privatisation, integration and globalisation

It is often stated that the main reason for the rise of the private sector, and the associated decline of public sector plant breeding, was the emergence of new technologies, particularly transgenesis, that could be better exploited in the private sector. This is not a new argument. As long ago as 1983, some agricultural economists were expressing concern about the future of the public sector: 'the high capital intensity of biotechnology-related plant breeding will render public breeding technologically antiquated vis-à-vis its more adequately funded counterpart in the private sector.'[323] However, things are a lot more complex than is implied by this statement. As we have seen in the previous two chapters, it was not just the new molecular biology and its technological manifestations that dealt a serious blow to public sector plant science during and after the 1980s. Other players, most notably governments and judiciaries, played at least as important a role in facilitating these processes as the scientists. In the legal domain, we have already noted the profound effects of the various forms of plant rights legislation enacted since the 1960s. For example, a recent journal review article begins as follows: 'In the past thirty years legal developments at national and international levels have completely reshaped the ways in which plant genetic resources are used in global agriculture.'[324]

Meanwhile, in the politico-economic domain, it was privatisation that had the greatest direct impact on public sector plant breeding, and this was most keenly felt in the UK. There seems little doubt that the recent decline of plant breeding research around the world can be traced back to those events in the UK, and specifically to the privatisations of the 1980s. The mass privatisation of all forms of state assets was pioneered in the UK in the early 1980s and gradually spread, albeit unevenly, across the globe. By 1991, the UK had completed the privatisation of assets worth almost $20 billion, Mexico $9.5 billion, Germany $8 billion, and Canada $3 billion.[325] In the UK, the privatisation programme was part of the economic dogma of the Thatcher administration of 1979–1991. One of the major architects of the privatisation programme, first in the UK and later overseas, was Madsen Pirie, who formerly worked for the House of Representatives in Washington DC and later went on to become President of the free-market leaning Adam Smith Institute in London.[326] The aim of privatisation is to remove the state from involvement in commercial activities that could be viewed more properly as the function of the private sector. This rationale applies particularly to large state-controlled ventures that had often occupied an inappropriate monopoly position in a particular part of the economy. Beginning in 1981, most publicly run industries in the UK, including telecommunications, major utilities and steel manufacture, were sold to private investors. By mid-decade the government began to look at its public sector research institutes, many of which were far from being viable units of wealth creation, but were sold off nevertheless.[327]

Amongst the public sector research establishments privatised and/or rationalised in the late 1980s and early 1990s were the National Engineering Laboratory, the Central Electricity Generating Board research centres at Leatherhead, Southampton and Barnwood, the Centre for Applied Microbiology Research at Salisbury, the National Seed Development Organisation (NSDO) at Cambridge,[328] and many research centres formerly belonging to the Agricultural Development and Advisory Service (ADAS). The drastic consequences of the ensuing wave of change that swept away so many of these institutions have been lucidly analysed by Rebecca Boden *et al.* in the book *Scrutinising Science – the Changing UK Government of Science.*[329] In the words of a reviewer of this book: 'Looking back over the last half century we have perhaps been too cavalier in discarding ... institutions for which Britain was once renowned. We may need to re-invent them if we are, as the Prime Minister wishes, to become the science capital of the world.'[330] While the consequences of the privatisations and restructurings of the past fifteen years have been at times regrettable, it cannot be doubted that, from the perspective of the late 1980s, the rather ad hoc jumble of publicly funded crop-related institutes that had emerged throughout the UK during the twentieth century (see Chapter 9) merited at least a modicum of rationalisation.

Unfortunately, however, when privatisation did occur in the research sector, it was often driven not so much by a desire to increase the efficiency and effectiveness of the research as by ideological factors. UK policymakers then developed the concept of 'near-market' research, which was deemed not to be the proper function or concern of the public sector.[331] Confusingly for those of us who were doing the actual research, we were simultaneously enjoined to refrain from 'near-market' projects and to pursue projects related to 'wealth creation'. At the same time, a series of technology transfer companies was set up to commercialise the fruits of our public sector research, which was supposedly in no way 'near-market'. These were indeed puzzling times for scientists like myself who were working in the UK research sector! As a result of the new policy, all public institutions deemed to be involved in 'near-market' research were ripe for wholesale privatisation. Or, if such work was only part of their portfolio, it was decreed that the research projects in question should be abandoned, or transferred to the private sector, or at least be funded by it.[332]

On a more pragmatic political level, privatisation also obviously involved a search for potentially lucrative public sector milch cows that could be sold off for the short-term windfall profit of the public purse. The proceeds of such sales, and savings from various reconfigurations and rationalisations, were then incorporated into general state revenues and could be deployed elsewhere, e.g. to pay off debts or finance electorally popular tax cuts. For example, in 1989, the Agriculture Minister announced to the UK Parliament that the cutbacks in 'near-market' agricultural research in public institutions amounted to a saving of $47 million over three years. Additional unspecified savings were generated by a wide ranging series of institute mergers and closures.[333] On a wider level, one can also question the economic wisdom of many of the science-related privatisations of the 1980s. In broader economic terms, it is well known that some activities are more suitable for privatisation than others. For example, one of the more serious potential flaws in the global privatisation agenda is that success is largely predicated on the presence of energetic and competitive markets, where efficiency gains can be readily realised. When such conditions are met, judiciously organised privatisations can indeed be beneficial. On the other hand, if a marketplace is unsuitable or unready for privatisation, the results can be more problematic. As stated in the classical study of British privatisation by Vickers and Yarrow:

Where product markets are competitive, it is more likely that the benefits of private monitoring systems (e.g. improved internal efficiency) will exceed any accompanying detriments (e.g. worsened allocative efficiency), a view that is generally confirmed by empirical studies of the comparative performance of public and private firms. In the absence of a vigorous product marketplace, however, the balance of advantage is less clear cut and much will depend on the effectiveness of regulatory policy.[334]

As we have seen in previous chapters, private sector involvement in the business of crop improvement was still at a relatively immature stage in the 1980s. Furthermore, the focus of many agriculture-related companies was becoming increasingly skewed towards business models that involved the use of a narrow subset of new and proprietary molecular-based technologies, i.e. agbiotech, rather than using the wider range of existing public domain breeding technologies. Development of the new molecular technologies was favoured by inconsistent state regulatory regimes in many countries. The factors that generated an immature agbiotech dominated marketplace have also contributed to what is, relatively speaking, a market failure of privatised plant breeding in the UK when compared with most of the other commercial sectors that were privatised at about the same time. For example, none of the immediate commercial beneficiaries of the privatisation of crop breeding institutes have pros-pered, and several have gone out of business altogether. Another measure of the immaturity of the agbiotech market is the way it became so skewed against the smaller companies established in the 1980s and early 1990s. Virtually all of these companies either went out of business or were bought up by multinationals in the mergers and acquisitions 'feeding frenzy' of the late 1990s. This period also witnessed a dramatic consolidation of the larger companies to create the current narrowly based oligopoly of the 'big four'. In terms of the thesis proposed by Vickers and Yarrow, therefore, the 'balance of advantage' in privatising the UK breeding centres can be seen to have been lacking. To examine this important process in more detail, we will consider the two most important UK crop-related privatisations, i.e. PBI and ADAS.

The Plant Breeding Institute

One of the most superficially profitable research institutes, and hence a prime can-didate for privatisation in the febrile atmosphere of the mid-1980s, was also the premier centre of crop breeding in the UK. This was the Plant Breeding Institute (PBI) in Cambridge (see Chapter 5 for a brief account of the early history of PBI). From 1987–1990, PBI was privatised and split up in a sale that failed to generate any direct income for the government.[335] The PBI privatisation set in motion a series of sales, mergers and consolidations of UK plant research centres that has continued until the present day. That first privatisation was unusual in several respects, but it is worth examining for some of the more general lessons that can be drawn from the experience. As stated in an analysis of the PBI privatisation as it was nearing completion in 1989:

It is suggested that some of the localised issues arising in the PBI context have already begun to be apparent in other areas of research science, such as energy, where privatisation is proposed. It is far from clear whether the policy encourages any greater competition or competitiveness. It is clear, however, that without a very careful assessment of the institutional changes wrought on public sector knowledge bases by privatisation, British science may find itself without access to those institutional structures on which future innovation in the new technologies will depend.[336]

The sale of PBI was the first example of government policy that not only involved a withdrawal from involvement in plant breeding research, but also ensured that existing plant breeding assets were transferred to an immature private sector marketplace that proved unable either to utilise them to their full potential or to develop them successfully for the benefit of the UK economy in the long term. The process of splitting up PBI at Cambridge in order to prepare the way for the sale of its most commercially attractive assets was initiated in 1987.[337] From 1987–1990, the Institute was gradually divided into a privatised entity, which remained in Cambridge, and a publicly funded, rump group (of which I was part) that relocated to the John Innes Centre site in Norwich. In the remainder of this section, we will look at the reasons for and consequences of the PBI privatisation.

PBI had always been the principal UK crop research centre, with a special focus on cereal, root, tuber and brassica crops. Since its establishment in 1912, PBI had pioneered the development of many of the most important UK crops, including semi-dwarf cereal varieties like those that were a key element of the Green Revolution of the 1960s. The Institute was unique in the UK and overseas in the way it combined strong research teams in both applied and basic aspects of plant science. New varieties developed by researchers at PBI were passed on to the National Seed Development Organisation (NSDO) for multiplication and commercial marketing of the seeds. Therefore PBI and NSDO worked together to take basic discoveries in plant science, incorporate them into new crop varieties, and make these available to farmers. Soon, more than three quarters of NSDO income was derived from varieties bred at PBI. Even before privatisation, the UK government was extracting a lot more from PBI than it was putting in. For example, in 1986, the total budget of PBI was about $7 million, but the net receipts for its seed sales (via NSDO) were almost $10 million.[338] During the 1970s, PBI had started an innovative programme of research into the organisation of DNA in wheat and its cereal relatives. In 1978, ten new staff positions were created at PBI to exploit the emerging science of plant molecular biology, and in 1985 a new Department of Molecular Genetics was launched under the leadership of Richard Flavell.

By the early 1980s, this work had led to some of the earliest steps in the emerging area of molecular genetics and DNA manipulation. Ironically, it was these innovations that exposed PBI to those who sought to dismantle and partially privatise the Institute. By the mid-1980s, PBI was made up of two successful groups of researchers, working in complementary but linked areas of plant science. The first group was the scientific breeders who, over the previous 75 years, had produced over 130 new crop varieties, including the most popular cultivars grown in the UK. This group had a pedigree second to none as outstandingly successful breeders of new elite crop varieties. And naturally, they were very much focused on field crops and applied research that was relevant to UK agriculture. Breeding successes at PBI continued apace well into the 1970s and 1980s. In 1972, PBI released the wheat varieties Maris Huntsman and Maris Nimrod, both of which increased yield potential by 10–15%. In 1980, they released a new wheat variety called Avalon; this widely grown variety combined high yield with good bread-making ability. The second group of PBI researchers comprised more recently recruited molecular biologists who pursued a very different type of research. Their molecular work was much more fundamental in nature and was very much at the cutting edge of basic biological science. For example, in 1980, PBI molecular geneticists reported the first example of the isolation and sequencing of a plant gene.[339] To a great extent, this second group of PBI researchers more resembled the sort of curiosity driven scientists to be found in most universities, than their more applied, crop breeding colleagues in the same Institute.

By 1985, PBI was much more than just a profitable producer of new seed varieties. During the late 1980s, a particular strength emerging at PBI was its combination of cutting-edge research on molecular biology, including transgene technology, with some of the most effective plant breeders in the country. The overwhelming importance of PBI in the breeding of the principal national crops, i.e. the cereals, is shown by the fact that in 1987 almost 90% of the total area of UK cereal crops was planted with PBI varieties.[340] It is especially ironic that, just as it was being broken up and sold, the PBI was being hailed by science policymakers in the USA as a model of interdisciplinarity to be emulated on that side of the Atlantic.[341] However, by the time I joined PBI at the end of 1989, much of this former 'jewel in the crown' of UK plant breeding had been sold off and the staff was in the process of being dispersed.[342] The proximate cause of the PBI privatisation was a new policy agenda proposed by the government research council responsible for funding much of the agricultural science in the UK. In 1985, the Forward Policy Review of the Agricultural and Food Research Council (AFRC) proposed that its many and varied portfolio of research institutes should be reorganised into eight new 'super' institutes. This was closely followed (in 1987) by the sale of the PBI breeding programme and its associated experimental farm to a private company (Unilever Plc.) under the government privatisation policy.[343]

In early 1990, those PBI scientists who had not been acquired by Unilever (myself included) were relocated to newly built facilities at the John Innes Institute where we formed the so-called 'Cambridge Laboratory'.[344] So, why had Unilever paid the not inconsiderable sum of $130 million for just a part of PBI and its scientific staff? Partly, PBI was purchased because, as we have seen, it was a good money-making venture. But another reason was that some of its breeding lines could potentially be exploited commercially via the emerging technologies of genetic engineering. In the event, Unilever did not profit greatly from its newly acquired PBI germplasm and, with some relief, it sold off the entire organisation in 1998 to the emerging agbiotech giant, Monsanto. However, not even Monsanto was able to benefit from this previously profitable enterprise and, in 2004, sold the PBI cereals business to the French-based seed company, RAGT-Semences. Today, little remains of PBI on its extensive site at Trumpington on the outskirts of Cambridge and there are now proposals to sell off this prime real estate for redevelopment as housing and retail units.[345]

The sale of the PBI cereals business in 2004 was the final link in a chain of events that started with the loss of the premier UK public sector plant-breeding centre when it was sold off in 1987–1989. The private sector then proved unable to use this asset profitably and by 2004 the country had also lost virtually all of its private sector plant breeding R&D. This seems like a lose-lose situation for UK plant science. Meanwhile, the former Head of Cereals Research at PBI and Director of the John Innes Centre, Mike Gale, has traced the origin of the decline of the public sector back to the regulatory changes, such as plant breeders' rights, that we examined in the previous chapter: 'Plant-variety protection was the death knell for public breeding programmes'.[346] While plant breeders' rights certainly marked the beginning of the decline of public sector breeding in the UK, we have seen in this section that the process was massively accelerated by the government policy of privatisation and rationalisation after the mid-1980s. We will come back to these points a little later, but meanwhile we will look at another major UK agricultural privatisation, that of ADAS.

The Agricultural Development and Advisory Service

The second major plant breeding asset to be privatised in the UK was the Agricultural Development and Advisory Service (ADAS). In the 1980s, ADAS was similar in many respects to the various Federal and State extension services in the USA (see Chapter 4). Interestingly, and in contrast with the UK, the US government decided in the 1990s that, far from privatising these services, their public sector functions would be expanded. Accordingly, in 1994, Congress decreed the merger of two USDA agencies (the former Extension Service and the Cooperative State Research Service), to create a much larger, better resourced body called the Cooperative State Research, Education and Extension Service (CSREES). The new CSREES can call on over 600 000 volunteers to facilitate State-level farmer extension services and public education.[347] Nothing in Europe comes close to CSREES in its size, scope of activities, or community involvement. Although it was the nearest UK equivalent to the US extension services, ADAS always had a much more modest remit. Prior to 1987, ADAS did applied research and provided free extension services to farmers. ADAS services included recommending crop varieties for specific regions, and advice on new varieties from public and commercial breeding programmes. ADAS also performed independent field trials of new varieties prior to their addition to the National Seed List. After 1987, as part of the privatisation agenda, the UK government obliged ADAS to charge fees for extension services and for some of its research. As government funding was reduced further, the cost of field trials was also increasingly borne by breeders and by farmers.[348]

The privatisation process continued such that, by 1989, ADAS had been split up into several self-funded, quasi-commercial, independent agencies that were required to compete with other public institutions and commercial organisations for contract research. The funding gap left by withdrawal of government support was only partially filled by industry levies. This meant that the new quasi-privatised agencies led a much more precarious existence than before, with no guaranteed continuity of funding or security of tenure. These conditions made it hard to pursue any but the most short-term and straightforward of research projects. One result was the steady departure of many highly qualified scientific staff. Another consequence of the ADAS privatisation was a sharp fall in expenditure on applied crop improvement. By 1991–1992, the total funding for such work had declined by 30% ($15 million), owing to the annual reduction of $23 million in government funding that was only partially matched by an additional $8 million from industry.[349] By 1997, ADAS had become a completely private company, providing commercial contract services, advice and information. By the early 2000s, the UK government was beginning to reconsider the privatisation of ADAS, as outlined in a paper from the Agricultural Research and Extension Network of the Overseas Development Institute: 'a greater government role in the provision of technical advice is now back on the agenda as one of the options being considered for tackling the *perceived negative consequences of fragmentation of services* and the slow take up of technologiess'[350] (author's stress).

Aftermath of the UK privatisations

An analysis of the consequences of the PBI and ADAS privatisations by Carl Pray of Rutgers University shows that the UK government saved very little money by privatising PBI and probably lost out in the longer term.[351] Although government made initial savings by cutting expenditure on plant breeding research, these were more than offset by the loss of revenue from sales and royalties for seeds.[352] According to Pray, government and taxpayers gained little, if anything, from the PBI and ADAS privatisations. Meanwhile farmers paid higher bills for (arguably poorer) expert advice and for seeds that were now only available from private companies. It can be argued that, in addition to farmers, the other major losers from privatisation are UK plant scientists who were deprived of much of their ability to link basic research activities with practical crop development. One of the main objections to the sale of PBI at the time was that it threatened to sever the connection between the more hands-on aspects of plant breeding from much of the innovative research work, including plant physiology, genetics and molecular biology. Ironically, the two halves of ex-PBI promptly sought to recreate such links after privatisation, but with new partners rather than with each other.[353] Hence, the Unilever-owned commercial plant breeding group that remained in Cambridge hired 25 new biotechnology scientists and built new laboratories for them. Meanwhile, the more research orientated molecular biologists and geneticists who had moved to Norwich set up a series of new joint ventures with more practically minded colleagues in a number of commercial plant breeding companies.[354]

During the financial year of 1987–1988, in the immediate aftermath of the privatisation of PBI, there was a rough equality of agricultural research funding from the public and private sectors in the UK, with a total value of about $950 million.[355] However, most of the private sector funds were spent on research into agrochemicals, food, and farm equipment ($400 million), with only about $100 million being spent on work related to crop breeding. Meanwhile, in the public sector, an increasingly large proportion of the $200 million funding for research institutes was being devoted to basic 'blue-skies' plant science research, instead of having any direct relationship to crop breeding. Already, at this relatively early stage, we can see that breeding research in the UK was being squeezed from both sides, as the private and public sectors alike withdrew funding in order to deal with other priorities, not all of which were necessarily scientific. In 1990, the UK government agriculture department (then called MAFF) withdrew all funding from 'near-market' agricultural research, with the loss of an annual $50 million to public sector institutions. There were further cutbacks in public funding for plant breeding and related research throughout the 1990s and well into the twenty-first century.[356]

These cutbacks in public funding were not matched by increases from the private sector. In fact, quite the opposite happened as the private sector too started to retrench and cut its research budgets after 2000. By 2006, therefore, the situation for UK crop breeding research as a whole had deteriorated considerably compared to the late 1980s. Almost all of the multinational companies that had initially bought up the privatised breeding centres had closed down their UK operations in this area.[357] Meanwhile, in the public sector, the switch to basic research had continued to its logical conclusion with most of the remaining institutes now operating as quasi-university departments, concentrating mainly on model plant species and academic questions, largely to the exclusion of crops and their more practical problems. At this point, the plight of the public sector had worsened to such an extent that even many private sector plant breeders were voicing concern about the adverse effects of public cutbacks. These concerns related mostly to activities such as germplasm development, especially in relation to the major UK crop group, the temperate cereals.[358] It was pointed out that, although the UK cereal gene bank still resided in the public sector, new research on germplasm screening and enhancement had been severely reduced in favour of more 'exciting' basic research. The result has been an inexorable drift towards less genetic diversity amongst UK cultivars, especially wheat. For example, in one case, seven wheat lines, each of which had been derived from the same cross, were marketed by four different companies as if they were completely separate varieties.[359]

It seems that by withdrawing support for public sector plant breeding in such a precipitous manner, the UK may have jeopardised its entire capacity for long-term crop improvement. This would indeed be a tragedy for the nation that produced Bacon, Fairchild, Darwin and Bateson. Here, we may recall the sentiments of an earlier English scholar as he bemoaned the fate of learning in his country thus: 'It has very often come into my mind what learned men there once were throughout England... and how foreigners looked to this country for learning and instruction ... and how we now have to get it from abroad.'[360] These words were written more than eleven centuries ago by King Alfred of Wessex. Although Alfred was not specifically discussing plant breeding, his sentiments still have much resonance in the present context. As it turned out, the UK breeders were not alone for very long in their predicament. It is said that misfortune loves company, and that was certainly the case in this instance.[361] Within a few years, the beleaguered UK plant breeding sector was joined by similarly unfortunate colleagues around the world, as we will now see.

10

Reaping the consequences

Have a care o' th' main chance, And look before you leap; For as
you sow, you are like to reap

Samuel Butler (1612–1680)

Hudibras

Introduction – privatisation exported

The wave of privatisations of plant breeding programmes gradually spread from the UK
to other industrialised countries after the mid-1980s. As we saw in Chapter 8, much of the
canola breeding programme in Canada was privatised with impressive dispatch during
this period.[362] In the USA, the funding of public sector plant breeding research remained
more-or-less constant during the period from the mid-1970s to the mid-1990s, at a time
when funding in the UK was already in steep decline. It was only after the mid-1990s that
the decline seen in most other industrial countries eventually spread, albeit in less virulent
form, to the USA.[363] For example, in a 1996 study, it was found that there had been a
reduction of about 12 scientist-years in plant breeding research in the US SAES system
from 1990–1994, and a corresponding growth of 160 scientist-years in the private plant
breeding sector.[364] In a recent review, breeder William Tracey of the University of
Wisconsin at Madison has noted the decline in the status and role of plant breeders in the
USA and depressingly opines that within a few years: 'Plant breeders will exist as
technicians for [genetic] engineering programs.'[365] However, on the plus side, it appears
that public sector plant breeding research has survived in far better shape in the USA
compared to other industrialised nations. For example, empirical studies of the impact of
mechanisms such as patenting and technology licensing at USDA suggest that these have
not significantly affected the overall public-good research agenda of the Department.[366]
Unfortunately for global agriculture, the USA is virtually alone in this regard.

The Anglosphere

One country that has surpassed even the UK in its zeal for privatisation is
New Zealand. Under a Labour administration in the 1980s, New Zealand was

especially vigorous in the implementation of a widespread privatisation agenda right across its economy. During the period from 1985–1992, its plant research centres were converted en masse into quasi-private bodies called Crown Research Institutes, a process involving large scientific staff reductions and a reorientation away from long-term research and towards more immediate, market focused projects. Prior to restructuring, much of the public research in New Zealand was carried out in institutes managed by the Department of Scientific and Industrial Research (DSIR). Beginning in 1985, DSIR and other state research providers were restructured to form Crown Research Institutes run as profit-making companies.[367] Across the Tasman Sea in Australia, the privatisation process started in the mid-1990s, as the Grains Research and Development Corporation (GRDC) switched its funding from state-based programmes to a small number of commercially focused operations.[368] During the 1990s, the main sponsor of public sector crop research in Australia, CSIRO (Commonwealth Scientific and Industrial Research Organisation), steadily reduced its applied programmes in favour of academic and strategic projects. The refocusing of CSIRO research away from varietal improvement towards high-tech, high-value projects was still causing political furore in late 2005 with publication of a new strategy that downgraded 'incremental agricultural research'.[369]

In Canada, while the general pool of funding for plant breeding research has declined, the fall has not been nearly as precipitate as in the UK or Australasia. For example, even as late as 2004, the vast majority of the major commercially grown cereal varieties still came from public sector programmes. Interestingly, no transgenic varieties of these crops have yet been released, so private companies can only rely on hybrid technologies (which have yet to be developed for most Canadian cereal varieties) or plant breeders' rights to protect their investments. The one major exception to this trend is canola, which along with wheat is the mainstay of prairie agriculture in Canada. Most new canola varieties are now provided by the private sector. This development may be related to the fact that, by 2005, almost 80% of the Canadian canola crop came from transgenic varieties.[370] Hence, once again, we have a link between the extent of private sector interest in a crop and the availability of a specific business-friendly type of breeding technology for that crop.

Europe

During the 1990s, similar bouts of privatisation and restructuring occurred throughout Western Europe, albeit not always as vigorously as in the so-called 'Anglo Saxon' countries, as discussed above. The continental nations of Western Europe have tended to have more statist traditions than 'Anglo Saxon' countries, such as USA, Canada, UK, Australia and New Zealand. This led to a markedly greater reluctance by the former to embrace the privatisation agenda pioneered by the UK. However, some European countries, such as the Netherlands, implemented far reaching privatisation programmes in the mid-1990s, and several other countries in the region have now moved in a similar direction. In the late 1990s, some of the most vigorous privatisation programmes have taken place in Eastern Europe, especially the Czech Republic and Slovakia, which privatised all of their state plant breeding centres. In contrast, the Baltic States, Poland and Hungary have retained most of their public breeding centres, albeit with reduced funding in many cases.[371]

Further to the east, the collapse of the former centralised government and the ensuing economic decline of the 1990s, have left most of the new nations of the former Soviet Union with a very poorly resourced and badly organised plant breeding infrastructure. The situation has been especially serious in the Ukraine, which was formerly one of the most agriculturally productive regions of Europe. For over a century, the Ukraine had an international reputation as the location of many of the plant breeding research and germplasm repositories in the former USSR. For example, as we saw in Chapter 4, it was breeders from the Ukraine who developed one of the important new commercial oilseed crops of the twentieth century, namely sunflower.[372] During the crisis years of the early 1990s, seed quality and productivity declined precipitately in many parts of the former USSR, including the Ukraine. Much of the public sector research effort collapsed as breeders no longer received salaries and their institutes went unfunded for years at a stretch. This was the period when some professional breeders and plant researchers were reduced to growing potatoes on the experimental plots at their institutes and universities, simply to feed their families, and in some cases priceless germplasm collections were left unguarded.[373]

Unlike the situation in the West when public sector breeding work was reduced, in the Ukraine there was no real private sector alternative to replace the now barely functioning public institutions. The rural economy was, and largely speaking still is, too weak to afford the purchase of (mostly imported) private sector seed. Instead, plant breeding reverted in some areas to a pre-twentieth century tradition of informal selection of landraces by farmers and localised exchange of germplasm between ad hoc groups of farmer-breeders. One stark statistic sums up the plight of Ukrainian agriculture: between the end of the 1980s and the late 1990s, crop production had almost halved in this exceptionally fertile region, a place that had once been dubbed the 'breadbasket of Europe'. Despite these immense vicissitudes, however, Ukrainian agriculture, its farmers, and its exceptionally talented breeders are beginning to recover and we can be a lot more optimistic about their prospects in the next decade than the last. We will come back to discuss the future potential of the Ukraine as a major food producer in Chapter 15.

Developing countries

During the latter half of the 1990s, the fashion for privatisation/restructuring of plant breeding assets began to spread beyond the industrialised world. In a succession of developing countries, more and more governments began to sell off or restructure their agricultural research and service centres.[374] In a few cases, such initiatives were positively beneficial, especially where some of the public sector organisations were overly bureaucratic and subject to political interference. An example of the potential benefits of privatisation is the case of national seed centres in some developing countries.[375] These organisations tend to be funded partially by international assistance and partially by the host government. Their mission is to accumulate, test, maintain and distribute seed stocks of the most appropriate cultivars for the country's farmers. In many cases, the publicly owned seed centres have tended to be large, monolithic institutions that were overly tied to the diktats of government policy, rather than run according to scientific and economic criteria. This has led to strong pressures from donor countries for an acceleration of the privatisation and diversification of seed supply in the recipient developing countries.

On the negative side, however, there are increasing concerns about the seemingly relentless retreat of the public sector around the world from agricultural research and development. More and more voices from across the scientific spectrum have expressed their reservations about this process. There is a widespread feeling that the over hasty withdrawal of the public sector is likely to give rise to ever more deleterious consequences for crop improvement, and that the situation is likely to become much more serious in the short- to medium-term future.[376] We will return to considering the current status of what remains of public sector crop research in Part V. In the meantime, we will analyse the consequences of the restructuring that we have just examined in terms of its impact on the kind of research that is now pursued in the field of plant science. As before, we will begin by looking at the situation in the UK, before moving further afield.

Academisation of public research

Following the withdrawal of UK government support for 'near-market' research in the public sector during the late 1980s and early 1990s, scientists were obliged, by default, to focus on more basic projects. In the case of plant breeding in particular, the obvious message from the privatisation of PBI and other government centres was that scientists should not get too applied in their research or they too would risk the removal of their funding. Similar trends were soon evident in other areas of public research in the UK, and in other industrial countries. Today, we see that the missions of publicly funded research institutions around the world have become ever more academically focused.[377] At the same time, however, researchers have been given confusing and inconsistent advice by governments about the need for greater engagement with the marketplace and for research that is focused on so-called 'wealth creation'.[378] While they apply globally, these dichotomies are probably more pronounced in the UK than anywhere else. And nowhere are they more evident than in the field of plant science research. By 2005, there was almost no remaining substantial public-good applied research related to agriculture in the UK. Each of the original score or more of publicly funded institutes had been privatised, merged, consolidated and otherwise disposed of, until their residue could be counted on the fingers of a single hand. Even this much reduced rump had undergone a profound change of its mission that has now become resolutely orientated towards basic plant science. I shall refer to this process, albeit somewhat inelegantly, as the 'academisation' of public research.

While academisation has been concentrated in research institutes, a similar process has occurred in many university departments in the UK. However, because universities have always specialised in basic research anyway, the paradigm shift here is less marked than in the public research institutes. In the public institutes, the academisation phenomenon has involved such a fundamental shift of mission and ethos that one can legitimately question whether their overall research role now differs in any meaningful way from that of universities. Moreover, the academisation of the research institutes has left an immense gap between what were the formerly well-connected areas of basic plant biology and its real-world application for agricultural improvement. The obvious and increasingly articulated danger of these developments is that many scientists in countries like the UK might well be doing some of the best plant research in the world, but such work is now largely deracinated from its original, largely agricultural, *raison d'être*.[379]

Academisation of public research has been accompanied by a seemingly inexorable rise in the scale of research, especially in biology. Whereas 'big science' in the past related mostly to physics, the latter half of the 1990s has seen the corresponding emergence of 'big biology'. But these labels are misleading: in reality 'big biology' is a very different creature to 'big physics'. The 'big physics' of the 1950s onwards was characterised by massive (in both size and cost) instruments, such as terrestrial and space-borne telescopes, subatomic super-colliders and super-conducting magnets. In contrast, the more recent emergence of 'big biology' has been characterised by large multidisciplinary research centres tackling *'grands projets'* such as genome sequencing, post-genomic studies, and the analysis of complex biological networks.[380] A major difference between 'big biology' and 'big physics' is that many biological resources are freely available across the entire research community, and many large research groupings are virtual, internet-linked entities spread across countries and continents, rather than physically housed in the same building.

In a (simplified) nutshell, one can say that whereas 'big physics' is very much hardware based, 'big biology' is more often based in software and informatics. Basic researchers into plant biology can nowadays call upon an unprecedented wealth of freely available physical and informatic resources that are held in various international research stock/data centres.[381] This makes it an extremely exciting time to be a biology researcher and the temptations to take up a career in basic science, in comparison with applied breeding, are indeed difficult to resist. The appeal of basic research has been enhanced by structural factors in the public sector that have militated against applied research. The carrot for a UK plant biologist considering a move into basic research was the prospect of joining one of the most dynamic areas of modern science. But there was also a powerful stick to propel the scientist towards such academic studies. This stick was the new mechanism of scientific funding that was based on a particular model of research assessment.

The thesis that will be examined below is that, in addition to the government policy against 'near-market' research in the early 1990s, a major force that has pushed UK institutes in particular towards basic science is the method by which they are assessed for their funding. This mechanism gives each institute a rating based on its research 'quality'. The crucial factor here is the definition of the term 'quality'. Because the assessment criteria are heavily weighted towards academic excellence, as defined by academic researchers themselves, one can see that it behoves an institute to score highly in areas of academic, rather than applied, research. As with many of the other developments that we have examined in this chapter (e.g. institute privatisations) such research assessment models were largely pioneered in the UK, before spreading to other countries.[382] We will now consider the process and the effects of these assessment exercises, especially in the UK.

Academisation and research assessments

For the past few decades, public sector research institutes in the UK have been formally assessed for the quality of their research about once every four years. Since the early 1990s, these assessments have used broadly similar criteria to those used for determining research quality in universities. The Research Assessment Exercise (RAE) is a regular process conducted nationally to assess the quality of university research and to provide a framework for the selective distribution of public funds for research by higher education funding bodies.[383] Since 1986, a large proportion of UK government funding for university research, totalling over $1.8 billion a year, has been allocated on the basis of RAEs.[384] Likewise, the assessments of UK research institutes have been used to reward those centres judged to be pursuing the best quality research, at the expense of those judged to be doing work of lesser merit. The key point here is that the main criteria used to judge the quality of a piece of research have tended to relate to its perceived academic standards rather than its utility.[385] The main output for the academically focused research scientist is publication in learned, peer-reviewed journals rather than more practical products such as, for example, improved seed varieties.

In the UK research assessment exercises, each scientist is typically asked to provide their four best publications from the appraisal period. These papers are then judged by a peer-review panel, consisting of senior scientists with expertise in the relevant discipline. The relentless pressure on academics and researchers to 'publish or perish' has had predictable consequences in the proliferation of journals and an almost exponential growth in the number of increasingly mediocre papers that are produced each year. It is increasingly rare nowadays to find a plant scientist who would happily comply with the exhortation of a practically minded breeder like Norman Borlaug to eschew the 'academic butterflies' of basic research. Instead, one has the relentless pressure to do the sort of research that can be published in academically highly reputed journals in order to score well in the next assessment exercise.[386] Other factors that contribute to the research quality score in the RAE include the number of research students, the amount of external grant funding, and so-called 'measures of esteem', such as honours, prizes and invitations to prestigious conferences, that accrue to a particular researcher. In many respects, these assessments of public sector research institutes in the UK are similar to the university RAE and hence, unsurprisingly, lead to similar outcomes regarding the ranking of different forms of research. The results of these evaluations also influence decisions on the possible restructuring, merging or closure of institutes.

The RAE process has reinforced still further a pre-existing tendency within the culture of biological research that places a greater cachet on certain fields like developmental and cell biology, while devaluing more applied aspects of biology (including, of course, plant breeding). As a consequence of these pressures, plant scientists in UK universities and institutes have tended to move away from practically orientated public-good research, in favour of more academic studies.[387] This change has been at least partially caused by the narrow performance criteria used in RAE and visiting group assessments, and an increasing pressure on individual scientists to publish in high-impact journals.[388] Another trend that has been fostered via the RAE in universities, and by direct intervention of research councils in research institutes, is the concentration of funding in an ever diminishing number of centres. The danger of such a trend was recognised in a report from 'Universities UK' that was released in 2003.[389] One of the most serious problems highlighted here is the lack of empirical evidence that the concentration of research activities in a few highly resourced, elite centres will actually lead to a commensurate improvement in quality. To quote from the Executive Summary of the report: 'There is no evidence that research concentration would necessarily create better research. Many relatively small research units ... have as high a research impact as larger units in their subject area. There is evidence that the only critical size threshold for research departments is at the smallest level, possibly equivalent to a single viable research group.'[390]

One of the main drawbacks of the concentration of research funding in a few favoured centres is the effective disenfranchisement from a research career of any scientist who happens to work elsewhere.[391] Ironically, this process of concentrating the research funding in a few elite centres is happening at a time when the availability of instant global communications, and (especially in biology) of global resource centres that are available to the entire community, are working together to render the old models of the physical concentration of research facilities increasingly obsolete. As we saw above in the discussion about 'big biology', it is now quite possible for a small but reasonably well resourced and intelligently managed laboratory to carry out work of the highest international standard, without necessarily having to be in the same building or campus as part of a large research centre. The disenfranchisement of many talented would-be researchers will also be to the ultimate detriment of the resource-rich centres of excellence, as the latter will be denied the input of colleagues who might otherwise wish to collaborate with them in order to take advantage of their facilities.

Interestingly, the viability of biological research as a relatively dispersed (rather than tightly concentrated) activity has always been implicitly recognised by funding organisations in the USA. Here, there is not nearly as much prejudice against small research groups who happen to work in relatively isolated areas. For example, many of the best US researchers in one of my own research areas (plant lipids) work in small teams in relatively little known State universities often located in out of the way places. You may not have heard of Clemson University in South Carolina, Williams College in Massachusetts, University of North Texas at Denton, or State University of Louisiana at Baton Rouge, but as plant lipid centres of excellence these institutions carry a lot more weight than Oxford, Cambridge, Princeton or Harvard. It is also rather telling that, in the supposed home of competitive free enterprise that is the USA, there is nothing like the sort of RAE system that is used so enthusiastically in the UK.[392] And yet, despite the absence of this somewhat ruthless UK-style 'winner-takes-all' approach to research funding, the USA remains the acknowledged global centre of scientific excellence, especially in biology.[393]

Due to the kinds of pressure outlined above, by the early years of the twenty-first century, the few remaining plant research institutes in the UK had largely turned away from crop-related work to more fundamental studies that tend to be focused on model organisms. This trend has been compounded by an increasing reliance by institutes on short-term competitive funding from research councils and government departments, rather than long-term 'core' funding.[394] Such external competitive funding is won by bidding against university groups for short-term projects (normally three years) that are assessed strictly according to academic criteria.[395] The vast majority of the more applied plant breeding research programmes would not fall into this category. Hence, we can see that there is a whole series of both structural and cultural factors conspiring against plant breeding research, both for the individual scientist and for the institute or university department.[396]

One might argue that researchers in more applied fields like plant breeding should try to work more with private companies, rather than expect to be funded by governments. There are several schemes for joint government/industry funding of some types of near-market research projects in most industrialised countries, including the UK. However, such schemes tend to be rather limited in their scope and in their potential to contribute to long-term crop improvement. Unlike some research council funding, these more near-market projects are only available in a few selected areas that are constantly redefined by policymakers. Applicants therefore have to fit their scientific proposal to the existing criteria of the grant awarder, rather than suggesting what may be a better project but one that unfortunately happens to be beyond the remit of that particular programme.[397] Other limitations are the need to secure matching funding from an often cash-strapped private sector, and the three-year funding limit, which is too short for most studies involving plant breeding. Another serious problem for UK researchers wishing to leverage government funding by collaborating with plant breeding related companies is that, during 2001–2006, most of these firms moved their R&D operations overseas.

In addition to these drawbacks, many government/industry schemes tend to reflect 'flavour of the month' preoccupations that may be of great public interest but do not necessarily coincide either with the most useful scientific opportunities or with the most important long-term crop-related challenges. For example, relatively large amounts of funding have been made available for politically popular biomass and biofuel projects, despite the poor economic justification for such ventures, not to mention an almost total lack of research on the breeding of suitable plant material for these uses.[398] Therefore, while this kind of government subsidised scheme might be suitable to fund narrowly defined, short-term projects, e.g. to investigate a new industrial use for a specific crop product, it is quite inappropriate for the funding of a more carefully thought out, long-term plant breeding programme. Because of these developments, it has now become almost impossible in countries like the UK to guarantee the long-term commitment required to sustain plant breeding programmes in the public sector.

Unfortunately, and despite what most people would accept as a good case for more resources being targeted towards more applied public sector plant research, we live in a world that has seemingly decided otherwise, at least for the time being. This has created a dilemma for the public sector institutions and has led research managers to look for alternative, non-governmental, sources of revenue. For example, some research directors of public institutes have sought closer partnerships with commercial companies. Another possible strategy is for an increasingly impoverished public sector to raise funds by more aggressive marketing of its goods, services and know-how.[399] Such strategies are fraught with pitfalls, however, as shown by events in industrialised and developing countries over the past decade or so.[400] The opportunities and risks of these strategies will be discussed in the next section. In Chapter 19, we will also compare the sporadic, short-termist, and often overly focused approach to industrial/academic research funding in the UK (and EU), with the more sustained and successful funding programmes available in the USA.

Dangerous liaisons – partnerships with the private sector

It has become increasingly common for public research institutions, normally (but by no means exclusively) in industrial countries, to seek commercial partners of various complexions.[401] A company might simply license a piece of technology from the institution, while in other cases the company might establish a longer term relationship, such as funding research in return for privileged access to the fruits of such work.[402] One of the more prominent and controversial examples of the latter arrangement was the 1998–2003 Novartis/Syngenta alliance with UC Berkeley. This deal caused considerable rancour within the University, and in 2004 it was eventually subjected to what turned out to be a severely critical (invited) review from a group of external academic experts led by Lawrence Busch of Michigan State University.[403] These reviewers argued that any benefits accrued by the host department at UC Berkeley or by the commercial sponsor, Syngenta, were relatively modest and not worth the adverse long-term consequences of the alliance, both within the institution itself and to its external reputation.[404] In the event, Syngenta terminated the alliance with UC Berkeley as soon as the contract had reached its initial five-year milestone. This implies that the company did not think it worthwhile to continue the arrangement, possibly owing to a combination of lack of benefits from the alliance itself and financial pressures that forced Syngenta to downsize its entire global research operation after 2002.

These events highlight another danger inherent in commercial partnerships, namely that the academic partner may be thereby exposed to the volatility and uncertainty of the marketplace, which may result in the sudden and unanticipated amendment or even termination of a contract. For example, several US State Agricultural Experiment Stations have experienced dramatic reductions in such private sector revenues. In 1989, the New Jersey Agricultural Experiment Station (NJAES) was making over $300 000 in royalties from its in-house developed asparagus varieties. However, following a legal dispute with its exclusive licensee, royalties declined precipitously to less than $8000 in 1990. In addition, NJAES was subsequently obliged to pay hundreds of thousands of dollars in legal fees. In California, local wine producers had historically provided UC Davis with an annual $800 000–900 000 for research on vine crops. This longstanding arrangement was abruptly terminated in 1991 after internal squabbling among the companies.[405] Such drastic and unplanned reductions in funding can force research managers to abort long-term projects and to lose well-qualified staff, neither of which can necessarily be retrieved, even if better times return at a later date.

Something rather similar occurred in the UK, when Syngenta unexpectedly terminated a research agreement on cereal genomics with the John Innes Centre in 2002.[406] This agreement had been launched with great fanfare in 1998 as a ten-year programme involving the investment of $80 million by Syngenta and the housing of 30 company scientists on the John Innes Centre site near Norwich.[407] Senior figures at the John Innes Centre described the abrupt decision as 'disappointing' and 'regrettable'.[408] The termination of the Syngenta agreement was linked to the decision not to proceed with renewal of the alliance with UC Berkeley, as discussed above. The root cause of both decisions was the major restructuring forced upon Syngenta by commercial pressures.[409] A further disaster befell the John Innes Centre in 2005 when its failure to win several short-term contracts from various sources resulted in a funding deficit of over $5.5 million, leading to dozens of staff redundancies.[410]

Overly restrictive legal agreements between public institutions and private companies can also jeopardise relationships with other public research organisations, as the Chinese government found out to its cost in the 1980s and early 1990s. During this period, the Chinese authorities licensed exclusive rights to its hybrid rice technology to two US-based multinational seed companies. As part of this process, the Hunan Hybrid Rice Research Center (HHRRC) also obtained a US patent on the process of producing hybrid rice. This was meant to deter other potential market entrants, whether from the private or public sector. Initially things went well for the Sino-US venture, as the seed companies paid an upfront fee of $200 000 plus an annual $50 000 for sales rights in the rest of the world. But after a few years the companies stopped the payments, because they claimed that it was not possible to produce successful commercial hybrids with the Chinese material. This agreement had not been very popular anyway with most of the scientists at HHRRC, because none of the money from the companies ever got as far as the researchers who had been involved in the original development of hybrid rice.[411]

To their dismay, as their income stream dried up, the HHRRC breeders were left, not only without commercial partners, but also without friends in the wider international breeding community. This was due to the original agreement signed with the companies, whereby HHRRC had agreed not to supply rice lines to other public breeding centres, most notably IRRI in the Philippines. The result was a poisoning of relations between the hapless Chinese breeders and the rest of the international breeding community. Eventually IRRI threatened to cut off all Chinese access to its public sector rice germplasm banks, with potentially drastic effects on future breeding programmes in the country. Finally, in 1994, the dispute was resolved and full mutual access to all germplasm was restored between China and the rest of the international public plant breeding community.[412] Meanwhile, a new scaled-down and much less exclusive joint venture was set up between HHRRC and a small research-based startup company in Texas. The new Sino-US venture quickly produced promising results and the first hybrid Chinese rice is now being commercialised in the USA, this time without disrupting other public-good breeding programmes.[413]

These examples highlight the inherent uncertainty involved in any alliance between a public institution and a commercial partner. Private companies that finance research projects at universities normally include clauses in their agreements that allow them to terminate funding at any time; and companies often do so at short notice.[414] As we saw above, even with the largest and seemingly most secure of companies, public institutions simply cannot depend on such arrangements to provide a reliable source of income in anything but the very short term. Over the past decade, we have seen dozens of examples of companies, both big and small, which have defaulted on research agreements with public sector bodies. Even in the best cases, the commercial sector is unable to guarantee long-term collaborative agreements with any degree of confidence. This relegates the utility of most commercial revenues to being somewhat risky, short-term palliative measures, rather than acting as a mechanism for the support of longer term scientific research.

Since most plant breeding research has timelines measured in decades, rather than months or years, it is a rather foolish and short-sighted manager or director who relies on private sector funding to support long-term, core projects in a public institution. This is a major challenge for the many plant research centres that are constantly being exhorted by their governments to raise ever more income from competitive bidding for grant initiatives and private sector funding. As we saw with the UC Berkeley/Syngenta case, it is also important to avoid the temptation to tie public centres too closely to a particular company. The take-home lesson is that the long-term interests of a public institution are far better served within the international research community, where free access to information and resources is the norm. There may be occasions when research managers are tempted by the lure of instant cash from the private sector, but while such lucre may not be exactly 'filthy', it is certainly ephemeral and not to be relied upon to fund such core activities as plant breeding or related long-term research programmes.

The penny drops

The parlous state of plant breeding research, especially in the UK, became increasingly evident in the early years of the twenty-first century. One of the most worrying recent developments has been the withdrawal of many major private companies from research, both in-house and collaboratively with public institutes. This has compounded the deficit left by the sale or adverse abandonment of much of the public sector breeding research effort during the previous 15 years. Not only has the UK government sold off or disposed of almost all of its plant breeding research capacity, the commitment to plant research as a whole has declined sharply since the late 1980s. This is evident from the funding patterns of the two main sources of money for plant research in the UK, namely BBSRC and DEFRA. The Bio-technology and Biological Sciences Research Council (BBSRC) is the major funding body for non-medical bioresearch in the UK. This agency funds the few remaining public institutes that carry out plant science research, including Rothamsted, the John Innes Centre, and the Welsh Centre (IGER). The BBSRC institutes collectively receive about $180 million per year in government funding.[415]

The second major public funder of UK crop research is the government agri-culture department (currently called DEFRA). In the past, DEFRA funded many long-term programmes related to crop improvement but this has been severely reduced because of two policy changes that were implemented in the 1990s. The first change was a steady annual reduction in the DEFRA funding of public research institutes. For example, between 1995–1996 and 2000–2001, DEFRA funding for the Horticulture Research Institute or HRI, fell by almost 30%.[416] In 2004, HRI was transferred to the University of Warwick, with the associated closure of two of its three research sites and redundancy for much of its scientific staff. At the same time, DEFRA guaranteed the remainder of HRI an eight-year funding package worth an annual $8 million. This was less than one third of the value of DEFRA funding to HRI in the mid-1990s.[417] In 2006, DEFRA also announced cutbacks in the funding of the Institute for Grassland and Environmental Research (IGER) in Wales to the tune of $2 million per year. This funding deficit resulted in the loss of at least 40 science posts and the further reduction in what remains of UK research into grassland and pasture crops.[418] In the second major policy change, any remaining funding from DEFRA was generally earmarked for short-duration, fixed-term, fixed-price contracts in specific and narrowly defined research areas that are offered out for tender via open public competition. Such projects have often been tendered in a reactive manner that reflects current public and/or political preoccupations, such as BSE (bovine spongiform encephalopathy) or FMD (foot and mouth disease [419]) rather than for more considered long-term strategic work. As quoted in a UK parliamentary report in 2004, the well-respected Biosciences Federation claimed that BBSRC's sustainable agriculture programme:

is being undermined by a lack of co-ordinated planning among the major funders, i.e. the BBSRC, Government departments and industry. For example, research funding is often redirected during crises such as the BSE outbreak, with consequent cut backs in longer-term strategic work... BBSRC has coped well despite budget cuts at DEFRA, but some research institutes have suffered ... and it has become harder for the BBSRC to provide the science to underpin policy.[420]

By the early years of the twenty-first century, it finally began to dawn on some scientists and policymakers that not all was well with plant research in the UK. Ironically, this growing structural deficiency in UK plant science was not always apparent to many of the plant researchers themselves, who were at their laboratory benches busily pushing back the frontiers of basic academic knowledge. The UK has long had a tradition for undertaking world-leading plant research. In recent years, largely due to its embrace of modern genomic and post-genomic technologies, the country has remained well in the forefront of this type of international research. From the sequencing of the first plant genome to the discovery of genes regulating flowering and dwarfing in plants, the UK has a string of recent achievements arguably second only to the scientific powerhouse that is the USA. However, this rosy picture hides a far from ideal reality as regards the balance of basic and applied plant research in the UK. Having disposed of much of its expertise in plant breeding, there is now a deepening chasm between the potential of the basic research and the reality of its application in the real world. This concern was articulated in a report published in June 2006 in which the House of Commons Science and Technology Committee criticised the UK Research Councils for a perceived lack of effectiveness in the promotion of knowledge transfer from basic research into its wider use in society.[421] As we will discuss further in later chapters, that void was not and cannot be filled by the private sector, a fact now also belatedly recognised by some of the companies themselves.[422]

UK researchers and policymakers finally started to realise the scope of the problem in 2003. Over the next few years, a series of consultations was undertaken, both within and beyond the plant sciences community. In the finest UK tradition, these were rather ad hoc affairs that emanated from some very disparate sources and often appeared to overlap in their remit.[423] One of the first reports, published in mid-2004, was from a working party of the BBSRC,[424] and identified four major weaknesses in UK crop research, namely: (i) no coherent strategy; (ii) lack of impact of basic plant science research on applied crop research; (iii) fragmentation of funding; (iv) shortage of suitably trained personnel. It was recommended 'that BBSRC should make the transfer of knowledge between plant and crop science a high priority.'[425] The clear implication was that, by 2004, such a transfer of knowledge was no longer regarded as a particularly important priority within BBSRC. As we saw in Chapter 1 and elsewhere, this linkage of basic and applied research is a perennial challenge that has confronted those interested in crop improvement since the early nineteenth century, when various European academies established special prizes and other incentives to achieve a similar goal. The immediate precursor of BBSRC was AFRC, which regarded the translation of basic research into crop improvement as a core area of its mission. The incorporation of AFRC into BBSRC in the early 1990s, and the more recent academisation process that we have just discussed, gradually eroded this linkage, which then withered on the vine of the privatisation agenda and RAE-driven priorities during the 1990s and beyond.

A further casualty of these policy changes was the kind of public-good plant breeding work done by for example PBI prior to its privatisation in the late 1980s (see above). The 2004 BBSRC report stated that it had: 'identified a widely perceived need for public-good plant breeding, in order to address crops and traits not emphasised by multinational interests and to *restore public confidence in plant breeding*' (my stress). This impression was strengthened by the stark conclusion of a review paper in 2004, by John Innes Centre (and ex-PBI) geneticist Phil Dale, that: '*There is virtually no plant breeding in the UK that is supported entirely by the public sector.*'[426] In view of the subsequent departure of all the major plant breeding companies from the UK, we could add that, unfortunately, this bleak verdict now applies to the private sector too. There are still a few isolated groups of plant breeders left in the public sector, e.g. the Scottish Crop Research Institute has an active and innovative group of barley breeders. However, these small bands of survivors are but a pale shadow of the once extensive public breeding effort in the UK.

In another discussion document, released in 2004 by the Agriculture and Environment Biotechnology Commission (AEBC[427]), the following observation was made in relation to UK agricultural research: 'Despite the focus on wealth creation, there is no sign of a move towards more near-market, experimental development type research. In fact, the Research Councils explicitly aim to continue to fund as much basic ... research as possible.' It is far from clear how UK science policymakers and the plant research community will respond to these reports in the longer term.[428] In mid-2005, BBSRC announced the provision of $14 million over three years for more targeted research on crop science, as distinct from basic plant science.[429] Although such funding is useful in the short term, it does not address the more deep-seated structural problems that now face UK plant research. These problems include, but are not limited to, the following:

- the relatively poor prestige accorded to research in applied plant science compared to its more academic counterparts;
- severing of sustained active links between plant scientists and the 'real-world' marketplace of practical varietal improvement, both commercially and in the international 'public-good' arena;
- the reduction in the size of the plant breeding community to below the critical mass threshold required to sustain activities such as long-term (decades) research programmes, added-value national and international collaborations, and the training of the next generation of breeders.

There are now very real concerns that much of the expertise in UK breeding will disappear once the present, much reduced, cohort of professional breeders retires over the next decade or so.[430] Very few students are choosing either to study or to do research into plant breeding and the UK is now left with a mere handful of university departments that specialise in agriculture.[431] This will make it very difficult for much of the excellent basic plant science research that is still being done in the UK to be translated into practical benefits in terms of improved crop varieties. This kind of process has been a classic failing of UK R&D in general for many decades. UK scientists are second to none in their ability to discover new and useful knowledge, but the means to translate such knowledge into economic goods is often signally lacking. The sad thing about the present state of UK plant breeding is that, unlike many other sectors of science and technology, there used to be a particularly effective knowledge transfer chain from discovery to exploitation, namely PBI and other crop focused centres, but this system was effectively destroyed in the 1980s and 1990s. To use a horticultural metaphor, UK plant science research is like a high-tech sprinkler system able to produce an ample stream of water (basic research) to sustain a profitable garden (crop improvement). Unfortunately, the state of the art hosepipe needed to carry the water to the garden (i.e. PBI etc.) was sold off for a few dollars at a garage sale. So now the water gushes away uselessly, while the parched garden gradually withers.[432]

The parlous situation of plant breeding in the UK cannot be viewed in isolation. Having acted as a pacemaker in moving forward the changes that have been brought about since 1985, the UK is now merely part of an international movement that has led to a worldwide decline of public sector plant breeding research. In 2000, a report addressing some of these issues was published by representatives of the national scientific academies of Brazil, China, India, Mexico, UK and USA. Regarding the balance of public/private sector research, the report makes the following points:

Whereas fundamental research is still being carried out in the public sector, the strategic application, in sharp contrast to the "Green Revolution", takes place largely in the private sector where much of the intellectual property is controlled. In these circumstances, research priorities are driven by market forces (e.g., price signals). Companies produce products whose costs are recoverable in the marketplace. There are also goods that benefit society as a whole rather than individuals and whose costs cannot be recovered in the marketplace (so-called public goods). Public sector funding is needed for such public-good work.[433] A classic example of a public good would be an improved plant that can be propagated by farmers with little deterioration, as with self-pollinated (e.g., wheat and rice), or vegetatively propagated (e.g., potatoes) crops. If such crop improvement research were left to normal markets for private provision, then it would be systematically under-supplied.[434]

In Part III, we have looked at the turbulent state of plant breeding, both public and private, since the 1980s. Private sector breeding was favoured by the coincidence of several factors, including new legislation, new technologies, and a new political climate. At the same time, the public sector underwent a considerable decline in many countries, especially the UK. Much of this decline was originally precipitated by the wish of governments to divest themselves of activities they felt more appropriate for the private sector, and the perception that the new biotech dominated companies might be more effective drivers of crop improvement. The coincidence of the privatisation agenda, which resulted in the depletion of the public sector, and the emergence of the powerful agbiotech paradigm in the private sector, dealt a severe blow to plant science in its more holistic sense as a provider of value-free knowledge that is meant to provide a genuine range of options for crop improvement. Meanwhile, within the much reduced public sector, the academisation process continues unabated. Even in developing countries, those fabulously plumed and gaudy 'academic butterflies', which were so derided by Norman Borlaug, are now tempting increasing numbers of scientists away from more applied areas of research. In Part IV, we will examine the new agbiotech paradigm in more detail and will ask whether its performance really matches up to its very considerable claims to represent a quantum advance in plant breeding.

Part IV
The agbiotech paradigm

Whoever could make two ears of corn, or two blades of grass, to grow upon a spot of ground where only one grew before, would deserve better of mankind, and do more essential service to his country, than the whole race of politicians put together.

Jonathan Swift (1667–1745)
Gulliver's Travels, Voyage to Brobdingnag (pt. II, Ch. CII)

11

Agbiotech: genes and dreams

Dreams are nothing but incoherent ideas, occasioned by partial or imperfect sleep.
Benjamin Rush (1746–1813), American physician and Congressman

Introduction

In the next three chapters, we will take a critical look at the past performance and future prospects of that putative emblem of twenty-first century agriculture: namely agbiotech. In Part III, we saw how this technology has come to dominate the thinking of many scientists, business people, policymakers, and others with an interest in plant breeding. The technology helped to spawn a dramatic resurgence of private sector interest in the commercial exploitation of crop breeding. But are the revolutionary claims of the agbiotech boosters really justified, or is this just another technology that has been artificially assisted by a fortuitous congruence of favourable patent protection and an ongoing decline in public sector breeding? Is the current agbiotech paradigm really a globally appropriate method of crop improvement? And are agbiotech strategies entitled to an ever increasing share of the limited resources available to international breeding programmes? In Part IV, we will examine these issues in detail, beginning in this chapter with a brief overview of the evolution of the agbiotech concept and development of the industry until the present day.

Agbiotech, or agricultural biotechnology, can be defined as the use of DNA-based technologies for crop improvement. Agbiotech is normally regarded as the development and use of transgenic crops. However, the term can also apply to use of DNA-based selection methods, such as marker-assisted selection (MAS) and TILLING, as discussed in chapter 3.[435] In this chapter, we will restrict discussion of agbiotech to the commercial uses of transgenesis, i.e. genetic engineering via the insertion of exogenous genes into plants. A great deal of hope has been pinned by many plant scientists,

157

companies and politicians on the potential of transgenic crops and other forms of agbiotech to usher in a new twenty-first century agricultural revolution. This is especially true in North America, where the major commercial developments have taken place hitherto, and there is much pressure from the US agbiotech industry, supported by the highest levels of government, to export the agbiotech paradigm to a global market. These attempts have met with considerable resistance, especially in Europe, leading to the first stages in a possible trade war between the European Union and the USA.[436] Meanwhile, the validity of the agbiotech paradigm itself is coming under increasing scrutiny, not only from anti-GM protestors, but also from more sober and impartial observers in business and academia.

Given these concerns, it is important for us to assess critically the current status of the agbiotech industry, and its possible future development, either as a solely private sector enterprise or as a more mixed public/private venture. During the past two decades, we have witnessed a dramatic growth in the agbiotech business sector, and the spread of transgenic crop varieties across many areas of the world. This has been accompanied by an overwhelming focus, often amounting to a seeming obsession, among both scientists and the general public, on this particular strategy of crop improvement.[437] The developments in plant science over the past few decades have also involved the conflation of broader research on crop improvement with the application of the particular and quite narrowly applied technology of transgenesis. This process has developed to the extent that it sometimes appears that crop improvement, particularly in a commercial context, has become virtually synonymous with the agbiotech approach.

This inappropriate fusion of two separate but interrelated concepts, i.e. the assumption that transgenesis is the best or even the only way forward, and that it is somehow quite different from the rest of plant breeding, has distorted the views of both scientists and the public on the future options for agricultural R&D. In reality, there is no such thing as a simple choice between transgenic or non-transgenic forms of plant breeding. There is just plant breeding, which will always rely mainly on field-based selection and trials, even after a new transgenic plant is created. All that transgenesis offers is a new method of enhancing variation, to be used or not as the case may be, alongside the many other technologies able to achieve similar results, such as mutagenesis and hybridisation. As shown in Figure 1, transgenesis gives the breeder a further set of options at the initial stages of a breeding programme, but has no impact on the remainder of the programme. To their abiding cost, several of the early agbiotech ventures suffered from a lack of this kind of realisation, namely that the successful addition of new genes to a crop is merely the beginning of the process of developing a new variety, and by no means marks its completion.[438]

The combination of the new transgenesis paradigm with the retreat of the public sector has meant that contemporary crop science is in a state of some uncertainty about the way forward. This phenomenon is not always apparent to the outside world, and indeed many academic and research focused plant scientists might claim to be unaware of what is happening in the more applied spheres of their profession. But, just because it has gone largely overlooked until recent years, the phenomenon is none the less serious. In the last chapter, we saw that many informed observers now believe that the decades-long decline in the public plant breeding sector in many industrial countries has gone much too far. Meanwhile, as we will discuss in Chapter 16, there are troubling signs that the carefully crafted network of international plant germplasm and breeding centres (CGIAR) is also coming under strain and is facing an ever more uncertain future.

One symptom of the strain being felt by some of the CGIAR institutes came in early 2005 when the two strongest centres, CIMMYT and IRRI, announced what amounted to a partial merger in what was, at least in part, a cost-saving exercise.[439] In 2003, practitioners of crop improvement were referred to in a *Nature* opinion article as a 'dying breed'.[440] In stark contrast to such pessimism, we are frequently told by those associated with the agbiotech industry that, thanks to genetic engineering, the world is on the verge of a new dawn in crop improvement. As if this were not confusing enough, we are also admonished on a regular basis that the prompt acceptance of transgenic crops globally is actually a *prerequisite* to feeding the increasing population of the world over the next few decades. Some of those who do not toe this line are even said by certain agbiotech proponents to have 'blood on their hands'.[441] We will look in detail at the arguments surrounding the issue of 'feeding the world' in Chapter 14, but suffice it to say here that the arguments are complex and the need for transgenic crops in this context remains far from proven.

To independent observers, the messianic fervour of some of the more prominent proselytes of transgenesis should strike a note of caution. As we will see later in this chapter, the 'life-sciences' business model of biotechnology that was so publicly championed by Monsanto CEO, Bob Shapiro, and others in the late 1990s has decisively unravelled. Despite the continuing chiliastic aspirations and upbeat pronouncements of its boosters, we cannot be sure about the long-term future of agbiotech as a sustainable business enterprise, at least in its current oligopolistic manifestation (see the section below, entitled *Domination by the 'big four'*). Since the turn of the millennium, the agbiotech sector has undergone a significant downsizing. As of 2006, there were only three or four major companies still involved in large-scale commercial cultivation of GM crops. The sector has also been noticeably slow to follow up from its initial rather basic portfolio of products (i.e. herbicide tolerance and insect resistance traits) and develop some of the more innovative, customer-friendly varieties that have been publicised by many of us (the author included) over the past two decades.[442] Moreover, although the global acreage of GM crops continues to grow annually, the few active companies in the sector still offer almost exactly the same narrow suite of traits that were around in the early 1990s.

Such restricted choice and sluggish pace of product development would have tolled the death knell for many companies in some of the more competitive and innovative modern high-tech industries, such as computing, photography and tele-communications. Agbiotech companies are naturally keen to develop consumer-friendly traits, but technical difficulties and, arguably, a lack of drive, focus and urgency means that they have manifestly failed to do so. Regarding the current GM crop traits, the market performance of the major multinational agbiotech companies has been decidedly mixed. For example, despite some good sales figures in the Americas, most larger companies have been shrinking and consolidating their research efforts over the past few years; and their wares remain largely excluded from the largest global trading bloc, the European Union.[443] Meanwhile consumers are still waiting for the kinds of products from agbiotech that will have direct appeal to them, e.g. cheaper, better tasting or more nutritious food. Having been told for two decades, that such products are just around the corner, the public is justifiably sceptical of these and many other claims from the agbiotech industry.

In the recent words of respected agricultural economist Lawrence Busch (the senior author of the report on the UC Berkeley/Syngenta collaboration, as discussed in Chapter 10): 'agricultural biotechnology, despite a few successes here and there, has thus far been a failure. It has failed to live up to the hyperbole, of course, but – more importantly – its proponents have failed to enroll citizens and consumers, and even the food processing and retailing industries around the world.'[444] Of course, despite what Busch says here, some aspects of agbiotech have been quite successful, otherwise it would not have spread as much as it has.[445] But in terms of the bigger picture, and the initial hype of the technology, he is quite correct in his judgement. In the rest of this chapter, we will discuss the often misunderstood role of transgenesis in plant breeding, and will assess its realistic potential for agricultural improvement. Readers who are interested in more detailed historical accounts of the agbiotech industry are recommended to consult some of the many and varied texts that are listed in the endnotes, with the caveat that they represent a distinctly mixed bag of greatly varying quality, accuracy and perspective.[446]

The artificial dichotomy of GM and non-GM

Origins of a false premise

One of the main reasons for many of the unrealistic expectations for transgenesis (or GM as it is popularly known) is the perception that it represents a qualitative, revolutionary change in crop improvement. This perception is also responsible for the vast majority of the fears about the technology, most of which are even more unfounded than the often hyped-up claims of the pro-GM lobby. Proponents and opponents of GM technology are equally guilty of fuelling this misapprehension, whether through ignorance, misunderstanding or opportunism. The roots of the perception of transgenesis as a radically different form of technology can be traced back to the 1970s, when researchers first developed the techniques of manipulation of DNA in vitro in a controlled, reproducible manner. As soon as it became possible in principle to isolate and copy genes from any organism and to insert these copies into a plant, it seemed as if there were no limits to the possibilities for the genetic engineering of crops. The main problem, for both research scientists and those with whom they communicate (e.g. politicians, media, business leaders and society at large), is that this optimistic conclusion about the potential of GM technology is, theoretically speaking, quite true: but it also happens to be immensely misleading in actual practice. So, how can we square these seemingly contradictory views about the potential of agbiotech?

We already know that most of the key traits that regulate the performance of crops are genetically determined.[447] Therefore, if we can identify and isolate the genes responsible for such traits and transfer them to a crop, we could create almost any desired genetic combination. It became apparent in the 1980s that this was indeed possible – at least in principle. For example, breeders could combine the best qualities of maize and wheat to create what would be a distinctly odd-looking, but immensely bountiful, crop that produced high-quality, bread-making grains as part of a large maize-like cob.[448] This would indeed be a revolutionary development in terms of increasing the yields and improving the nutritional quality of cereal crops around the world. A second example relates to my own research area, namely oil crops. For the past fifteen years, many of us who do research in this area have been advocating the engineering of so-called 'designer' oil crops to produce virtually any type of carbon-based product, ranging from pharmaceuticals and biodegradable plastics to vitamin-rich edible oils.[449] The trouble with these aspirations was that the translation from theoretical possibility to real-world application took far longer than any of us had ever imagined. In the meantime, such attempts to engineer super cereals or designer oil crops remained largely irrelevant to the actual needs of farmers on the ground, whether the latter were high-volume soybean producers in Kansas, or subsistence cassava cultivators in Kenya. Nevertheless, during the 1980s, many of us talked up the potential of the technology, through a mixture of ignorance about the (non-) immediacy of its application, and/or as a way of publicising our research and securing additional funding for its continuation – as we were exhorted to do by government policymakers, faculty/department heads, and institute directors.

This was indeed a golden era for many molecular biologists. In my own case, I was approached by several foreign governments, multinational oil companies, chemical manufacturers, and some rather dubious businessmen.[450] All had substantial funds to offer and we discussed many projects, most of which were based on totally unrealistic expectations of GM technology. By the early 1990s, many scientists believed that the new GM/transgenesis paradigm had more or less replaced conventional plant breeding, and that almost anything was achievable within the next few years. Our infectious enthusiasm and boundless optimism were soon communicated to the public via the various media, who were often only too willing to compound our own modestly hyped-up aspirations with their own brand of sensationalism. During these early days of transgenic crop development, the talk was all about radical alterations to crop traits, and very few researchers mentioned such workaday options as herbicide tolerance. For a start, it was much easier to communicate the idea of revolutionary changes in crop appearance or nutritional content than something as mundane as herbicide tolerance or insect resistance (which had already been achieved anyway in some crops using conventional breeding approaches). At this stage, the perception of GM technology in the media and among the public at large was rather positive in most countries, including the UK.

The more serious backlash against GM crops came much later, especially after the mid-1990s, by which time public sector scientists had been more or less removed from a stage that was increasingly dominated by a few multinational corporations.[451] As it turned out, these companies had a completely different agenda to our original notion of radically different and improved crops for all. The new 'dumbed down' commercial version of genetic engineering was used to manipulate some of the most basic and scientifically simple production traits, such as herbicide tolerance and insect resistance. As we all knew, these particular traits had already been successfully manipulated by non-transgenic methods. This meant that, in breeding terms at least, there was little qualitative novelty involved in the new developments.[452] Therefore herbicide tolerance and insect resistance traits tended to be of little interest to most researchers. However, despite their lack of any particularly innovative qualities (in scientific terms), these new transgenic crop varieties were much more easily patentable, simply by virtue of being transgenic (see Chapter 7). This was the decisive factor that led to production of varieties with transgenic herbicide tolerance traits, rather than development of the same traits using less patentable alternative technologies like mutagenesis.

It should be emphasised that not all agbiotech companies made this step. Several of the larger companies, such as Dupont, decided to stay away from the simple input traits and work towards longer term targets relating to more consumer focused output traits, such as taste or nutritional quality. Unfortunately for the agbiotech sector as a whole, it was the actions of companies like Monsanto and Novartis in commercialising simple input traits in the mid-1990s that set the agenda for transgenic crops over the next decade. The trouble with this particular manifestation of transgenesis was that, although it was being used in a relatively trivial and non-innovative manner, it still carried all of the hyped-up baggage that had been generated by the earlier ambitious expectations of radical and revolutionary manipulations to crops. In a sense, therefore, we had collectively created a kind of monster of expectations; and when the decidedly feeble creature was eventually released, many people still greeted it with a fear of Frankenstein-like proportions.

Spot the difference?

At this point, it is germane to ask: *what is really so different about genetic engineering that requires it to be placed in a special category from all other forms of plant breeding?* The public has been told by many scientists, politicians, agbiotech companies and anti-GM campaigners that transgenesis is qualitatively different from other forms of plant breeding. Indeed, in most countries, this difference is even enshrined in law. Unlike other advanced breeding technologies, the application of transgenesis is subject to a confusing and inconstant set of regulations that differ markedly from country to country. Also, unlike other forms of variation enhancement in crops, transgenic technology is seemingly so special that it has stimulated a massive new corporate sector, with a host of proselytising boosters who seek to spread the Word around the world. GM is so special that it has stirred some people to protests that have involved the destruction of property, physical threats, legal injunctions, ballots, demonstrations, and a veritable torrent of opinion in print, online, and in the broadcast media. GM is even so special that hitherto law-abiding matrons have helped to lay waste experimental trial plots of innocuous transgenic plants, risking imprisonment and a permanent criminal record in the process. One could not imagine such people getting into such a state and risking their liberty for the sake of opposition to other plant breeding strategies, such as wide hybridisation or induced mutagenesis.

But, maybe we (including some plant scientists) have got things wrong. Perhaps we have exaggerated the differences between transgenic and non-transgenic breeding for our own various purposes? In order to address these points, we need to consider what, if anything, is so radically different about transgenesis compared to other methods of genome manipulation. An obvious answer might be that transgenic organisms contain new genes of 'foreign' origin. However, one thing that modern genetics has taught us is that genomes are plastic, malleable, and rather inconstant entities that are in a continual state of flux.[453] Genes are constantly being exchanged between the main nuclear genome and the smaller mitochondrial and plastid genomes. Huge quantities of foreign DNA, often of viral origin, have entered almost all plant and animal genomes, including that of humans. Some of our crops, like maize, seem to be able to tolerate and even make use of this foreign DNA, such that it has been allowed to grow and multiply until it makes up over 95% of the genome. Other crops, like rice, appear to be more fastidious and have removed a large proportion of their exogenous DNA during their recent evolution.

Gene transfer within and between plant genomes

Recent evidence from genomic research shows that DNA is constantly travelling to and from the genomes of both plants and animals.[454] This process is referred to as horizontal gene transfer between species.[455] There are several mechanisms by which genes from one species can be transferred to another in plants: these include host–parasite gene exchange,[456] gene transfer via a plant virus,[457] genes transfer via mycorrhizal fungi,[458] gene transfer from a biting insect,[459] and non-standard fertilisation involving more than one pollen grain.[460] Some newly discovered examples of interspecies gene transfer in plants include movement of an isomerase gene from a member of the genus *Poa* (which includes meadow-grasses and bluegrasses) to the genome of an unrelated species, sheep's fescue, *Festuca ovina*,[461] movement of a transposon gene from rice to members of the *Setaria* genus,[462] and the transfer of a mitochondrial gene from an asterid (a group of flowering plants that includes the Solanaceae) to members of the gymnosperm genus, *Gnetum*.[463] As more plant genomes are sequenced and analysed in detail, it is becoming ever more apparent that inter-organism gene transfer between unrelated species is much more common than was suspected hitherto. There are even cases of genes being transferred from plants to animals. For example, the simple aquatic animal, *Hydra viridis*, has been found to contain a fully functional ascorbate peroxidase gene that was transferred to it from a former symbiotic partner, the photosynthetic green alga, *Chlorella vulgaris*.[464]

As a result of DNA transfer over millions of years, well over 90% of the DNA in most plant and animal genomes appears to be of exogenous ('foreign') origin. Much but by no means all, of this 'extra' DNA is probably non-functional, and some of it may even be parasitic in certain instances. This may be the reason for the removal of the 'extra' DNA in some crops like rice, where most of it has been steadily deleted from the genome over the past six million years. But, if rice can rid itself of much of its 'extra' DNA, why has maize or wheat (or, indeed, humans) not acted likewise? This is one of the most interesting questions facing molecular geneticists as they seek to unravel the many remaining enigmas of our genomes. It may be that rice is just better at removing its exogenous DNA than most other plants and animals, or the 'extra' DNA might be performing some useful function in those organisms in which it has not been removed.[465] Hence, we find that in other crops like wheat and maize, the exogenous DNA has persisted and even increased so that it now makes up about 80–90% of the genome. Remarkably, the maize genome has doubled in size over the past three million years, since its divergence from the related cereal, sorghum, and is still increasing to this day.[466]

Genetic manipulation or manipulation of genomes?

During the course of modern, *non-transgenic* plant breeding, we have learned to manipulate genomes by artificially transferring genes from other species into crop plants, i.e. by wide crosses (Chapter 2). Sometimes genes are even transferred from relatively unrelated species with which the crop could not possibly interbreed under normal circumstances. As a theoretical example, consider a high-quality, high-yielding elite variety of bread wheat. Let us assume that our wheat crop is threatened by a virulent new fungal pathogen. Luckily, we hear that colleagues in Asia have discovered an unrelated wild grass species that is resistant to this crop-destroying fungus. All we need from this wild grass is the single gene that confers resistance to the fungus; any of its other genes would probably just diminish the performance of our elite bread wheat. The challenge is to transfer a single gene from the wild grass to our unrelated domestic wheat plant. One solution is to perform a wide cross, followed by a backcross/selection programme. First, we add pollen from the wild grass to fertilise the bread wheat. Normally, the mother plant would quickly abort the resulting grossly abnormal embryos. To prevent this, the strange and unnatural hybrid grass/wheat embryos have to be removed from the female plant and put into artificial tissue culture in the laboratory, where they will be coaxed into growth by a complex cocktail of chemical growth regulators.

Sometimes this does not work, so an even more radical approach is required. The nucleus of a cell from the wild grass is chemically or mechanically disrupted, and some of its contents are injected into an intact cell from a bread wheat plant. This results in transfer of some chromosomes or chromosome fragments from the wild grass into the wheat nucleus. Because the wild grass chromosomes are so alien to the wheat genome, the new hybrid plants are normally infertile. The solution is to add a DNA-disrupting chemical toxin, such as sodium azide or colchicine, in order to force each chromosome to divide. Some of the resulting doubled-haploid plants will be fertile, but they are still useless as crops because they now contain a lot of unwanted genes from the wild grass. By backcrossing large numbers of the doubled-haploid plants with the elite wheat, the wild grass genes are progressively diluted from the genome. The breeders ensure that the desired fungal resistance gene remains, however, and the eventual result is a plant that is more or less identical to the original elite wheat variety, except that it now contains a new resistance gene from the wild grass. In this case, the breeder has used a rather roundabout and long-winded method to transfer a single desired gene from one species to another.[467]

As we saw in Chapter 2, this sort of breeding strategy has been used to create hundreds of new varieties of crops around the world, especially in developing countries. Such strategies may not involve transgenesis, but one could hardly call them anything other than radical and technologically sophisticated forms of genome modification. In principle, such non-transgenic methods could attract many of the same sorts of concerns as are normally only associated with transgenic technologies. For example, there are those who worry about the transgression of the 'species barrier' by moving genes from one species, or genus, to another.[468] There are also worries about the fate of DNA transferred from one genome to another and possible disruption of genes by the newly inserted DNA.[469] All of these concerns apply in equal or greater measure to the kinds of non-transgenic breeding methods, such as wide crossing and embryo rescue, that we have just examined.

Now, let us imagine an alternative scenario in which the breeder already knows exactly which gene from the wild grass is responsible for the disease resistance trait. Rather than bothering with a complex series of wide crosses, not to mention a lengthy series of tissue culture steps involving powerful genome-disrupting chemicals, the breeder could instead choose to isolate the desired gene directly from the wild grass. The gene could then be copied and inserted into the bread wheat via transgenesis. The result in both cases is essentially the same: one or more genes have been transferred from one species to another. In some of the progeny, the new gene(s) may not be in an ideal position in the genome. This means that more screening may be needed to select only those plants in which the new gene(s) are fully functional, do not interfere with existing genes, and are inherited in a stable manner over several generations.

To the farmer, it does not matter how the new crop variety was created. In both cases, it is the product of highly intrusive manipulations of a crop genome, of a sort that could not possibly occur in the absence of human intervention. Both procedures carry risks and have their own advantages and drawbacks. Both approaches involve genetic alterations of a magnitude that would have been impossible to achieve by pre-scientific breeding. Both approaches can, in principle, give rise to objections based on concepts like 'naturalness' or 'species integrity'. And yet, despite their similarities, these two approaches to crop improvement are treated by scientists, companies, lawmakers and anti-GM campaigners alike as if they were radically different. The non-transgenic route is essentially unregulated and is regarded as slightly old-fashioned (by scientists), unprofitable (by companies), non-patentable (by lawmakers), and as 'natural breeding' (by anti-GM campaigners). In the same vein, the transgenic approach is very highly regulated, and is deemed novel, patentable and profitable, by agribusiness; it is also considered to be not just unnatural, but also positively dangerous by anti-GM activists.

We are all genetic manipulators

There is a broad perception amongst the interested public,[470] and even amongst many scientists, that most researchers in fields related to plant biology and its application to crop improvement are automatically supportive of GM technology and its application. This is a misleading and simplistic impression that conflates the current manifestation of this new technology with the multitude of possible future directions that it might take, both scientifically and as a commercial enterprise. The current state of commercial agbiotech is almost a caricature of the true nature and potential of the technology. For a start, virtually everyone who has worked in the field of plant biology recognises the immense contribution that transgenesis has made as a research tool in the study of plant growth and development. Over the past twenty years, this use of transgenesis has revolutionised our understanding of plant genetics and development. Transgenesis is now enabling us to tackle more complex problems involved in discovering how plants determine their body shape, how they deal with environmental stresses, and how they tackle pests and diseases. However, to a great extent, much of what we have learned over the past decade or so about plants has merely shown us how much more still lies undiscovered about these apparently simple, but in reality very complex, organisms.

As many of us have been saying for the past two decades or more, most researchers also recognise the seemingly unlimited theoretical potential of agbiotech to enable the manipulation of plant traits in some quite radical ways. But this does not necessarily relate to the major concerns of plant breeding today. Despite the much proclaimed successes of agbiotech in manipulating a few simple input traits by transgenesis, it is almost certainly the case that the more significant, and normally quite unremarked, achievement of modern high-tech breeding has been in the use of marker-assisted technologies. In the words of Jorge Dubcovsky, a wheat molecular geneticist at the University of California, Davis: 'Fortunately, biotechnology has provided additional tools that do not require the use of transgenic crops to revolutionize plant breeding.'[471] Major M. Goodman, a well-known maize breeder from North Carolina State University has also stated that: 'Plant breeding is unlikely to be radically altered by genetic engineering despite progress in genomics.'[472] At this point, it is also worth quoting a longer passage from Goodman in which he seeks to redress the balance away from the current inappropriate and ultimately damaging bias towards agbiotech:

Once the euphoria over the promise of transgenics fades, the closing of so many quality breeding programs, the loss of valuable sales staff, and the centralization of decision-making at company headquarters are almost certain to be regarded as tragic, even by stockholders interested in short-term profits. There are few good investments that are more long-term than rational plant breeding. Repeated studies have shown that very high returns on investment are available from expenditures on [non-transgenic] breeding ... but the returns are not the instantaneous sort favored by the five-year funding plans currently in vogue. The usefulness of a breeding program is probably more dependent on continuity than ingenuity. The probability of great success by any one breeder is small, but the odds of success of a group of reasonably competent breeders working independently and continuously [and, I might add, sharing seed] is high. At present, the evidence that these same rules apply to biotechnology is almost nonexistent.[473]

It is a pity that the sober judgements of such highly respected independent scientists as Goodman, Dubcovsky and many others, who have nothing against agbiotech per se but who recognise its current limitations, seem to have been drowned out by the many shrill voices from those vested interests that seem to dominate all sides of the public discourse about agbiotech. We must also realise that one can be a firm believer in the future potential of agbiotech for crop improvement, without automatically being a fervent supporter of each and every aspect of its current commercial manifestation. This point has, alas, eluded most of those involved in the rather sterile and polarised public debates about GM crops, where anyone with an opinion must be either pro or anti with nothing in between. The danger of this sort of fundamentalism is that the neutral or disinterested scientist finds it very difficult to stay away from the polarised extremes. The GM crops issue then becomes conflated with other unrelated issues, such as anti-globalisation and anti-capitalism, distorting the dialogue even further by forcing people into a see-mingly simple choice between the 'bad' US-dominated, globalised, high-tech GM crops and the 'good', 'traditional' non-GM crops. Such intellectually lazy and irrational oversimplification is by no means unique to the GM crops issue: indeed it is regularly mirrored in the wider political sphere.[474] But it should nevertheless be resisted by anybody capable of informed and independent reasoning.

Agbiotech today – the worst of all possible worlds?

We have seen that there has been a sharp decline in the size and effectiveness of public sector plant breeding research and its application over the past couple of decades. To a great extent this decline has supposedly been compensated for by a much increased effort from the private sector. Although this was partially the case during the 1990s, it is becoming ever more apparent that it is no longer true. I have already argued that much of the private sector involvement in crop improvement during the twentieth century has, for quite understandable commercial reasons, been rather narrowly focused compared with the perspective of most public sector researchers. This situation is especially acute in agbiotech where, even after well over a decade of commercialisation, only a couple of simple input traits have been developed, and only four major transgenic crops are grown. We may therefore wish to ponder whether, by decimating public sector plant science and relying on an immature and increasingly biotech focused private sector, we have not ended up with the 'worst of all possible worlds' for the future of agriculture.[475]

A headless chicken?

My first confrontation with a real-life 'headless chicken' came as we celebrated the *fête national* of 14 July in a small village in Southwestern France. To my surprise, the recently decapitated fowl really did run around madly for several minutes, scattering bystanders and causing chaos at the barbecue, before expiring in a mass of bloodied feathers. We also use this term to describe somebody who sets off recklessly on a venture without due consideration for the future consequences. This individual, like Aesop's hare, soon tends to run out of steam and is eventually overtaken by the remainder of his slower, but more considered, companions. So, is the agbiotech industry to be thought of as a sort of headless chicken, running amok with its new transgenesis technology, and leaving havoc in its wake throughout the rest of the plant breeding sector? And will it ultimately suffer the fate of the over-hasty hare in being overtaken by a far less glamorous, and less well resourced, but more broadly based set of tortoise-like public sector breeding technologies?

Surprisingly, given their vast financial and intellectual resources, some of the agbiotech companies sometimes appear to be their own worst enemies, both via their actions and via the words of some of their most senior officers. For example, public sentiments about agbiotech seeming to be getting out of control, as increasingly articulated in the mid-late 1990s, were not exactly soothed by the following statement in a 1998 newspaper interview by Monsanto director, Phil Angell: 'Monsanto should not have to vouchsafe the safety of biotech food. Our interest is in selling as much of it as possible. Assuring its safety is the F.D.A.'s job.'[476] Unfortunately such statements (and note that this is from the person in charge of corporate communications for his company) were all too typical of a cavalier disregard for public opinion and an abdication of corporate responsibility, which is inappropriate in today's highly sensitive, consumer focused and risk-averse marketplace. During the 1990s, the culture of 'trust us, we know best' was coupled with the widespread perception that this was the new way forward for plant breeding. As we have seen, public expectations and concerns about genetic engineering had already been heightened by the exaggerated optimism of many researchers in the area.

It is certainly the case that most agbiotech companies have tended to focus on transgenesis, sometimes neglecting alternative technologies for crop improvement. This is not to say that private sector breeders do not regularly use alternative methods of creating genetic variation, such as mutagenesis, wide crossing and hybrid technologies. However, there has been something of a culture shift since the 1980s, especially in the larger companies, whereby the advocates of transgenesis have gradually gained more influence and power over company policy and research strategy. Moreover, companies rarely accord new crop varieties developed by non-transgenic methods the same sort of prestige and publicity that is granted to new transgenic varieties. The former therefore tend to remain relatively invisible, while the transgenic varieties gain the spotlight of both company and media attention. One of the best exemplars of the seeming obsession with the agbiotech paradigm in a leading company is former Monsanto CEO, Bob Shapiro. It was Shapiro who seized upon the technology with an almost religious fervour and duly launched the first major wave of transgenic crop varieties in the mid-1990s. According to former Pioneer Hi-Bred CEO, Tom Urban: 'Shapiro has this messianic sense about him. If he said it once, he said it three or four times: Put us together and we'll rule the world. We're going to own the industry. Almost those exact words. We can be a juggernaut. Invincible.'[477]

Shapiro was by no means alone at Monsanto in his enthusiasm for agbiotech as the new crop improvement paradigm. In 1996, Robert Fraley (co-president of Monsanto's agricultural sector) commented as follows on his company's acquisition policy: 'What you are seeing is not just a consolidation of seed companies, it's really a consolidation of the entire food chain.'[478] However, this fascination with agbiotech and the penchant for buying up other companies did not bear the promised fruit. Instead, the corporate overreach and ultimate failure of the 'life-sciences' policy at the company led to the ousting of Shapiro as CEO and the near demise of Monsanto itself in the early years of the twenty-first century.[479] What seemed to some as corporate 'arrogance', as exemplified by the earlier quotation from Phil Angell, was later admitted by Shapiro's successor as CEO at Monsanto, Hendrik Verfaillie, in a remarkable, if belated, apologia in November 2000. In his inaugural speech, entitled 'A New Pledge for a New Company' Verfaillie said:

Monsanto focused so much attention on getting the technology right for our customer – the grower – that we didn't fully take into account the issues and concerns it raised for other people. ... We didn't understand that when it comes to a serious public concern, that the more you stand to make a profit in the marketplace, the less credibility you have in the marketplace of ideas. When we tried to explain the benefits, the science, and the safety, we did not understand that our tone – our very approach – was seen as arrogant. We were still in the 'trust-me' mode when the expectation was 'show me'.[480]

Monsanto has subsequently improved its PR policy in some areas, although many observers would agree that it, and the rest of the agbiotech industry, still

have a long way to go before they regain any substantial measure of public confidence and trust, especially in Europe.[481] One example of how far Monsanto has
yet to go in the battle for public 'hearts and minds' came in early 2006, when the
company used legal injunctions to block unloading of three cargoes of Roundup
Ready® soybeans from Argentina at the ports of Santander and Bilbao in Spain
and Liverpool, UK.[482] The Monsanto embargo was a bizarre mirror image of the
Greenpeace blockades of Monsanto derived Roundup Ready® soybeans from the
USA in the late 1990s (see Chapter 17). Monsanto took this drastic action because
the Argentinian soybean producers had allegedly failed to pay royalties due for
their use of the GM seeds. In the end, Monsanto only succeeded in alienating even
more Europeans, as well as thousands of Argentinian farmers who wanted to grow
GM soybeans, and even the giant US-based multinational seed company, Cargill,
who owned the 5900 tonnes of soybean flour that was blocked from unloading at
Liverpool.[483]

Meanwhile, PR problems apart, the agbiotech sector is still hampered by its
narrow technology focus on transgenesis and input trait modification. Even
within this area, the industry continues to use outdated and consumer-unfriendly
techniques, as we will discuss in the next chapter. The over reliance on transgenesis still evident in some of the present-day agbiotech companies would be bad
enough for the long-term future of plant breeding, even if the private sector
continued to grow at the expense of the public sector. But what may make a bad
situation much worse is that the private sector may, as some suspect, have already
passed its peak, and might even be in decline. Many industry observers argue that
the boom years of the private sector may be over, and that its investment in crop
improvement is on the wane. If true, this really would leave us in the worst of all
possible worlds. Having disposed of much of our public R&D effort, we may also
lose the interest and investment potential of a retreating private sector (see
Chapters 1 and 19).

Moreover, a cash-strapped agbiotech industry might be tempted to milk its
existing transgenic products for all they are worth, rather than invest in costly new
research on improved transgenic varieties with more customer appeal and wider
applications. This would increase further its isolation and reduce its chances of
helping to develop the true long-term potential of transgenic crops. So, what is the
actual evidence of a decline in the private sector plant breeding effort? We have
already seen that several of the largest agbiotech companies downsized their research
operations across the world from 2000–2004. But is this symptomatic of a larger
industry-wide trend, or is it a transient blip in its progress? To answer this question,
we need to review the operations of the commercial agbiotech industry over the past
decade and assess their effects on the marketplace.

Rise and fall of the 'life-sciences' biotech business model

At the height of its growth phase, in the mid-late 1990s, the agbiotech industry went through a remarkable series of mergers and acquisitions. This process started when several of the major multinational pharmaceutical companies merged with each other and then acquired agbiotech companies in order to reconfigure themselves as vertically integrated life-sciences businesses. An idea of the extent of these multibillion dollar mergers and acquisitions can be gleaned from the following examples.

- Sandoz, which had already bought Northrup King, Sluis & Grut, and Hilleshög, merged with Swiss pharmaceutical and agrochemical company Ciba-Geigy to form Novartis in 1996.
- Zeneca, itself a biotech-based spinoff from UK-based chemicals giant ICI, bought several smaller agbiotech companies (such as Mogen and Ishirara Sangyo Kaisha) before merging in 1999 with Scandinavian-based multinational pharmaceutical company, Astra, to form Astra-Zeneca.
- German and French agrochemical and pharmaceutical combines, Hoechst-Schering-AgrEvo and Rhône Poulenc, both of which had previously acquired several smaller agbiotech companies, merged in 1999 to form a super agro-pharming combine called Aventis.
- Meanwhile, Monsanto, which over the past few years had already bought up numerous smaller seed and agbiotech companies (such as Holden's, Calgene, Agracetus, Incyte, Asgrow, PBI Cambridge and DeKalb) was itself purchased in 2000 by the recently merged Swedish–US pharmaceutical giant, Pharmacia Upjohn.

The rationale behind these mergers was that the resulting companies would use similar sets of expensive and sophisticated bio-based technologies, such as molecular genetics and automated screening of large sample populations. These core technologies were to be used to create a wide-ranging portfolio of value-added products, from human and animal pharmaceuticals for medical applications to transgenic plants and animals for agricultural use. The basic methods of DNA and protein manipulation are relatively similar regardless of whether one works with *E. coli*, maize or cattle. Therefore, it was argued, considerable economies of scale would be achieved by combining the R&D operations of the diverse companies involved in these mega-mergers. It was also expected that significant research synergies would be created by bringing together scientists from such different backgrounds. However, these massive conglomerates proved too unwieldy, and the economic basis of the integrated life-sciences business model soon unravelled. In particular, the mismatch between high-value, low-volume, very profitable pharmaceuticals and much lower value, high-volume, and less profitable agricultural products became starkly apparent.

The adverse impact of the various controversies surrounding the introduction of GM crops from 1998–2002 probably also played an important role in persuading senior executives to abandon their short-lived strategy of combined life-sciences mega-companies. At the time, CEOs and company boards were faced with poor profitability across the agbiotech sector, coupled with the emerging threat of the adverse consumer reaction to GM crops possibly spilling over to affect their relatively buoyant pharmaceutical divisions. The eventual decision was therefore something of a 'no brainer'. Within a couple of years, from 2000–2002, all of the large integrated mega-companies had divested themselves of their agbiotech and agrochemical components, and had gone back to the business of making serious money from pharmaceuticals. Hence, in 2000, Astra-Zeneca and Novartis spun off their plant science divisions to form a new solely agbiotech-based company named Syngenta. In 2001–2002, Pharmacia Upjohn, the parent company of Monsanto, sold off what had become an ailing agbiotech division of an otherwise profitable pharmaceutical company. In 2001, Aventis also decided to focus on its far more profitable pharmaceutical division by selling off its agbiotech group, which was renamed Aventis Crop Science.

This series of spinoffs created three smaller, but more focused, agbiotech companies, i.e. Monsanto, Syngenta and Aventis Crop Science. However, the troubles of these companies did not end here. Aventis Crop Science went on to suffer several GM-related disasters, including the StarLink case, which alone probably cost it over $500 million, and was eventually absorbed by Bayer in 2001.[484] During the next two years, both Monsanto and Syngenta retrenched and downsized their R&D operations around the world. For those associated with the industry, the beginning of the twenty-first century seemed to be as turbulent and unsettling as the massive shake-ups that had hit the public sector in the 1980s and 1990s. By 2003, former John Innes Centre Director, Mike Gale, was saying that: 'The industry is in turmoil.'[485] The company with the largest investment in commercial agbiotech, Monsanto, was still posting quarter year losses of as much as $42 million in late 2004, although recent performances have been better.[486] By the end of 2005, even the mighty Dupont/Pioneer conglomerate was forced to announce substantial R&D cutbacks as part of an effort to save $1 billion over three years.[487]

Domination by the 'big four'

As of mid-2006, the agbiotech industry was dominated by Monsanto, Syngenta and Bayer, plus the former chemical company, DuPont. US-based DuPont invested heavily in crop biotechnology during the 1990s, gradually selling off other assets such as the Conoco Oil Company and its pharmaceutical division. In order to secure its new crop-based focus, DuPont bought the original US seed giant, Pioneer Hi-Bred, for close to $10 billion during 1998–1999. Although they are much smaller than the major global pharmaceutical concerns, these four agbiotech companies are still multinational giants. Collectively, they control most of the world seed market and plant breeding industry. The 'big four' are especially dominant in the arena of agbiotech IPR, where they owned over 77% of all US utility patents in 2005.[488] While four companies may seem like enough of an oligopoly, the commercial agbiotech sector, i.e. that part of the industry that is actively growing transgenic crops, is even more concentrated as a virtual monopoly. In this sector, a single company, Monsanto, owned 90% of the seed used in transgenic crops in 2005.[489] But is this oligopoly/monopoly situation really having a deleterious effect on the market? After all, there are several areas of most industrialised economies, e.g. the automotive and oil industries prior to the 1970s, where oligopolies endured for many decades without unduly distorting particular markets.

There are two major questions that we should consider here. First, does the presence of a few large firms lead to the imposition of unnecessarily high prices in the market? Second, do these firms exert a stranglehold that is constraining the future development of crop improvement, whether via agbiotech methods or via alternative approaches? Starting with the price issue, in theory a monopolistic company should be able to use its position to extract high profits from the market because there is virtually no competition to enforce price discipline. However, the evidence is inconclusive as to whether this is really happening. Monsanto did indeed increase its technology fees and earnings from transgenic crops sixteen-fold from $235 million in 1996 to $3.8 billion by 2001. But over the same period, the area planted to such crops expanded more than thirty-fold, from 1.7 to 52.6 million hectares. So, if anything, Monsanto was earning about half the amount of fees per hectare of transgenic crops in 2001 than it had in 1996, despite its supposed stranglehold on the market. During this period, Monsanto lost as much as $8 billion.[490]

More generally across the agbiotech sector as a whole, a study published in 2003 estimated that the industry was only just breaking even and that it would take another 3–5 years to earn a market rate of return on its investments.[491] More recent figures show a mixed picture with Monsanto reporting a return to healthy profitability while other agbiotech companies report more modest incomes for the period 2003–2005. The relative success of Monsanto may be behind an unprecedented alliance between two of the other members of the 'big four' that was announced in April 2006.[492] This involved a new 50:50 joint venture between DuPont and Syngenta, branded as 'GreenLeaf Genetics'. Much of the collaboration is focused on the sharing or cross licensing of existing input trait technologies for herbicide tolerance and insect resistance. This partnership will challenge the domination of Monsanto in the agbiotech arena, possibly converting a near monopoly into a duopoly, but otherwise it seems unlikely to alter the fundamental structure of the agbiotech sector as a whole.

One of the problems for all of the large agbiotech companies is that they cannot rely on their highly patented and highly priced transgenic seed as a source of reliable income for more than a few years. This is because any new transgenic variety will itself become obsolete after several years as new elite cultivars are developed across the industry. Therefore, companies must continually cross their transgenic lines with the current best available germplasm in order to keep up with the breeders. For example, if a new (non-transgenic) maize cultivar with greatly improved disease resistance and yield is developed, it might rapidly eclipse the existing transgenic cultivars whose only distinctive agronomic asset is herbicide tolerance or insect resistance. The reason is that the former, more complex (non-transgenic) traits are often far more valuable to the farmer than the simpler (transgenic) traits. Hence, a high-yielding non-transgenic crop that was resistant to the major diseases might be a much better income earner than a modest yielding transgenic variety that happened to be herbicide tolerant.

So why does the agbiotech industry not develop these more valuable traits? The answer is that, of course, they would love to do this, but genetically complex characters like yield and fungal resistance are still proving very difficult to manipulate by transgenesis, whereas they have been successfully modified by alternative breeding techniques for many decades. In other words, despite a great deal of hype, commercial transgenesis is at present a pretty clunky and restricted technology that can only address a few relatively minor traits out of the dozens of others that are of interest to farmers (these topics will be discussed in greater detail in Chapters 12 and 13). In order to keep their customer base intact, the companies now realise that they need the whole panoply of other crop improvement technologies as well – and that is expensive. The upshot of this may be a retrenchment of the R&D operations as companies can no longer afford to keep developing expensive new transgenic varieties, while also keeping abreast of developments in the remainder of plant breeding. To quote former Pioneer Hi-Bred breeder (and Senior Vice President), Donald Duvick, talking about agbiotech R&D:

The payoff from this research will be long-range and will require consistency of application (and of funding) from either public or private sources if it is to succeed. ... One could also imagine that private industry would be unable or unwilling to devote sufficient funds to this research over the long term, thinking in particular of those firms whose management had supposed that biotechnology alone (or nearly alone) would be sufficient at this time to generate a continuing stream of improved cultivars.[493]

The second issue that confronts the private sector in the longer term is whether the dominance of a few large companies that own most of the IPR (i.e. patents) and PBR (plant breeders' rights) will stifle the entry of new players into the market and therefore act as a brake to innovation. According to the USDA, the mergers of the 1990s resulted in a concentration of patent ownership in the agbiotech sector whereby the top ten patent assignees controlled over half of agbiotech patents issued before 2000.[494] Several market solutions to this challenge have been proposed, including the alteration of patent or antitrust regulations, or the creation of new institutions to pool IP or to use inventions to circumvent monopoly positions.[495] Although there are numerous examples of big companies refusing to share or licence their technology either to competitors or to the public sector, such restrictive practices do not seem to have helped their profitability, while they have certainly contributed to their demonstrably poor public image.[496] On the other hand, it has also been argued that while most (but not all) of the proprietorial technologies are in principle available across the industry, the transactions costs, e.g. in setting up licensing arrangements, are so large that they effectively prevent the entry of small entrepreneurial startup companies and most public research organisations into the marketplace.[497]

It may also be the case that what is seen to be the rather stuttering start of commercial agbiotech will continue to act as a disincentive to the entry of more companies into the market in order to challenge the dominance wielded by Monsanto and others.[498] The apparent adverse effects of the 'big four' concentration in the USA, especially on overall innovation and the activities of new companies, have been discussed recently in the context of a possible investigation by the Federal Trade Commission (FTC).[499] This report concluded that the 'big four' agbiotech companies have the capacity to decrease total market investment in R&D and that the industry is on the cusp of sliding into a situation of 'research inefficiency and/or noncompetitive behaviour in the plant biotechnology innovation market'. However, any intrusive FTC involvement in the agbiotech business sector should be regarded as a last resort, especially given that the available evidence suggesting that markets can and will continue to adapt and can redress the current imbalances without external intervention.[500]

In the medium term, there are several large companies waiting on the sidelines to enter the agbiotech market, and possibly challenge the current dominance of Monsanto and others. Multinational companies, such as BASF and Dow Agrosciences, have developed the capacity to produce transgenic crops, but are awaiting the arrival of more customer-friendly traits before entering the market on any large scale.[501] For example, in 2006, BASF Plant Science announced the investment of $320 million in a variety of agbiotech traits over the three-year period until 2009.[502]

This could mean that the current quasi-monopoly enjoyed by Monsanto will only persist for several more years, and that within a decade the market will have opened up to include as many as six relatively large companies. Hence, a more competitive market might emerge, opening up the prospects for more widely acceptable transgenic crop varieties. But the mere presence of a few more companies will do little to address the fundamental issue of whether transgenesis is really the most sound and economical investment for crop improvement at the present time. This is something that will be resolved in company boardrooms and the marketplace, rather than in research laboratories. But in the meantime it is by no means certain that the 1990s paradigm of commercial agbiotech will survive in its present form into the second decade of the twenty-first century. To a great extent, the future success or otherwise of the agbiotech industry will lie in its capacity both to innovate its products and to deliver these to its customers.[503] As we will see in the next chapter, its past performance in this regard has been decidedly mixed and future prospects remain uncertain.

12

The future of transgenic crops I: improving the technology

Become addicted to constant and never-ending self-improvement
Anthony J D'Angelo (1997)
The College Blue Book

Introduction

Transgenic crops will almost certainly have a viable future, providing they deliver what is really desired by farmers and consumers. It is useful to remember that the only benefits that any new crop varieties ever deliver in plant breeding terms are agronomically useful traits. If the same traits can be delivered by other, more accessible, breeding technologies, especially at lower cost and with less regulatory encumbrance, then, all things being equal, transgenesis may not necessarily be the favoured route. For example, during the Green Revolution, semi-dwarf crop traits were developed in the absence of transgenesis. Now that we know the identity of the genes involved in these traits, we could in future use transgenesis to create semi-dwarf varieties of other crops, including many of the orphan crops in Africa. Indeed, Japanese breeders have recently identified a new set of semi-dwarf traits in rice, plus the corresponding genes, and the use of these genes in transgenic varieties could result in a doubling of yields in the crop.[504] However, even in this case, it is not absolutely necessary to use transgenesis to modify these genes. Because the position and nature of the genes are known, molecular markers have been developed that will enable breeders to select mutants carrying the desired semi-dwarf phenotype. In this case, a company or pubic sector breeder could, in principle, choose between transgenic and non-transgenic methods. Such choices should be dictated by criteria such as cost, timescale, feasibility, utility, and in the case of the private sector, potential profitability.

If transgenesis is to secure a viable long-term future as a well accepted and widely adopted strategy for crop improvement, three key issues should be addressed by the agbiotech industry, and by public sector plant scientists. First, the core technology itself, which is over twenty years old, is arguably now outmoded, relatively 'clunky', and has unnecessarily given rise to public concerns: it should be updated by adopting alternative methods as soon as possible. Second, management (including, where appropriate, segregation) of transgenic crops by companies, farmers and down-stream processors needs to be considerably improved. Third, much more focus should be placed on some of the consumer-friendly output traits and high-value traits that should be available to breeders and farmers in the medium term.[505] In this chapter, we will examine the first two of these issues, i.e. the technology itself, and the management/segregation questions. The third major issue (better products from GM crops) will be discussed in the following chapter.

Obsolete technologies?

It is a remarkable feature of virtually the entire current generation of transgenic crops that they still use technologies dating from the mid-1980s. This is even more puzzling because some of these technologies have caused public concern for well over a decade, although more acceptable alternatives have often been available to industry. Failure to address these issues reflects a malaise in parts of the agbiotech industry, whereby market signals have been repeatedly ignored or derided. There are many examples of such self-defeating behaviour, most notably in the attempts in the mid-late 1990s to import transgenic soybeans into Europe despite increasingly sceptical public opinion. Such attempts occurred despite warnings from scientists and others about the real likelihood of a backlash amongst European consumers already sensitised by a series of highly publicised public health scandals, ranging from the BSE and Salmonella scares to the Pusztai affair involving a study on rats fed transgenic potatoes.[506]

Although the consensus among scientists is that many of the concerns about GM technology, such as antibiotic resistance genes or 'superweeds', have little scientific basis, public concerns about the technology still need to be addressed by any company seeking to sell such material to its customers. This is true for any market, but is especially so for a high-profile, consumer-driven sector such as food, where perception can play at least as important a role as reality. For example, if objective scientific criteria were the sole arbiters of consumer behaviour, one might argue that there would be no such thing as organic foods.[507] As it is, there is widespread, if unfounded, public belief that organic food is invariably superior and more nutritious.[508] Hence, organic foods sell well despite costing as much as 50% more than conventional alternatives. The agbiotech industry therefore needs to address all consumer concerns, rational or not, particularly if there are readily available solutions to hand. We will now look at three well-known examples of such concerns about transgenic crops, namely: the presence of antibiotic resistance markers and other unnecessary DNA in GM plants; the effects of random insertion of foreign genes into a plant genome; and the risks of transgene movement to other plants, possibly leading to the creation of 'superweeds'.

Selectable markers

Selectable markers are used to distinguish between cells that carry transgenes and those that do not. In many cases after a transformation procedure has been carried out, only a tiny proportion of cells in a tissue or culture may be transgenic. Relatively powerful selection methods are required to eliminate the large number of non-transformed cells, while preserving the small proportion of cells that have success-fully incorporated a functional transgene(s). One of the most effective techniques developed in the early days of plant genetic engineering was based on antibiotic resistance genes. Following transformation, tissues or cultures were treated with a selected antibiotic that would kill all non-transgenic cells; only the small number of transgenic cells that carried the antibiotic resistance gene (plus the gene(s) for the trait of interest to the breeder) would survive. The issue of using antibiotic resistance genes in transgenic crops was widely debated among scientists from the early 1990s, as soon as it became clear that there was a (minuscule) theoretical possibility that such genes might be transferred to bacteria in the gut of a person eating food in which such DNA was still present.[509] If this bacterium were pathogenic, its acqui-sition of resistance to an antibiotic would obviously limit some of the clinical options for treatment of any resulting disease.

Among scientists, opinions were divided about whether the exceedingly minute risk of gene transfer justified the abandonment of an efficient and relatively inex-pensive selection method for transgenic plants. An example is an article from 1992 entitled 'Removal of selectable marker genes from transgenic plants: needless sophistication or social necessity?'[510] Even at this early stage of agbiotech develop-ment, it was recognised that there might be significant public concern about the presence of antibiotic resistance genes in edible crops, however small the risk might be. Moreover, the issue had been raised by the industry itself back in 1990 when Calgene asked for a ruling from the US Food and Drug Administration about the safety of such genes in its transgenic tomatoes.[511] At the same time, there were several strategies available to companies that would allow them to avoid the presence of antibiotic resistance genes in crops, either by removing them at an early stage of breeding, or by using less controversial types of selectable marker.[512] Despite the availability of alternative selectable marker systems, agbiotech companies have been very slow to eliminate antibiotic resistance genes from their transgenic crops.

As recently as the spring of 2005, it emerged that an unapproved transgenic maize variety, developed by Syngenta and mistakenly supplied to farmers from 2001–2004, contained a gene conferring resistance to the antibiotic, ampicillin. This antibiotic is commonly used for treating human and animal infections. Although the risks of transmission of functional ampicillin resistance to pathogenic bacteria are negligible,[513] the fact that information on its presence in the unapproved Bt10 maize was initially withheld from the public led to deep anxieties about the conduct both of the company and of the US regulatory agencies.[514] Eventually, in April 2005, the USDA fined Syngenta $375 000 for their infringement of the regulations in allowing the release of Bt10 maize into the food chain. In the UK, this affair was played out within the context of heightened public concern about a coincidental rise in hospital-acquired infections due to antibiotic resistant bacteria, and especially the much feared methicillin resistant strain of *Staphylococcus aureus* (MRSA). While the MRSA outbreaks in the UK and elsewhere had nothing whatsoever to do with transgenic maize carrying ampicillin resistance genes, it is not difficult to imagine the subconscious links that can be created in the minds of the public; not to mention the way that such events are readily manipulated for their own purposes by opponents of agbiotech. To an outside observer, therefore, it is puzzling that agbiotech companies should still be using this kind of technology, given the known public concerns (however misguided), and the existence of alternatives that would avoid the need for any such controversy.[515] A sign that at least one company is belatedly recognising the importance of the selectable marker issue came in 2006 when BASF Plant Science and French biotech spinoff (from the Pasteur Institute) Cellectis announced the licensing of meganuclease recombination technology for the removal of marker genes from transgenic crops.[516]

The use of antibiotic resistance genes in transgenic crops is not the only case of an outdated technology being needlessly employed in a way that only serves to fuel the concerns of opponents, and even some neutral observers, of the industry. As we can see from the MRSA scares mentioned above, the whole area of food and health is not just highly newsworthy, it is also much influenced by subjective sentiment.[517] Unsurprisingly, the steady stream of controversies around antibiotic resistance has had a corrosive effect on overall public confidence in agbiotech. Another selectable marker system with adverse effects on public opinion is herbicide tolerance. This system has been used from the earliest days of genetic engineering and is still present in some of today's commercial transgenic crops. Once again, concerns were raised, as long ago as the late 1980s and early 1990s, about the possibility of the escape of such traits via pollination with other cultivars or with sexually compatible wild relatives of the crop.[518] Such risks do not apply to hybrid crops but for open-pollinated crops, like oilseed rape or oats, this type of gene flow is a real, if low-likelihood, possibility.[519] Once again, industry has been slow to respond to such concerns and several transgenic crops still carry herbicide tolerance marker genes.

Despite it being in its own self-interest for the agbiotech industry to take a vigorous lead in developing marker-free transgenic crops, the public sector seems to be at the forefront of many such innovations. For example, one of the few remaining examples of public sector involvement in transgenic technology development in the UK, a programme called Biotechnology Resources for Arable Crop Transformation (BRACT), aims to produce transgenic plants free of selectable marker genes. Interestingly, these are referred to as 'clean gene technologies',[520] which carries the perhaps unfortunate connotation that the older transformation technologies are less than clean. Similar terminology was used in a *Nature Biotechnology* commentary on a new method that avoids the use of bacterial antibiotic resistance markers for the selection of transgenic plants. The article was subtitled 'A plant gene that confers antibiotic resistance provides a "cleaner" selectable marker for plant transgenesis'.[521] I raise this point because scientists and industry alike often complain about inappropriate use of emotive terminology by anti-GM campaigners. Examples are terms like 'Frankenstein foods' and 'terminator genes', which have now passed into the everyday language of the GM crop debate. However, we should recall that it was scientists themselves who invented dramatic labels like 'genetic engineering' and 'gene revolution'. It is therefore difficult for the same scientists to criticise others for using similarly emotive terms, especially if the former are then prepared to imply that certain aspects of the technology involved in transgenesis might be 'unclean'.[522]

Transgene insertion

The second area where the agbiotech industry has been slow to innovate technically is in the mechanism of transgene insertion into plant genomes. As we saw in Chapter 3, the two principal methods of choice are biolistics and *Agrobacterium*-mediated gene insertion. In both cases, exogenous DNA is inserted randomly into the genome and several copies might be inserted either in the same place or into different locations in the genome. The transgenic DNA may also fragment into numerous smaller pieces that become inserted elsewhere in the genome.[523] None of these outcomes are necessarily deleterious for the functioning of the desired transgene. And even if they result in unwanted phenotypes, the latter can be screened out by subsequent selection as the transgenic plantlets are propagated and backcrossed to elite cultivars. However, both biolistics and *Agrobacterium*-mediated methods of transgene insertion are relatively inefficient, and the technology would be greatly improved if transgenes could be targeted to a single predefined location in the genome, in order to give minimal interference with the expression of endogenous genes.

It is also becoming increasingly evident that many of the most desirable traits for manipulation via transgenesis will require the insertion of several, and perhaps over a dozen, genes. With the existing technology, and use of similar or identical regulatory sequences (promoters, terminators etc.) for each transgene, there is a risk of interference with expression between the various transgenes. Attempts have now started in several academic laboratories to develop new methods, including co-transformation and use of polycistronic transgenes, to address the issue of multiple gene insertion into plants.[524] Methods of targeted transgene insertion have existed for many years, and are well-established tools in mammalian and yeast systems.[525] One of the major techniques is homologous recombination. Transgenesis via homologous recombination has not seen the same success in plants as in animal and fungal models, because targeted integration tends to occur at very low frequencies in plant systems.[526]

A significant advance in achieving targeted gene insertion into rice plants was reported by a Japanese public sector group in 2002, but more work is needed before such techniques can be applied routinely.[527] One noticeable feature of the development of many of these basic technologies, such as transgene insertion methods, is the very low profile of the private sector in the research process. This is curious because ownership of a significantly improved technology platform like gene insertion would give a company a useful competitive edge, and would certainly be a public relations coup. Note also that the existence of an agbiotech oligopoly is a further factor that will tend to act against such technological innovations. It is also the case that this kind of technology development has historically been of less interest to academics than the more exciting challenges of basic research. This means that, as public research becomes ever more 'academicised', and companies fail to develop new technologies, mainstream commercial agbiotech tends to become ossified into dependence on relatively outmoded methods.

Ironically, the main reason for the recent resurgence of public sector academic interest in targeted gene manipulation has come from a desire to achieve knockouts of specific genes in order to investigate their function as part of fundamental scientific research, rather than any burning desire to develop more efficient forms of transgene insertion for commercial agbiotech purposes.[528] The need for researchers to do targeted knockouts of genes arises from the runaway success of genome sequencing programmes in describing tens of thousands of genes that still have no known function. Indeed, it is a telling comment on the current priorities in plant science research that tens of millions of dollars are being spent on sequencing programmes that have produced many terabytes of data, while we still only know the function of about one hundred of the 60 000 genes in the genome of our major global food crop, namely rice.[529] So it seems that, thanks to the success of genomics, new plant transformation technologies may well be developed in the near future. However, they will most likely be developed by the public, rather than the private, sector and for reasons that are completely unrelated to the betterment of the agbiotech industry per se.

Biological confinement

Another contentious technology-related issue that has long plagued the agbiotech industry is how to reduce or confine undesirable gene flow from a transgenic crop to other organisms. Of course, genes are constantly being exchanged between organisms, and spontaneous gene flow between crops and their wild relatives has been of great importance in the evolution of many crops. Gene flow between different plant species has been especially important in the evolution of domestication-friendly crops, such as cereals, the Solanaceae and brassicas. In the case of simple, single-gene traits like herbicide tolerance, the consequences of unwanted gene flow are very similar whether the crop is transgenic or not. In such cases, gene flow becomes an issue of agronomic management rather than transgenesis itself. In other cases, however, gene flow from transgenic crops can create unique challenges that should be of direct concern to the agbiotech industry. For example, engineering crops to express pharmaceutically active compounds (see *Biopharming* section in the next chapter) might create problems if these crops could transfer such genes to weedy relatives.[530] There are also sound commercial reasons for preventing the possibility of gene flow, since this would enable a company to confine its proprietorial genes to a single variety, thereby minimising the chances of their accidental or deliberate transfer to other varieties of the crop not owned by the same company.[531]

The issue of the so-called 'biological confinement' of transgenic organisms is not restricted to plants, as noted in a comprehensive report from the US National Academies in 2004.[532] However, plants have been a major focus of concern because they were the first transgenic organisms to be released into the open environment on a large scale: as of 2006 GM crops were grown on over 100 million hectares around the world. As with the examples of selectable markers and transgene insertion methods, the agbiotech industry has seemingly placed little priority on developing strategies of biological confinement for transgenic crops. Once again, it has been largely left to the public sector to develop such technologies, and once again progress has been relatively slow because such work is not necessarily regarded as cutting-edge research. Despite this, there are now several potentially applicable gene containment systems that are being developed, some of which are quite ingenious.[533]

One of the many possibilities is to transfer genes to the plastid, rather than the nuclear, genome. This means that the transgene is not present in the pollen of the plant and can only be transmitted via maternal inheritance in the vast majority of crop plants. A further merit of plastid transformation is that the transgene can be inserted into a predefined position in the plastid genome, rather than the random gene insertion mechanism dictated by current technologies. Progress on plastid transformation has been relatively slow, but significant technical improvements now make this an increasingly feasible option for crop engineering.[534] Among other bio-containment options being explored are apomixis, male sterility, cleistogamy, seed sterility, and inducible gene expression (see note for further explanation of these methods).[535] In the next few years it is probable that several of these biological confinement systems will become available for use for future generations of trans-genic crops. But these techniques will not have been developed by the agbiotech industry itself. As with the development of improved gene insertion technologies, the failure of the agbiotech industry to develop biological confinement systems can be seen as a reflection of its narrow focus and, perhaps, its restricted capacity for technical innovation in comparison to public sector scientists.

Other technologies

In this critique of the agbiotech industry, I have focused up to now on the major global players who are involved in large-scale commercial farming of transgenic crops. Most of the smaller and sometimes more innovative agbiotech companies disappeared or were acquired during the late 1990s. It is frequently the case in rapidly moving areas of science and technology that large multinationals can appear less effective at innovation than smaller companies, despite often purchasing the latter, along with their IPR portfolios and research staff. However, this does not mean that today's private sector is necessarily lacking in the ability to develop technological innovations. In the next chapter, we will see examples of a new breed of specialist small companies who are doing some particularly creative R&D in the area of biopharming. Unlike many of the previous small agbiotech startups, these newer companies tend to be more narrowly, and realistically, focused as R&D service providers, rather than seeking to get into the much more uncertain and messy business of crop improvement per se.

As we saw with selectable markers and transgene insertion, development of improved technologies has mainly come from the public sector. This is an unusual situation in the commercial sphere and can be levelled as a criticism of the overall innovatory capacities of the agbiotech industry. However, providing market forces eventually prevail, and R&D costs continue to fall, we should soon begin to see the emergence of other companies specialising in development of improved commercial transgenic and non-transgenic technologies. There are already signs of this with the non-transgenic technology called TILLING (see Chapter 3). Another interesting approach is the use of engineered zinc finger repressor proteins (ZFP TFs) to regulate genes in plants. As we will discuss in the next chapter, one of the greatest challenges to conventional transgenesis is the manipulation of complex multi-gene traits by simply inserting one or two transgenes. A small US company, Sangamo

BioSciences, with funding and technical assistance from the public sector and some agbiotech companies, has developed ZFP TF technology to manipulate large numbers of genes via the insertion of a single modified transgene.[536] There are other efforts to open up multigene traits to transgenic manipulation, although as stated in a recent review: 'One of the major technical hurdles impeding the advance of plant genetic engineering and biotechnology is the fact that the expression or manipulation of multiple genes in plants is still difficult to achieve.'[537] Looking further ahead, the agbiotech giants might emulate other areas of the biotechnology and information technology industries by outsourcing some 'frontier-orientated' R&D to the increasing number of innovative research-based service companies (see Chapter 18).[538] In the meantime, recent bibliometric data show that published output of research into plant transgenic technology has failed to grow over the past decade.[539]

Management, segregation and other challenges

In addition to the recent rather lacklustre performance of the agbiotech industry in technology innovation, there are several management issues that merit attention. Such issues include the imposition of prescribed agronomic practices on growers, and implementation of reliable, effective systems to segregate non-compatible crops, and their downstream products. Before tackling segregation, we will consider another management area where agbiotech companies have had a reasonable degree of success, namely refugia. Insect resistance in transgenic maize, cotton and potatoes was conferred by insertion of a gene encoding a protein toxin from the gram-positive soil bacterium *Bacillus thuringiensis* (Bt).[540] The obvious danger of relying on a single class of toxins is that it tends to establish a strong selection pressure favouring survival of insects able to sequester or detoxify the toxin. To combat development of insect resistance to Bt toxins in the field, farmers are advised to set aside refugia; these are areas adjacent to the main transgenic crop that are sown with non-transgenic seeds of the same crop. In the refugia, non-tolerant insect populations continue to thrive and hopefully out-compete any conspecifics that develop Bt tolerance. This strategy relies upon the cooperation and enforcement of good management practices by growers and can fail if not implemented rigorously. Following almost a decade of increasingly large-scale cultivation of Bt crops, the evidence suggests reasonably good farmer compliance and a high level of effectiveness of the refugia strategy in preventing (or, more realistically, delaying) the acquisition of Bt resistance in the target insect pests.[541]

   While refugia are a modest success story, the identification and segregation of transgenic crops from other crops remains one of the most contentious management issues for the agbiotech industry. As discussed below, a series of well-publicised failures to segregate certain GM crops or their downstream products has already severely impacted the industry, both financially and in terms of public confidence. While the series of often costly management failures in the recent past, such as the StarLink affair (see Chapter 13), have fortunately had no significant public health implications, they should nevertheless remind the industry that segregation and identity preservation (IP) procedures need to be critically re-examined as it moves into a new generation of transgenic crops with modified output traits.[542] As discussed above, the latest highly publicised episode in a growing list of management failures involved a transgenic maize variety, called Bt10, that was widely grown for several years in the USA (and exported for human consumption to Europe) although it had not been granted regulatory approval.[543] In this case, failure to segregate occurred at the beginning of the supply chain, when Syngenta accidentally multiplied up and supplied the wrong seed to farmers over a period of several years.[544] As the Bt10 and StarLink examples demonstrate, the issue of segregation needs to be tackled all the way from the company research laboratory, through its breeding and multiplication operations, through to labelling and selling seed to farmers via local merchants. And this is before the seed gets to the farm, where even more challenging segregation issues arise, both on-farm and throughout the rest of the supply chain to the crusher, processor, manufacturer and retailer. Not only will successful segregation require more stringent and wider ranging management oversight, it will be much more expensive than current practices. This means that new products from such crops need to be so attractive to consumers that the latter will, as with organic produce, willingly pay the considerable price premiums required to fund the necessary IP measures.

To look at the possible impact of the segregation issue on output trait modified crops, we can consider some transgenic oil crops already under development. Varieties of oilseed crops with different oil compositions will obviously need to be strictly segregated from one another. For example, we would not wish to mix seed containing premium edible oil with similar looking seeds containing an inedible oil designed for polymer manufacture. Unless a transgenic variety completely replaces all non-transgenic varieties of the same crop, it will require segregation and identity preservation at every stage from seed production and distribution to farmers, through on-farm storage and planting to harvesting, transport to mills, crushing, and downstream processing. Such measures can add at least 10–20% to product costs and, given the complexity of the supply chain, from breeder to consumer, via growers, crushers, processors, and retailers, IP also creates formidable new management problems. The difficulties in ensuring strict segregation of otherwise indistinguishable transgenic crops have often been highlighted, but are still underestimated by industry.[545] It is estimated that over a dozen new varieties of oilseed rape (the major European oil crop) could be engineered over the next few years, all looking identical, but producing completely different oils, some edible and some not. Imagine the scope for farmers to mix up bags of identical looking seeds and then grow mixtures of industrial and food oils in the same field. Or the consequences if a truck driver delivers seed to the wrong hopper at a processing facility, and contaminates a batch of therapeutic-grade oilseed for hospital use with an inedible seed variety. Segregation poses formidable challenges for an industry that, from recent performance, has not always appeared ready and willing to confront such issues.

Although they are by no means trivial, such management problems can and should be resolved by companies and farmers. After all, for more than three decades, large numbers of farmers in the UK have been growing two chemically very different but identical looking, non-transgenic oilseed rape varieties, in some cases on adjacent farms. These two varieties respectively produce an eminently edible high-oleate oil, and an industrial-grade, high-erucate oil prohibited for human consumption.[546] Despite the potential for cross-pollination or other forms of mixing between the two varieties, and the fact that over 400 000 hectares of rape is grown in the UK every year, careful management by farmers, seed crushers, and oil processors has ensured that there have been no significant instances of cross contamination in the past thirty years. So, segregation can and does work, even with large-scale commodity crops, but everybody in the food chain must cooperate to ensure that strict standards are observed. This becomes even more of an issue when we are confronted with not two, but maybe as many as twenty, different varieties that require such segregation. The recent StarLink and Bt10 incidents do not inspire confidence in the ability of sections of the agbiotech industry (including their downstream clients, such as farmers and processors) to regulate and manage themselves. And they certainly call into question the wisdom of growing crop varieties that contain therapeutically active compounds alongside major commodity varieties of the same crop (see the next chapter). As with the issue of outdated gene insertion techniques, the segregation challenge is eminently solvable, but this will require a sustained effort by the industry and a greater recognition of its wider responsibilities and interests.

13

The future of transgenic crops II: improving the products

To improve is to change; to be perfect is to change often.
Winston Churchill (1874–1965)
attributed

Introduction

For almost twenty years, the appearance of a large number of sometimes radical new transgenic crop traits has supposedly been just around the corner, but none of these 'wonder products' has yet made it to large-scale commercial cultivation. One of the commonest criticisms of the agbiotech industry is that a few companies precipitously commercialised a narrow set of relatively trivial input traits in the mid-1990s, rather than waiting to develop a wider range of consumer-friendly product traits. In this chapter, we survey the new generation of transgenic traits now being developed, their prospects for success, and some possible alternative non-transgenic strategies to generate the same traits in commercial and subsistence crops. We will consider two categories of traits, namely: input traits, which affect how the crop is grown without changing the nature of the harvested product; and output traits, which change the quality of the crop product itself, e.g. by altering starch, protein, vitamin or oil composition.

Input traits

Almost the entire current portfolio of commercial transgenic crop varieties contains modifications to either or both of just two input traits: herbicide tolerance and/or insect resistance. However, these two examples are far from the most important yield-limiting input traits in most farming systems around the world. More important yield traits relate to other types of biotic (from other living organisms) or abiotic (from the non-living part of their environment) stresses encountered by plants.

189

We will now survey some of the GM and non-GM strategies to enable better tolerance by crops of some of the many biotic and abiotic stresses they routinely encounter.

Biotic stress tolerance

Although insects and weeds can on occasion exert significant biotic stress on crops, the most important organisms in this context are viruses, bacteria, fungi and nematodes. These are all major pathogens of crops and there has been much research aimed at producing resistant varieties using transgenic approaches. While many chemical fungicides and nematicides are available to help farmers control these pathogens, there are no equivalent virus-control agents, so the combating of viral diseases normally relies on the endogenous resistance of the plant itself. In the absence of such resistance, viral infections can be particularly devastating to a crop. This has stimulated efforts to engineer viral resistance into transgenic crops. The commercial cultivation of several transgenic potato, squash and papaya varieties with virus resistance genes has already been approved in some countries.[547] So far, these efforts involve relatively small crop areas and have not yet led to commercial transgenic modification of any of the major crops, especially the cereals. In the medium term, however, the use of transgenesis to produce virus resistance in crops is a very promising area, and one where the agbiotech approach may well be the best option for combating many crop diseases.

Engineering of endogenous resistance to bacterial, fungal and nematode[548] pathogens has been much more problematic to address than viral or insect resistance, although several promising approaches have been demonstrated, at least in principle.[549] One possibility is to insert resistance genes, such as the Xa21 bacterial blight resistance gene that was recently transferred to five Chinese rice varieties.[550] Anti-fungal agents like phytoalexins or chitinases have also been expressed in plants[551] and a gene encoding an anti-fungal protein from alfalfa was transferred to potatoes.[552] The transgenic potatoes became resistant to the soil-borne fungus *Verticillium dahliae*, which causes some $70–140 million in damage to US potato crops each year. However, the resistance was quite specific and the potatoes were still susceptible to the late-blight fungus *Phytophthora infestans*.[553]

This highlights a major problem in engineering fungal resistance into crops, namely the difficulty in producing broad-spectrum resistance, without transferring numerous resistance genes. Also, fungal resistance often evolves spontaneously in the field and can sometimes be found in other varieties of a crop, or in sexually compatible wild relatives, from which it can be transferred to an elite crop cultivar by conventional breeding. In the transgenic potatoes mentioned above, a variety called

Russet Burbank was transformed with an alfalfa gene so as to confer resistance to the fungus, *Verticillium dahliae*, but there is another non-transgenic potato cultivar called Russet Ranger that is already resistant to this fungus. Hence, rather than adopt a GM approach in this case, the resistance trait from Russet Ranger could have been transferred, or 'introgressed', into Russet Burbank by simple sexual crosses. It is possible that in the longer term, more useful forms of transgenic fungal resistant traits may be developed but, at present, the non-transgenic approach seems to be the more realistic option for most crop breeding programmes.

Abiotic stress tolerance

The most important abiotic constraints encountered by crops are due to factors such as nutrient limitation, salinity, drought and thermal (both heat and cold) stresses. In most conventional farming systems, nutrient deficiencies are overcome by use of supplemental fertilisers, such as nitrates, phosphates, calcium and other minerals. However, the remaining abiotic stresses are a serious concern in many parts of the world, especially in regions where climate change or increasing soil salinisation threaten crop yields. Drought and salinisation are already the most common causes of sporadic famine in arid and semi-arid regions, and are the most significant threats to agricultural productivity in many regions. As much as one third of the world's arable land may become degraded by salinisation over the next 25 years.[554] The almost inevitable resulting human conflicts over water resources at both local and international levels could exacerbate food shortages still further in the affected regions. Although abiotic stress is often regarded as primarily an external (environmental) component of crop performance, physiologists have shown that there is a great deal of untapped genetic variation for stress response in all the major crops.[555]

Much has been made of the potential for genetic engineering for the improvement of stress-related input traits and this prospect is a major aspect of the biotech industry case for more extensive use of agbiotech in developing countries.[556] However, as many researchers in the field have pointed out, our limited knowledge of stress-associated metabolism in plants still constitutes a major handicap to effect such manipulations in practice.[557] Another problem that farmers and breeders have long been aware of is the synergistic effect of different stresses on crop performance. It is often the combination of such stresses that is so deleterious to the crop in an agronomic context, rather than the effect of any single stress. However, molecular biologists have tended to focus (for understandable reasons) on single stresses. Unfortunately for this piecemeal approach, recent studies have shown that simultaneous application of several stresses gives rise to unique responses that cannot be predicted by extrapolating from the effects of stresses given individually.[558] Because the co-presence of several stresses is the norm in the open environment, the success of a molecular approach to stress remediation in crops will require a broader and more holistic approach than we have seen hitherto.

Salt tolerance has been a particular focus of claims for significant results from transgenic approaches. One of the key prerequisites for the success of a gene insertion strategy to combat salt tolerance is that it should be regulated as a simple genetic trait, i.e. one involving a very small number of genes. Although such simple genetic regulation has been claimed in some cases in experimental studies,[559] it seems more likely that salt tolerance in most crops in the field is in fact a rather complex multi-gene trait.[560] Meanwhile there have been some successes at engineering salt tolerance in laboratory situations. One example is a transgenic tobacco line, expressing an *E. coli* mannitol-1-phosphate dehydrogenase gene that accumulates elevated levels of mannitol, and can withstand high salinity.[561] Laboratory and small-scale field studies have also shown that the accumulation of other compounds, including betaine or trehalose, in transgenic plants may enhance salt tolerance.[562] In a UC Davis study, rapeseed plants expressing an *Arabidopsis* vacuolar transport protein tolerated as much as 250 mM sodium chloride (about half the concentration of sea water and enough to kill most crops) without significant impact on seed yield or composition.[563]

However, it is not clear whether such relatively simple modifications will lead to a sustained effect on crop yields in the much more complex real-world cropping systems, where osmotic stress is often linked with a combination of other factors such as periodic aridity, mineral/salt build-up and/or erosion. This means that the jury is still very much out on the amenability of salt tolerance *in the field* (which is the only type of interest to breeders) to modification by genetic engineering.[564] Unfortunately, attempts to improve salt tolerance through conventional breeding have also met with very limited success, largely owing to the complexity of the trait. In the meantime, we know that salt tolerance must be an especially complex trait, physiologically speaking, because there are so many tolerance mechanisms in salt-adapted plants in the wild. This should lead to some caution when interpreting claims in the scientific literature that the transfer of one or a few genes can increase the tolerance of field crops to saline conditions.[565] The way forward here is to investigate as many realistically promising strategies as possible, but if I were a practical field breeder with limited resources, and our present state of knowledge, I would probably focus most of my resources on non-transgenic approaches to salt tolerance.

Drought tolerance, like salt tolerance, seems to be controlled by a complex set of traits that may have evolved as separate mechanisms in different plants. In the near future, it is likely that aridity will increase around the world. This will be caused by factors such as lower rainfall due to climate change, and the diversion of upstream water supplies from rivers, e.g. for dams or irrigation, leaving farmers in downstream regions bereft. It is surprising therefore that there have been relatively few attempts to produce transgenic drought tolerant crops, even by publicly funded organisations. An Australian group has recently reported that a single gene, called *erecta*, might regulate much of the genetic variation for drought tolerance in the model plant, *Arabidopsis*.[566] This approach merits further attention, but as with salt tolerance, it may turn out that in a practical field situation many other genes are involved in addition to *erecta* or its equivalents. Instead of transgenesis it is now possible to use advanced breeding methods to improve the agronomic performance of existing drought tolerant crops in arid regions. One of the most important such crops is pearl millet, which is grown on over 40 million hectares in Africa. The similarity in gene order, or synteny, between the pearl millet genome and that of the other major cereals[567] means that drought tolerance traits could be introduced into local varieties via marker-assisted conventional breeding. Another option is to use wide crossing and tissue culture methods to cross millet with one of the other high-yielding cereal crop species to create a new drought tolerant, high-yielding hybrid species. As discussed in Chapter 2, breeders have used such a strategy to create the new rye/wheat hybrid species called *Triticale*.

The final type of abiotic stress that we will consider here is that caused by certain soil-borne metals. Soils may contain metallic minerals due to geological processes, but in many cases toxic metals or their derivatives have been deposited as a result of a range of human activities, from mining to poor irrigation practices. We already know that some plants can tolerate relatively high levels of otherwise toxic heavy metals. The metals are often absorbed but, instead of harming the plants, they are chemically bound, or chelated, to specific proteins or other compounds, which allows them to be sequestered in a non-toxic form. The ability to express such traits in crop plants could extend their range of cultivation. Other ideas are to use such plants for 'bioprospecting' and 'bioremediation'. In bioprospecting, a metal sequestering crop is cultivated in an area that contains valuable mineral deposits that are not sufficiently abundant to justify direct mining. The possibilities of such an approach are illustrated by a report that a brassica species has been used to accumulate gold.[568] In bioremediation, engineered plants could be used to remove toxins from degraded soils so that they can be returned to productive use.[569]

One of the best-studied classes of proteins involved in heavy-metal tolerance is the metallothioneins, which bind metals such as zinc, cadmium and copper. Experimental trials have shown that transgenic rapeseed or tobacco plants expressing mammalian metallothioneins can tolerate elevated levels of heavy metals.[570] Although these approaches are yielding promising preliminary results, it will be necessary in future to carry out a thorough analysis of the field performance of such transgenic plants in order to assess possible unintended, or pleiotropic, effects of the transgene. This is particularly relevant to the overexpression of metallothioneins in plants because the precise biological function(s) of these proteins is still unclear, and it seems likely that they are involved in numerous physiological processes in addition to metal tolerance. Examples of such processes include fruit ripening, leaf senescence, and infection by viruses, bacteria and fungi.[571] Some of our own research showed that one class of metallothionein was the most actively expressed gene during the ripening of oil palm fruits.[572] Presumably this particular metallothionein plays an as yet unknown, but very important, role in this key developmental process. Indeed it is quite possible that the primary role of metallothioneins in plants relates to activities such as redox regulation rather than metal sequestration. This is a good example of how our emerging, but still imperfect, knowledge of many aspects of plant physiology and biochemistry is revealing that assumptions informing some of our genetic engineering strategies may require revision. It is still worth persevering with metallothionein studies, but it will be important to assess thoroughly all developmental stages of any resulting transgenic plants for any unexpected phenotypic behaviour.

Output traits

Modification of output traits will be necessary if transgenic crops are to deliver anything new to their ultimate customers, i.e. domestic consumers. Examples of such attributes include improved nutritional composition, enhanced taste, longer shelf life, plus the ability to make completely novel types of food product, thanks to a modified seed or fruit composition. There are two major challenges to manipulation of output traits via transgenesis, namely segregation/identity preservation and the complexity of the genetic traits themselves. The first issue is essentially a management problem, as we saw in the previous chapter. The second challenge is more technical and relates to the large numbers of genes that might need to be manipulated to effect a change in a particular trait. This genetic complexity is coupled with our still relatively poor understanding of many underlying biochemical and physiological processes that link expression of such genes with the phenotypic manifestation of the trait in question. These are significant challenges and they are one of the reasons for the relatively slow progress in bringing transgenic output trait modified varieties to the marketplace.

One of the most obvious ways that food can appeal to the modern consumer is by having a significantly better nutrient profile than many of the products currently on offer. Some of the many nutrient-related characteristics under consideration for use in transgenic crops include: improved vitamin content, lower saturated fat, increased omega-3 oils, presence of certain antioxidants (some of which are also vitamins), better essential amino acid balance, and higher levels of essential minerals like iron or zinc. In addition, there are significant sectors of the population with special dietary needs that can be met by producing foods in which certain problematic proteins have been modified or eliminated. Examples include decreasing gluten levels in some cereals for those with celiac disease, or eliminating allergenic proteins in seeds, especially nuts (including peanut). These allergens cause often violent responses in the increasing numbers of people who have various forms of nut allergies. The list of possible products is extensive and the prospect of creating such novel foods was one of the main drivers of the original research on agbiotech in the late 1980s. We will now briefly survey a few examples of these products and assess their prospects of commercial success over the coming years.

Enhanced vitamins

The so-called 'golden rice' developed in the late 1990s by a Swiss-based group is probably the best-known example of a nutritionally enhanced transgenic crop.[573] The grains of this rice variety are yellow, rather than white, because they were engineered to accumulate the pigment, β-carotene (provitamin-A), which is normally absent from rice grains. The transgenic rice contains three new genes encoding enzymes responsible for conversion of geranylgeranyl diphosphate to β-carotene. It is claimed that consumption of this rice by at-risk populations may alleviate the chronic vitamin A deficiency (leading to xerophthalmia and night blindness) that currently afflicts some 124 million children worldwide. Such claims are hotly disputed by anti-GM groups and the 'golden rice' has yet to prove itself in large-scale field and nutritional trials in target developing countries.[574] One of the main reservations expressed about the original varieties of golden rice was the relatively low content of provitamin-A which might entail the daily consumption of several kilograms of rice to meet dietary requirements for vitamin A.[575] If true, this would clearly be impractical, especially for children who would be the most important target group for the rice. More recently, this particular problem has apparently been solved by replacing the daffodil phytoene synthase gene with a similar gene from maize. Use of the maize transgene in rice led to a 23-fold increase in provitamin-A levels. This means that grains of the improved version of 'golden' rice are now bright orange, rather than the insipid yellow of earlier varieties.[576]

Although this was an exciting technical achievement, it remains the case, as ever, that the development of golden rice plants in the laboratory is only the start of a lengthy process of breeding and feeding trials that may take anything from five years to a decade. Many years of backcrossing into local varieties and field tests lie ahead before we will know whether golden rice varieties could become viable crops. Not the least of the challenges ahead is to ensure that the newly expressed provitamin-A is in a form that can withstand processing, storage and cooking of the rice, while remaining in a form that is completely bioavailable following consumption by people. Bioavailability of nutrients in foods is an often overlooked limitation on their dietary efficacy. There are many cases of vitamins and mineral nutrients that are lost during post-harvest treatments, such as processing and cooking. Alternatively, nutrients in a particular food product might pass through the human digestive system without being absorbed, e.g. due to chelation or other forms of chemical complexing. Probably the best-known example of this is spinach, where only 2% of the iron in the leaves is actually bioavailable, because most of it is strongly bound to oxalates and passes right through our digestive system without being absorbed. Notwithstanding the many challenges that face golden rice, the development of various cultivars is well under way and this crop may eventually have a modest impact on human nutrition in some parts of the world.

While golden rice is a promising development, there are many alternative ways in which plant breeding can enhance vitamin A levels in food crops without recourse to the expensive and time consuming transgenic approach, with all of its many complications as regards regulation, IPR and public perception. Several such initiatives are now under way and, although they have not attracted the massive media attention (both good and bad) that has been lavished on golden rice, some of these conventionally bred crops are much closer to release to farmers, most particularly in sub-Saharan Africa. For example, there are two programmes aimed at enhancing vitamin A levels in the staple food crops, maize and sweet potato. The first of these is the High β-Carotene Maize (HBCM) Initiative, organised by the HarvestPlus consortium. The latter includes the United States Agency for International Development (US-AID), the International Food Policy Research Institute, Washington, and the International Center for Tropical Agriculture, Colombia (ICTA, which is part of the international network of CGIAR centres).[577] While one aim of the HBCM Initiative is to develop vitamin A enriched varieties of maize, an equally important goal for researchers at Iowa State University is to assess the bioavailability of the vitamin A in the kinds of maize products actually eaten by the target population in Africa.

The second programme is called Vitamin A for Africa (VITAA), and is jointly funded by the Macronutrient Initiative and US-AID. As part of VITAA, the dietary impact in school-age children who were given high-carotene varieties of sweet potato, *Ipomoea batatas*, has been evaluated, with promising results.[578] Sweet potato is globally the fifth most important crop on a fresh weight basis and is especially important in Africa, where it is traditionally cultivated by women. Plant breeders from the International Potato Centre (CIP, which is another CGIAR centre) at

Lima, Peru have recently produced new varieties of orange-fleshed sweet potatoes. The new sweet potatoes are highly enriched in a readily bioavailable form of vitamin A. These new clonal varieties are being mass propagated in Peru and sent for a 12-month quarantine period to Kenya, before being distributed to centres in Ethiopia, South Africa, Uganda and Tanzania for further evaluation and on-farm testing. The final stage will be to provide the sweet potatoes for cultivation by needy farmers throughout sub-Saharan Africa. VITAA researchers from Michigan State University and the International Potato Centre have concluded that the new orange-fleshed varieties could replace white sweet potato varieties, and thereby directly benefit over 50 million children under six years old currently at high risk from vitamin A deficiency.[579] It will be interesting to compare the impact over the next decade or so of VITAA sweet potatoes and HBCM maize in Africa, which few people have heard of, with the much more heavily publicised golden rice varieties in Asia. Hopefully all three crops will make a contribution to human welfare, but for the pragmatic breeders in developing countries, the VITAA and HBCM strategies seem to have the virtue of greater simplicity and practicality under current regulatory and technical constraints.

Another important nutrient group attracting the attention of breeders is the vitamin E complex, which includes four tocopherols and four tocotrienols, all having antioxidant properties. These compounds are implicated in reduction of blood cholesterol levels, plus protection against ageing and several forms of cancer. Vitamin E group compounds are found in most non-processed (i.e. cold pressed or virgin) vegetable oils, but are often lacking in the vast range of prepared foods made from the more common, and cheaper, processed oils. There is interest in trying to increase the levels of this group of fat-soluble vitamins in plant oils using a variety of approaches. For example, transgenic plants accumulating 10–15-fold higher levels of vitamin E compounds have been engineered by adding genes encoding homogentisic acid geranylgeranyl transferase (HGGT) from several cereals to *Arabidopsis* plants.[580] The same report, which was from a group at DuPont, indicated that overexpression of a barley HGGT gene in maize led to a six-fold increase in tocotrienol content in the grain of the transgenic plants. The prospect of boosting vitamin E content in grain crops by the insertion of a single gene means that this might be one output trait that is realistically amenable to transgenic manipulation. Whether health-conscious consumers would accept such a product, and pay a considerable premium for it, in the present climate is another matter.

Once again, this raises the issue of alternatives to the transgenic approach for improving the vitamin content of foodstuffs. In the case of vitamin E, it turns out that there are several other choices for the consumer who is interested in better dietary sources of this desirable antioxidant. For example, many non-transgenic plants already contain high levels of various members of the vitamin E group, including many coloured fruits and seed oils. One of the most interesting sources of vitamin E is the fleshy, oleaginous fruit of the oil palm. In recent years, breeders have examined several African varieties of oil palm that produce an edible oil, which is highly enriched in a range of vitamin E group compounds.[581] The unprocessed, or 'virgin', form of this oil is a marvellous and visually striking burgundy-red colour, due to the additional presence of nutritionally important antioxidant pigments including yellow/orange carotenoids and deep-red lycopenes.[582] This nutritious oil is already used by people in tropical Africa, but has not yet been taken up widely elsewhere, possibly because of a feeling by retailers that Western consumers already accustomed to the more common bland (in taste and colour) light-yellow oils might be put off by the unusually vivid colour.

However, there is growing interest in red palm oil as people begin to recognise potential benefits, and it is now available in some US and European health food stores. Because this high-yielding vitamin E rich variety of oil palm is an existing mainstream crop, the development of a robust production, supply chain and retail market for red palm oil would be a lot cheaper (especially for consumers) than engineering alternative transgenic crops. Such a strategy would also be of immense benefit to farmers in the potential producer countries in West Central Africa, such as Sierra Leone, Liberia, Ivory Coast, Ghana, Togo, Benin, Nigeria, Cameroon, Congo and Angola. It remains to be seen if vitamin E rich foods are really desired by consumers and, if so, whether they would prefer to purchase a familiar-looking oil from a transgenic crop, such as soybean, or the strikingly coloured, non-transgenic, virgin red palm oil from the tropics. For understandable commercial reasons, the agbiotech industry is backing the former option, but it would be good for customer choice if the latter alternative were available as well. Perhaps both products could find a market niche, but most importantly, if both were available, there would be a free choice and the consumers would be the final arbiters.

Biofortification with essential minerals

Many crops are relatively deficient in key minerals required in the human diet. People subsisting mainly on cereal crops, especially if these are grown in mineral-deficient soils, are at risk of deficiencies in such minerals as iron, zinc, calcium, copper and selenium. For example, rice is highly deficient in iron, and almost one third of the world population, mostly women, suffers from some form of iron deficiency, much of it related to a lack of dietary iron. The resultant anaemia is especially serious during pregnancy and can result in impaired physical and mental development of neonates. Iron-rich rice is, therefore, another potential transgenic product that could help combat a widespread nutritional deficiency, in an analogous manner to golden rice. As we saw above with spinach, a major problem with many iron-rich plant foods is the low bioavailability of the iron owing to the presence of anti-nutritional compounds like oxalates. Many grains contain similar iron-sequestering compounds, such as phytates. In a trial study, rice has been genetically engineered to have higher levels of iron, coupled with a reduced amount of phytates.[583] So far this preliminary work, which was done by the same Swiss-based university group that produced the original 'golden rice', has not been developed into a crop breeding programme. At present, therefore, it seems that the development of transgenic iron-rich rice is a rather distant prospect. It may also be possible to improve the content of other essential minerals, such as zinc, via transgenic methods.[584] Zinc is essential for a healthy immune system and its deficiency in children is also associated with poor growth, reduced motor and cognitive development, and increased susceptibility to infectious diseases. It is noteworthy that most of these transgenic studies on iron and zinc are being carried out by university groups in industrial countries.

In contrast, plant breeders in developing countries have focused almost exclusively on non-transgenic strategies. Genetic variation has been found for traits, such as high iron and zinc contents and low phytate levels, in major cereal crops such as rice.[585] In particular, IRRI breeders in the Philippines have found that some existing aromatic varieties of rice contain double the normal amount of iron, as well as greatly enhanced levels of zinc. For this reason, crop breeding institutes like IRRI are tending to concentrate more on this simpler and more practical approach to what they call the 'biofortification' of staple crops such as rice. These non-transgenic varieties are being developed at IRRI and elsewhere as part of the CGIAR Global Challenge Program on Biofortification, with funding of $90 million over 10 years. The target is to use conventional breeding strategies to produce improved grains that are enriched in iron, zinc, vitamin A, selenium and/or iodine, depending on the dietary needs of various target populations in different parts of the world. Breeding programmes at IRRI have shown that rice varieties can be produced with five-fold higher amounts of iron in the post-milled grains. Furthermore, in double-blind trials, a joint US–IRRI study showed that this biofortified rice was readily accepted as a foodstuff by a group of young Filipino women (trainee nuns) and that it boosted their bodily iron status to a significant degree.[586]

Improved oils

The above examples of crop traits related to vitamin and mineral nutrition demonstrate that, although transgenesis may have a role in some crops, it is far from being the only option, and it is by no means a prerequisite to the breeding of nutrient enhanced crops in general. The situation is very similar with regard to the production of several types of nutritionally enhanced oil crops.[587] Although epidemiological evidence on the optimal composition of dietary fats can seem confusing, there are some fairly simple facts that have emerged over the past decade or so. In general, saturated fats, especially stearic acid, should be kept to a low level in the diet and *trans*-fatty acids should be avoided, if possible.[588] Polyunsaturates may be taken in modest amount, but should ideally be accompanied by fat-soluble antioxidants such as vitamins A and E. Overall, the most desirable edible fatty acid is the mono-unsaturate, oleic acid, which is the main constituent of olive oil, as well as being one of the key elements of the much-touted 'Mediterranean diet'. Breeders have been trying for several decades to use transgenic and non-transgenic methods to create oil crops enriched in monounsaturates, especially oleic acid. Such oils have the additional benefit that they do not need to be highly hydrogenated in order to produce margarines and other spreads. Since hydrogenation can be either reduced or avoided altogether, fewer or no undesirable *trans*-fatty acids will be produced, meaning that the oil is of much higher quality and value for food production.

Over the past fifteen years, there have been many attempts to produce very high oleic varieties of all the major oil crops.[589] In the early-mid 1990s, such efforts were very much focused on the transgenic route and high-oleic cultivars of soybean and oilseed rape were produced, but not commercialised. However, it seems that, more recently, attention has switched back to the use of non-transgenic breeding, with the successful introduction of soybeans with 83% oleic acid, oilseed rape with 70–80%, and sunflowers with over 80%.[590] The high oleate sunflower variety, NuSun, has been particularly successful. Its oil does not require hydrogenation and works well in commercial frying applications. By 2001, over 200 000 tonnes of NuSun oil were being produced and the hoped-for commercial breakthrough came in the same year with the announcement that Proctor & Gamble would be using NuSun oil exclusively in its popular Pringles line of potato chips.[591] In the present uncertain climate about GM foods, it is unlikely that this kind of success in the highly competitive added-value processed food market would have been possible if NuSun had been a transgenic variety.[592]

In our final example of nutritionally enhanced foods, we will consider the very long chain polyunsaturated fatty acids, or VLCPUFAs. Unlike the previous examples that we have looked at, these fatty acids are completely absent from all higher plants. This means that, in this case, conventional breeding is not an option and the only way to produce them in oil crops is via transgenesis.[593] There are numerous reports about the importance of dietary supplementation with these fatty acids for human health and wellbeing. For example, dietary VLCPUFAs have been shown to confer protection against common chronic conditions such as cardiovascular disease, metabolic syndrome and inflammatory disorders, as well as enhancing the performance of the eyes, brain and nervous system.[594] It is worth mentioning here that none of these VLCPUFAs are strictly 'essential' in the diet in the same way that vitamins are. The only unequivocally essential fatty acid is *cis*-9,12-linoleic acid, an omega-6 fatty acid that mammals are unable to synthesise. This essential fatty acid is common in most grains and is present in all leafy vegetables. All of the VLCPUFAs can be synthesised from linoleic acid in a well-nourished and healthy individual. Unfortunately, many Western diets do not deliver a balanced spectrum of fatty acids intake, so VLCPUFAs and other fatty acid supplements are required to maintain optimum health.[595]

Consumption of fish is currently recommended in most Western countries, in order to provide a balanced diet. Much of the nutritional benefit of the fish actually comes from the VLCPUFA content of the fish oils. These fatty acids are not usually synthesised by the fish themselves, but are derived from microorganisms, especially photosynthetic microalgae, that are ingested as part of their diet. As an alternative to fish consumption, therefore, it is possible to purchase VLCPUFA dietary supplements that are produced from cultured microalgae or fungi. However, low oil yields and high costs of oil extraction have limited the scope for this production method, while ever dwindling fish stocks are also threatening supplies of the primary dietary source of marine oils. This situation has led to renewed interest in the possibility of producing transgenic oilseed crops that are capable of producing significant quantities of VLCPUFAs in their storage oils. Attempts to engineer transgenic VLCPUFA rich oil crops are currently under way in several industry/academic collaborations in various countries, including Australia, Germany and the UK.

There are numerous technical challenges involved in this kind of metabolic engineering but several reports during 2004–2005 indicated that the goal of VLCPUFA-producing crops might be within sight.[596] In one rather heroic experiment, no fewer than nine different genes from various fungi, algae and higher plants were inserted into the oilseed, *Brassica juncea* (also known as Indian mustard and a close relative of oilseed rape). The resulting plants accumulated up to 25% arachidonic acid and 15% eicosapentaenoic acid in their seed oil.[597] Of course, as with the earlier examples described above, it is a moot point whether such nutritionally enhanced transgenic crops will be acceptable to consumers. In general, shoppers with the most developed awareness of nutrition also tend to be among the most prejudiced against any product of transgenic origin. Nevertheless, it may be the case that, once such products can be marketed as nutritionally superior to comparable foodstuffs, they may prove to be popular, especially with people who are averse to the taste and/or expense of fish.

In addition to food products, transgenic crops have been proposed as sources of a huge range of non-food items, ranging from biodegradable plastics and starches to novel oleochemicals and fibres. We do not have the space to consider such products here, but as with the other traits examined above, progress has been mixed over the past decade, and in many cases non-transgenic approaches may prove more fruitful in the medium term.[598] However, one group of non-food products that we should consider is the very high-value therapeutic agents, including antibodies and vaccines, that can be made in transgenic crops via so-called 'biopharming'. Such crops represent a radical departure from the normal low-value, high-volume business model of conventional agriculture, and could become a new and highly profitable manifestation of agbiotech, as we will now see.

Biopharming – the killer app?

The vision

In order to grow and prosper, the commercial agbiotech sector needs to produce crops that have a decisive advantage over non-transgenic crops. At present, it is questionable whether output trait transgenic versions of relatively low-value, high-volume commodities, such as oilseeds or starchy cereals, will ever recoup their full investment costs, or be worth segregating for identity-preserved markets, unless a significant price premium can be charged to consumers (and not just the farmers). But, owing to the relatively poor margins in markets for such bulk commodities, it may be difficult to extract significant premiums from end users. Even in the case of medium-value products, such as industrial enzymes, nutraceuticals and biopolymers, I have previously argued that the economics of such transgenic crops are questionable.[599] However, there is one remaining area in which transgenic crops could yet find their elusive 'killer app'; and that is biopharming. Also known as molecular farming (or pharming), biopharming refers to the engineering of plants to express high-value therapeutic products, such as vaccines, antibodies, blood-clotting factors, and anti-cancer drugs.[600] Biopharming has the potential to initiate a new era of very high-value production from plants that could be profitable to participating growers, as well as potentially providing a cheaper source of medications for developing countries.[601]

Biopharming has both advantages and drawbacks compared to the traditional fermentation-based production systems for pharmaceuticals. Advantages include reduced production costs,[602] easy scale-up from sub-gram to multi-tonne quantities, and a more versatile range of protein products compared to alternative prokaryotic systems like *E. coli*.[603] Some drawbacks include the following: ensuring rigorous containment of plants grown outdoors in the field; the complex and expensive purification methods often needed to recover active therapeutic products from plant extracts; low expression levels of recombinant proteins in the plant; requirements for expensive and time consuming clinical trials; batch production, rather than the more efficient continuous process; and the unfamiliarity of many in the mainstream pharmaceutical industry with this new technology.[604] Biopharming was first mooted in the early 1980s and expression of pharmaceuticals in transgenic plants was demonstrated in 1989.[605] Early attempts to produce therapeutic proteins in plants foundered owing to low yields and difficulties in extracting active compounds from plant tissues.

Biopharming generated renewed interest in the late 1990s, as technology improved and low-value commodity transgenic crops ran into increasing problems of consumer acceptance and lack of profitability. At that time, commercial biopharming was, and mostly still is, dominated by a disparate group of small research focused companies, rather than the multinational agbiotech giants. In some cases, these small companies have formed joint ventures with various larger firms, including mainstream pharmaceutical businesses. In many other cases, they act independently and are funded by venture capital or other speculative investment vehicles. Some of these companies are developing scientifically exciting, innovative strategies for contained production of therapeutics in plants, while as we will now see, other companies are moving in a much more controversial direction.

Flawed strategies

Unfortunately for the wider image of biopharming, several of the relatively new breed of smaller agbiotech companies are taking what seems like a potentially self-defeating approach to biosafety issues in their quest for high-value products.[606] For example, over recent years a series of contamination scandals has needlessly tainted this nascent industry.[607] The most notorious such episode was the ProdiGene affair of 2002, in which a few transgenic maize plants expressing a pig vaccine contaminated a soybean crop destined for use in human food and/or animal feed.[608] ProdiGene is a Texas-based biotech company seeking to produce a variety of vaccines, antibodies, therapeutics and enzymes in transgenic plants. The most serious problem in the strategy pursued by ProdiGene, and by several other biopharming companies, is the use of mainstream food crops as production vehicles for their medical products. Moreover, they routinely use Midwestern maize for many of their trials, even though it is grown in the same areas where tens of thousands of hectares of food-grade maize is being grown. Maize plants are well known to pollinate other maize fields up to 800 metres away.[609] This means that transgenic maize expressing pharmaceuticals could theoretically end up contaminating food supplies via several different routes, including pollen transfer (which has a low likelihood), mixing of seed on- and off-farm (as happened in the 2000 StarLink affair, described below), and via transgenic maize volunteers in subsequent crops (as happened in the 2002 ProdiGene affair).

Even before ProdiGene carried out its controversial field trials in 2001–2002, it was already well known that effective segregation of transgenic maize crops, both on-farm and post-harvesting, was far from easy to achieve in the US agrifood system. For example, in 2000, an unapproved (for food use) transgenic variety of maize called StarLink was discovered in the human food chain throughout the USA. Millions of tonnes of maize seed, flour, and food products such as tortillas were recalled for disposal. The disaster cost Aventis CropScience an estimated $500 million, as it was required to buy back the contaminated maize and to compensate various injured parties.[610] There were international consequences, as US trade with maize purchasing countries like Japan and Korea was adversely affected, and the food industry as a whole suffered a severe shock.[611] Aventis CropScience never really recovered from this setback: within a few months most of its senior management team in the USA had been replaced, and it was subsequently sold by its parent company to Bayer CropScience.[612]

Despite the salutary precedent of StarLink, ProdiGene went ahead with a series of what turned out to be inadequately managed field trials of its transgenic pharma-maize, resulting in at least two episodes of crop contamination in Nebraska and Iowa in 2003.[613] The transgenic maize was expressing a vaccine intended to prevent *E. coli*-induced diarrhoea in pigs. Although the vaccine protein is non-toxic to people, it is not something that one would wish to have in one's tofu or soymilk, and the crop was certainly never meant for food use. Discovery of the contamination led to incineration of the fields and destruction of more than 12 000 tonnes of soybeans, at a cost to ProdiGene of about $3.5 million. In an unprecedented action, the company was also fined $250 000 by the US government.[614] This affair embarrassed many in the agbiotech and biopharming industries, and resulted in calls for tighter guidelines on the confinement of plant-made pharmaceuticals by the Biotechnology Industry Organisation (BIO). It is worth noting that BIO is normally very supportive of any agbiotech venture and is generally against what is sees as the over-regulation of the industry as a whole.[615] One might think that these very public and costly episodes of contamination would have made biopharming companies somewhat chary about using major food crops as production vehicles for their compounds. Indeed, a few companies, most notably Monsanto, abandoned this area of research in the wake of the ProdiGene affair.[616] But such was not the case in the industry as a whole. Not only did ProdiGene and several other companies continue to use maize for biopharming,[617] many of them went on to express their pharmaceuticals in other important food crops, including rice, barley, wheat and soybean.[618]

This policy alienated many in the food industry, who had nothing to gain from biopharming crops, but stood to lose a great deal, both financially and in terms of public confidence, if their soybean or rice crops became contaminated. Once again, in 2004–2005, biopharming slid towards another industry-damaging public confrontation. The new controversy involved California-based Ventria Bioscience, which had developed transgenic rice varieties engineered to express anti-microbial proteins. Ventria proposed to test the transgenic rice by growing over 50 hectares of the crop in trial plots located in the Central Valley of California, which is also home to a thriving rice industry worth over $500 million per year. Understandably, this proposal caused great concern to rice farmers despite assurances from CEO Scott Deeter that 'Ventria will work hard to keep its rice isolated'.[619] After much controversy in California, Ventria bowed to pressure from the US Rice Federation and other bodies and the proposed test site was relocated to what was hoped would be a more welcoming location in Missouri. However, this promptly led to a further storm of protest among the rice growers and users of that State. At one point, the brewing giant, Anheuser-Busch, threatened to stop buying Missouri grown rice for use in its beer, because of the fear of possible contamination.[620] Following these delays, Ventria was unable to carry out any field trials in 2005, but it still plans to grow transgenic rice elsewhere in the USA in the future.[621]

The most puzzling and disappointing aspect of the ProdiGene and Ventria affairs is that such controversies are completely unnecessary and are readily avoidable. The primary issue of concern to most people in the industry, and amongst the general public, is not so much the concept of biopharming per se, to which many would extend a qualified approval. Rather, it is the use of major food crops as production vehicles for potentially problematical pharmaceuticals or other bioactive compounds, should they ever enter the human (or animal) food chain.[622] In the worst case scenario, if a company is determined to use a food or feed crop for biopharming, it would at least seem sensible to grow the biopharming crop as far as possible away from the major areas of cultivation of the food/feed varieties of the same crop. And yet, it seems that exactly the opposite is happening in crop after crop. Plans are being made to cultivate several new biopharming varieties of maize in Iowa, which is home to hundreds of thousands of hectares of food/feed maize.[623] And most recently, a company called Agragen Inc. has sought to grow commercial biopharmed flax crops in North Dakota, which is the centre of the largest edible flax growing region in the USA, producing a food/feed crop worth over $150 million/year.[624] The problem for existing growers of conventional varieties of these crops is that the new pharming varieties will only ever take up a small fraction of the total crop area, so very few farmers will be able to grow these crops. However, all of the maize/rice/flax growers will share the risks from any contamination incidents, and may possibly lose markets (especially in sensitive food sectors), without the prospect of ever deriving a cent-worth of benefit from the new biopharmed crops. The short-sighted actions of some biopharming companies are therefore alienating the very farmers who have historically been some of the strongest supporters of agbiotech, by making the elementary mistake of forcing them to take on what may be a significant risk for no equivalent gain. This situation is particularly regrettable because it is so unnecessary to use major food/feed species at all for biopharming.

Improved strategies

It is neither necessary nor desirable to use major food crops to express pharmaceutical products. There is a host of alternative species that can be used in pharming, including minor food crops, non-food crops, or even plants that are not used as crops at all. One of the best candidates is tobacco, *Nicotiana tabacum*, a widely grown but inedible crop that is increasingly frowned upon as a source of an addictive and life-threatening cocktail of drugs and toxins. In technical terms, tobacco is one of the most amenable plants for genetic engineering, and its leaves can express therapeutic proteins at impressively high yields.[625] There are also several ingenious strategies to ensure that pharmaceutical-producing tobacco plants cannot be misused or escape into the wider environment. For example, the transgene responsible for production of a pharmaceutical can be linked to a promoter, ensuring the gene is always switched off unless it is induced by a specific external signal. Hence a transgenic tobacco crop an be harvested and taken to a secure storage area without containing any trace of the pharmaceutical product because the gene is still inactive. This ensures that there is no possibility of escape or contamination of the wider environment. Only when the harvested leaves (which are still alive) are sprayed with a chemical inducer, will the novel compound be synthesised and extracted in the normal way.[626] A wide range of transgene-inducing compounds is available, including various detergents, the steroids dexamethasone and estradiol, copper, ethanol, herbicide safeners, and the insecticide methoxyfenozide.[627] Other recent studies have highlighted a host of additional mechanisms available for effective bio-containment of biopharmed crops.[628]

Thanks to these and other strategies, biopharming can potentially be as safe as any of the many other age-old methods of producing and extracting useful drugs from plants. Several small biotech firms, such as Icon Genetics and Large Scale Biology Corporation, and some university groups, have been using such novel strategies.[629] One of the approaches taken by Icon Genetics in Germany is especially interesting in that it does not rely upon insertion of a transgene into the genome of the plant.[630] Instead, a viral vector is used to deliver the gene into the cytoplasm of the plant cells from where the gene product is transiently expressed. This results in very high expression levels of the transgene without it ever entering the nucleus and becoming integrated into the genome.[631] As a consequence, the transgene is only present temporarily (possibly for as little as a few days) in the plant cells and cannot be inherited by the next generation, e.g. if some of the plants were to escape into the open environment.[632] This type of transient, viral-delivery system is being trialled in non-food crops like tobacco in order to minimise even further the chances of contamination of the food chain. Tobacco was also the transient, virus-mediated production system used by Large Scale Biology Corporation in a large bio-manufacturing unit for contained production of antibodies, vaccines and other therapeutic proteins in the heart of the Kentucky tobacco growing country.[633] Among these products were an antibody to control dental caries, a lipase to treat atherosclerosis, and the protease inhibitor, aprotinin, to treat inflammatory response during cardiac bypass surgery.

Another, even more secure, strategy is to grow non-food plants in contained culture, e.g. in glasshouses or polytunnels (inexpensive, tubular plastic versions of glasshouses). Many conventional food crops, including hitherto seasonal fruits like strawberries, are already grown year-round in this way. Similar culture methods, including hydroponics, have been used to grow transgenic varieties of species such as duckweed, *Lemna gibba* and *Lemna minor*, in order to express high-value proteins.[634] The particular advantage of duckweed is that the recombinant proteins are secreted from the plant into the culture medium from where they can be readily purified, hence avoiding all the problems associated with extraction from plant tissues. Moreover, because clonal *Lemna* plants are vegetatively propagated and do not produce pollen or seeds, the production system has a very high degree of bio-containment. Transgenic *Lemna* hit the news in 2004–2005 when Biolex Therapeutics in the USA bought up both French biotech startup, Lemnagene, and US protein therapeutics company, Epicyte Pharmaceutical, Inc. Newly enlarged Biolex is currently developing large-scale production of recombinant alpha-interferon (used for treatment of hepatitis C) and has other *Lemna*-based products in the pipeline. Plant-based production of pharmaceuticals can be taken a stage further, in terms of bio-containment and safety, by using in vitro cell cultures rather than whole plants as expression systems. The drawbacks here are technical (it has been difficult to achieve reliable scale-up) and economic (cell cultures are expensive to maintain). The advantages are much improved biosafety, which may be important in the case of some highly active and potentially toxic compounds, and easier maintenance of purity standards. In recent years, encouraging progress has been made in the use of plant cell cultures to produce such compounds as the anti-cancer drug, taxol,[635] and the versatile anticholinergic drug, scopolamine.[636]

Looking to the future, the successful application of biopharming has the potential to transform production of many therapeutic products by reducing production costs and improving biosafety. It is estimated that vaccines against many endemic diseases in developing countries could be produced in plants at a fraction of their present cost. Similar techniques could be used to produce other low-cost medications for some of the most serious diseases of poorer countries, including AIDS and hepatitis, where current treatments are prohibitively expensive for many of those afflicted. This is an entirely laudable vision and one that has a good prospect of realisation within the next decade.[637] In principle, there need be very little risk in the cultivation of crops that contain pharmaceutical products. After all, at least a quarter of all prescribed medications, including such potent drugs as the painkiller, morphine, and the new anti-malarial medication, artemisinin, are already derived from plants that are grown in the open environment.[638] Many of these compounds cannot be synthesised chemically, so we still need to grow dozens of crops for the production of drugs. Indeed, medicinal plants are currently enjoying something of a renaissance as efforts are made to cultivate and breed them systematically, rather than simply gathering them as wild plants, which can lead to their extinction. Modern biotech methods like marker-assisted selection and tissue culture are also increasingly being applied to improve the yields and reduce production costs of these eminently useful plants.[639]

For many millennia, people have been growing and collecting a vast range of pharmaceutically active medicinal plants with little or no risk to their food crops. Ideally the same principles should hold true for medicinal compounds from bio-pharmed crops. However, the actions of a few companies are in danger of jeopardising the future of biopharming. A combination of short-termism and a somewhat lax regulatory regime in the USA (as acknowledged by the USDA Inspector General in 2005) has allowed biopharming to go somewhere that it should never be, namely into the arena of the major commodity food crops.[640] We may have to wait for yet another major contamination scandal, possibly this time directly affecting human health, before parts of the private sector come to their senses and move to safer, and in the long run potentially more profitable, plant systems for biopharming. Meanwhile, we should note that biopharming is based on a totally different kind of business model than conventional large-scale agbiotech. Different companies are involved, that are more akin to earlier small-scale agbiotech startups. These companies tend to be much more science based, innovative and entrepreneurial than the agbiotech giants. The departure of Monsanto from this sector in 2003 implies that they recognised the mismatch of trying to maintain a high-value, low-volume operation alongside their current low value, high-volume seed and agrochemicals business.

So, does biopharming have a future, and will it really be the killer app of agbiotech? The answer is 'yes and no'. Biopharming could be a real success for a few players, but each product might only need a few hectares of land for its annual production, so very few farmers will be involved. Biopharming might end up being carried out by relatively small contract companies on behalf of larger pharmaceutical concerns, who then test and market the products using existing supply chains. While the conventional agbiotech giants may not fit into this scenario, the smaller biopharming companies could be very successful indeed. However, it behoves the biopharming sector to remember that their name may sound like 'farming' but the 'ph' means that they are in the pharmaceuticals business. As noted by Kirk *et al.* in their recent analysis of plant-made vaccine production: 'Successful use of this technology is highly dependent on stewardship and active risk management by the developers of this technology'.[641]

In this new business model, we are no longer talking about feeding the world or growing millions (or even tens) of tonnes of plants. Rather, we have the high added-value manufacture of some potentially very profitable and very useful pharmaceuticals in plants. Given the anticipated profits from this business, it is regrettable that some biopharming companies are taking a short-sighted 'cheap and easy' route by using major food crops as production vehicles. As we have seen from for example Icon Genetics, a little additional upfront investment, and a lot of clever innovation, could virtually eliminate contamination problems and make biopharming no more controversial than growing sweet wormwood, *Artemisia annua*, as a source of the precious new anti-malarial drug, artemisinin. Meanwhile, the current confused and inconsistent regulatory regimes around the world for biopharming production, testing, and use in either human or veterinary contexts is impeding future development of the technology both for profit and for public-good applications.[642] In conclusion, therefore, biopharming could and should have a bright future, but it will be one that is very different from the current forms of agbiotech. In the meantime, one hopes that, despite the unnecessary risks involved in some of the current industry practices and the confused regulatory environment, the biopharming goose will be permitted to survive long enough to lay its golden eggs.

Part V

Increasing global crop production:
the new challenges

A seed hidden in the heart of an apple is an orchard invisible.
Welsh proverb

14

Feeding the world – fallacies and realities

I will venture to affirm, that the three seasons wherein our corn has miscarried did no more contribute to our present misery, than one spoonful of water thrown upon a rat already drowned would contribute to his death; and that the present plentiful harvest, although it should be followed by a dozen ensuing, would no more restore us, than it would the rat aforesaid to put him near the fire, which might indeed warm his fur-coat, but never bring him back to life.

Jonathan Swift (1667–1745)
Famine[643]

Introduction

It is frequently opined in the popular and scienctific media alike that crop production may have serious difficulties in coping with projected increases in the global human population over the next fifty years. It has also become commonplace to hear statements, from biotech companies, politicians and even some public sector scientists, that this putative crisis in food production can only be fully resolved by the global deployment of transgenic crops.[644] Of course, over the past few centuries, we have repeatedly heard various Malthusian predictions about imminent famine, all of which have proven to be misplaced.[645] As we saw in Chapter 4, it was forecasts of future famine that prompted the USDA to begin its programme of worldwide germplasm collections in the late nineteenth century. The spectre of overpopulation was a recurring theme during the twentieth century, but the most egregious instances of misguided predictions of imminent apocalypse occurred in the late 1960s and early 1970s. At that time, numerous 'experts' informed us that global famine was just around the corner.[646] In reality, of course, even as they were being uttered, these dire warnings had already been gainsayed by the achievements of the Green Revolution crop breeders such as Norman Borlaug and his colleagues at CIMMYT (see chapter 6). So is there really an imminent food crisis in the twenty-first century, or are these concerns yet another false alarm?

213

As we will see below, we currently produce more food per capita than ever before, and more than enough to feed the world for many years to come. It is therefore puzzling and worrying to hear warnings about imminent global starvation from many shades of opinion, including respected scientists. It is even more troubling to hear powerful and influential political leaders arguing that the best way of addressing the supposed food crisis is to invest massively in agbiotech on a global scale. The 'saving the world with GM crops' lobby has already convinced important decision makers, including US presidents Clinton and Bush, of the rectitude of their crusade.[647] Ironically, this case for an imminent food crisis is rejected by those in anti-GM environmentalist groups. But, is the pro-GM lobby actually correct this time; do we really face a food crisis? In the next two chapters, we will begin by examining the reality of the food supply situation in the light of population growth over the next 40–50 years. We will seek to address the following questions. What is the most likely magnitude of the expected population growth? How will the extra people be fed? What are the prospects for increasing crop production to cover increased food demand? And, finally, are transgenic varieties really required as part of a sustainable global crop improvement programme?

The case for a global food crisis by 2050 is based on three premises, one of which is sound while the other two are decidedly suspect, as we will now discuss. The first premise is that human populations will rise appreciably over the next half century. This premise is based on well-characterised population trends. There is widespread agreement among demographers that the world population will increase, possibly by as much as 50%, by 2050. The second premise is that we are now growing food on virtually all of the potentially arable land around the world. However, we will see that recent evidence on land availability, from South America and elsewhere, seems to contradict this almost universally held assumption. The third premise is that conventional, non-transgenic, breeding techniques have been exploited almost to their limit and that we cannot possibly expect the yield gains of the last fifty years to continue.[648] We will see that this argument is based on a narrow analysis of a few of the major commercial crops. It ignores or understates the immense and barely tapped genetic potential of dozens of other staple crops around the world.

Even concerning the major crops, the experienced and respected breeder, Donald Duvick of Pioneer Hi-Bred, has recently stated that 'yield gains can continue at the same pace for several more decades' and that 'conventional plant breeding will continue as the essential foundation',[649] while Nobel laureate Norman Borlaug has argued that the world already has the technology available to feed 10 billion people on a sustainable basis.[650] In the rest of this chapter, I will argue that, although we must not allow ourselves to become complacent about population growth, we nevertheless have an amply sufficient capacity in our production, management and

crop improvement systems to avert a serious food crisis. This conclusion is supported by the most recent findings from respected international bodies.[651] For
example, in 2004 an FAO report titled *World Agriculture: Towards 2015/2030*, stated
that, between now and 2030:

> ... growth in food production will be higher than population growth. By the year 2015/2030 per
> capita food supplies will have increased and the incidence of undernourishment will have been
> further reduced in most developing regions ... The world population will be increasingly well-
> fed by 2030, with 3050 kilocalories (kcal) available per person, compared to 2360 kcal per person
> per day in the mid-1960s and 2800 kcal today. This change reflects ... the rising consumption in
> many developing countries whose average will be close to 3000 kcal in 2030.[652]

The projections for food production in this FAO report are based on current
breeding technologies and make no assumptions, either for good or ill, about the
contribution of transgenic varieties to future food yields. In short, this and other
robust evidence from impartial and reliable sources suggests that it is most unlikely
that we will face anything resembling the catastrophic mass hunger that has been
repeatedly predicted from the time of Malthus until the present day. With regard to
the best way to improve food production, I will argue that the yield gains realised
from readily implemented measures, such as better management and infrastructure
in developing countries, bringing more land into cultivation, and abolition of per-
nicious subsidies and tariffs in industrial countries, will achieve vastly more than
even the most optimistic dreams of the proponents of transgenic crops. We will now
begin by examining the evidence for population increase and the frequently overlooked role of economic growth in augmenting future food production.[653]

Population, economic growth and food production

Projections of global population growth over the next fifty years are in general
agreement in estimating that, while the rate of growth is now slowing down
appreciably, there will still be an additional 2.6 billion people by 2050.[654] Virtually
all of this population growth will take place in developing countries, especially in
Asia and Africa. The projections are based on current trends that include an
increasingly dynamic and productive economic outlook for most of the developing
regions that will experience significant population growth. Factors that curtail
economic growth will, ironically, lead to much higher rates of population increase,
because less well-developed societies tend to have far higher ratios of births per
woman.[655] Indeed, it is largely the increasing affluence of countries like China and
India, due in part to the Green Revolution of the 1970s, that has facilitated the
steady decline in global birth rates over the past few decades.

For the past fifty years the world has increased its food production at a much faster rate than the population increase. Excess food production has been especially pronounced in industrialised countries, where surpluses have been accentuated by agricultural subsidies from governments. The oversupply of food has been reflected in a steady decline in the prices of edible commodities over the past fifty years. For example, in 1950 palm and soybean oils cost about $2 000 per tonne, whereas current prices are more like $400 per tonne.[656] This represents a staggering 80% decline in the value of these internationally traded staples. Similar declines have occurred in the value of other major crops such as rice and wheat. Higher food production has far outstripped increases in demand, whether due to rising populations or increased affluence. In many parts of Europe and North America, crop surpluses have even contributed to a significant depression in agricultural sectors, leading to the setting aside or abandonment of good arable land for food production. This agricultural overcapacity is one of the resources that can be tapped to increase food production in the future.

Targeting disadvantaged regions

In a sense, we already produce enough food to cope with the expected population increase until 2050. This is because, although the world population has increased by 2.2-fold since 1950, food production increased by 3.1-fold over the same period. We now produce 40% more food per capita than in 1950 and most of the world's population now consumes significantly more food per capita than previously.[657] So, even if the world population does increase by 40% in the next few decades, it could still be fed from our current levels of agricultural production, provided that we all reverted to a 1950s-like diet, i.e. less meat and fat and a lot more plant carbohydrates. However, while such a diet might be beneficial for the chronically overfed in industrialised countries, it is unlikely that we will all need to go back to what would undoubtedly be a less-than-universally popular 1950s dietary regime of bread, potatoes and turnips.[658] In reality, we have several options for increasing global food supplies in addition to breeding higher yielding crops. Other options that have been overlooked by many in the food debate include expansion of arable land area and improved management of existing crop systems (see below).

Although current world agricultural production provides a modest overall surplus of food, for a variety of mainly economic and/or political reasons, this food does not always reach the most disadvantaged groups. Given the political will to distribute them, sufficient food reserves already exist to cope with periodic localised shortages due to events such as drought, disease, flooding or warfare. Moreover, the FAO predicts that improved agricultural and economic development will cause the number of hungry people to fall from 770 million to 440 million by 2030, despite a predicted population increase of well over one billion people during the same period.[659] In the past, the region that always seemed to miss out on advances in agriculture was sub-Saharan Africa. According to a series of FAO reports, the number of hungry people in sub-Saharan Africa will not decline appreciably between now and 2030.[660] In contrast, as discussed above, the numbers in the rest of the world are predicted to fall sharply over the same period.[661]

Economic growth and sustainability

It is obvious that one of the primary concerns of global agriculture over the coming decades should be to provide sufficient food to sustain increasing human populations. Two other concerns about the 'mission' of agriculture are that it should be both economically rewarding for the farmer and sustainable in the long term. The former concern is far from being realised in the present unfree global marketplace, with large government subsidies being paid to certain farmers, most notably in the richer industrialised countries of North America, Europe and Japan. To make things even worse, much of the trade in agricultural commodities from developing countries is restricted by a combination of quotas and tariffs in order to further protect uncompetitive farmers in industrialised countries.[662] This artificial rigging of the market against developing country producers obviously affects them in the short term by keeping them in poverty. It also impedes their ability to address long-term sustainability issues because it suppresses their potential for economic advancement. We will discuss subsidies further in the next chapter.

The issue of economic development and its relationship to sustainability is of great importance for developing countries, especially in view of potentially trade-damaging criticisms of their sustainability and environmental practices from people in the comfort zone of the rich industrial world.[663] There is good evidence that the capacity of a nation to address issues like sustainability and the environment is closely linked with its economic advancement. One of the best-known manifestations of the link between sustainability and economic wellbeing is the so-called ECK (or environmental Kuznets curve) hypothesis. The ECK hypothesis posits an inverted U-shaped relationship between average per capita income and unsustainable activities, such as uncontrolled pollution from excess fertiliser runoff or industrial emissions. In other words, environmental degradation tends to increase sharply during the early stages of industrialisation, as it did previously in much of Europe and North America. Pollution rates then tend to level off and finally to fall as average incomes rise, and environmental awareness and the wherewithal to address such challenges also increase. Most post-industrial economies have already passed through this cycle. The overall hypothesis has been neatly encapsulated by economist Theodore Panayotou as follows:

At low levels of development both the quantity and intensity of environmental degradation is limited to the impacts of subsistence economic activity on the resource base and to limited quantities of biodegradable wastes. As economic development accelerates with the intensification of agriculture and other resource extraction and the take off of industrialization, the rates of resource depletion begin to exceed the rates of resource regeneration, and waste generation increases in quantity and toxicity. At higher levels of development, structural change towards information-intensive industries and services, coupled with increased environmental awareness, enforcement of environmental regulations, better technology and higher environmental expenditures, result in leveling off and gradual decline of environmental degradation.[664]

Most economists would agree, however, that we cannot simply wait for market forces to enable most people in a particular region to get (relatively) rich, before such countries begin to tackle their most serious environmental problems. It seems that a judicious measure of government intervention is normally warranted in order to supplement and catalyse market forces towards more environmentally benign practices.[665] These arguments appear to be borne out in the case of some of the more economically advanced developing nations. As such countries become more affluent, they tend to start taking agro-environmental and sustainability issues more seriously. A good example is the case of oil palm cultivation in Malaysia.[666] In recent years, the Malaysian oil palm industry has started to take greater note of environmental and sustainability issues, thanks to a mixture of self-interest and pressure from government and other advisory bodies. For example, in the late 1990s the Malaysian government instituted a 'zero burning' policy on all plantations.[667] Prior to this, many plantation managers had burned old or diseased palm trees before planting new seedlings.[668] The new policy means chopping down trees, mulching, and stacking the chips, and a delay of six months while the material rots in the ground. Although these methods are both more expensive and less efficient at controlling diseases than traditional burning, they are deemed to be more environmentally sound; and this has enabled oil palm to maintain and expand its markets in eco-sensitive regions such as Western Europe. Therefore, although an over-naïve interpretation of the environmental Kuznet's curve hypothesis is not warranted, there is much empirical evidence to commend the notion that economic growth can often greatly facilitate environmental amelioration in the longer term.[669]

A major challenge in addressing agricultural development in needier regions is to look beyond a facile reliance on palliative measures, such as food aid or redistribution, in order to alleviate local deficiencies in food production. A far more satisfactory and durable long-term solution, for all concerned, is to enable food producers in such regions to generate their own surpluses, by a combination of biological and economic measures. Biological measures include access to improved seed and agronomic advice, while economic measures can encompass everything from trade liberalisation to improved market infrastructure (access to loans, reduced taxes, better communications and transport etc). Advice on sustainability issues to developing countries is also important, but this must always be given in the context of their stage of economic development. The most effective long-term mechanism to ensure that the ever shifting targets of agricultural sustainability are met is to facilitate the development of a thriving agrarian economy, with well-educated and financially secure farmers who will then tend to act more sustainably in their own long-term interests. One of the most important challenges for agricultural managers, including farmers, will be to ensure that improved yields do not come at the expense of long-term sustainability of farming systems.[670]

As we have seen above in the case of oil palm, increased awareness of sustainability issues goes hand in hand with increasing affluence. This is relatively straightforward for a major crop, like oil palm, that is produced mainly on large plantations, but is less easy to facilitate in the case of smallholder cropping systems. One of the guarantors that agricultural sustainability at all levels of production is maintained as a key objective in a country is the continued participation of public sector scientists as major players in the overall enterprise of both national and international crop improvement. In his book *The Doubly Green Revolution*, President of the Rockefeller Foundation Gordon Conway argues that many of the failings of the first Green Revolution came from a lack of attention to issues of long-term sustainability in the kinds of crop growing systems that were introduced so quickly in the 1960s and 1970s. For example, the over-reliance on fertilisers and lack of attention to soil structure led to soil erosion, nutrient depletion, falling water tables and salinisation, even in the immensely fertile region of the Indian Punjab.[671] These challenges are now being addressed by initiatives from CGIAR and national crop research centres in the affected countries. We will now examine the question of land use, and in particular the prospects for utilising potentially productive, but currently unused, land for arable farming.

Expanding the area of crop cultivation

In 2005, the global area of arable (cultivated) cropland was about 1380 million hectares, or just over 10% of the total land mass. This may seem like a rather small proportion to use for such a vital purpose as food production, but much of the world's land is unsuitable for farming owing to climate, topography or various other forms of use by humans, including as urban and industrial centres. We should also remember that the global arable area of 1.4 billion hectares is dwarfed by the 3.4

billion hectares that is used for pasture, i.e. raising animals for meat and dairy production.[672] Although not all pastureland is suitable for cultivation, a significant proportion could certainly be brought into arable use. Moreover, this could often be done without reducing livestock numbers, e.g. by making wider use of feedlots and other intensive animal husbandry regimes. In addition to converting pasture, there are several other options for continued expansion of arable cultivation around the world, as we will now see. Over the past 150 years, new land has been brought into arable production on an immense scale. For example, in the sixty-year period from 1860–1920 a remarkable 440 million hectares was cultivated for the first time, more than half of it in the newly occupied Great Plains region of North America. A similar scale of transformation took place over the next fifty years until, by the third quarter of the twentieth century, the vast majority of the potentially productive temperate land of the northern hemisphere had been brought under the plough. More recently, a considerable amount of new land has been brought into cultivation in developing countries, while in the industrialised nations the arable land area has actually declined. Between 1961 and 1997, the arable land area of developing countries rose from 610 to 750 million hectares (a 23% increase), while that of the industrialised countries fell from 650 to 630 million hectares. This decrease of over 20 million arable hectares in developing countries, which is almost as large as the entire land-mass of the UK, is largely the result of setting aside ever more potentially productive land as fallow, owing to surplus food production.[673]

The received, or conventional, wisdom from many scientific experts, and from pundits of every hue, states that we have now more or less reached the limit of usefully cultivated land on the planet. This is despite published evidence that there has been no slowdown in the rate of conversion into arable land in the developing world as a whole. The most recent data from the late 1990s show that the conversion rate is still at 4 million hectares/year, which if anything is slightly higher than the average rate of 3.9 million hectares/year for the entire period from 1961–1997.[674] This suggests that there is no shortage of land for conversion to farming and that arable expansion is still proceeding at the same steady rate as over the past fifty years. We should therefore ask the simple question: where is all this new land that is still being settled and farmed so assiduously? The short answer is: mostly in South America. It is a curious omission of those who advocate the prevailing 'limits of cultivation' argument that they have largely ignored the enormous untapped potential for arable expansion in places such as South America. Their argument also ignores the existence in industrialised countries of comparatively large areas of abandoned, but potentially productive, arable land. As we will now see, the abandoned and the new lands both carry the promise of huge additional capacity for food production over the coming decades.

Reclaiming abandoned and set-aside land

The abandonment of farmland is nothing new. Compared to today, many regions in Northern Africa and the Near East were farmed much more intensively two millennia ago. For example, Northern Africa was the main grain growing area for the hungry imperial metropolis of Rome from about 200 BCE until 400 CE. Much of the best arable land in these regions was abandoned following a catastrophic series of invasions, plus possible climatic changes, at about the fifth century CE.[675] Most of this area is now so agronomically degraded that it would be a formidable task indeed to bring it back into the cultivation of annual crops such as cereals. Despite these challenges, there is a lot of useful research into such reclamation. Among the techniques being investigated are the use of seawater desalinisation for water provision, new forms of irrigation, and the breeding of salt and drought tolerant crop varieties. It is possible that these efforts may eventually bring some of this land back into cultivation. However, to be realistic, the reclamation of such climatically and environmentally degraded land, and its capacity to be a major contributor to increasing food production, remains a distant prospect.

Over the past century and a half, tens of millions of hectares of productive arable land have been abandoned for more benign reasons than the series of manmade or environmental disasters that befell post-Roman North Africa. This more recently abandoned land is still potentially productive and so could be reclaimed for farming much more quickly and easily than the degraded Northern African soils. Some of most recently abandoned land has only been fallow for a few years because of relatively short-term factors, such as the policy of set-aside in the EU since the 1990s. For example, in 2005, the EU set aside 6.6 million hectares from food production because its heavily subsidised farmers were generating such large, and politically embarrassing, surpluses. According to the terms of the Common Agricultural Policy, this area will increase to 7.7 million hectares by 2009, and then to as much as 12 million hectares.[676] Many other industrialised countries have similarly vast regions of recently set-aside land that could be reconverted to arable use with relative ease.

Meanwhile in other regions, mostly also in industrialised countries, a great deal of arable land has been abandoned over a longer period, but could still be recovered for farming should the need arise. One example of the almost complete abandonment of a whole region was the exodus from New England and Hudson Valley farms in the USA, once the more productive lands of the Western frontier were made available for settlement, in the nineteenth century. New England was once largely covered by primeval forest, with small clearings that were managed by the Amerindian inhabitants.[677] Following the post-Columbian European incursions, immigrant farmers cleared this land for agriculture at such a pace that, by the middle of the nineteenth century, crops or pastures covered nearly three quarters of the non-mountainous land in southern and central New England. Much of the New England farmland was abandoned just a few decades later when more attractive areas were opened up in the West. Many of these Western farm plots were offered freely for settlement by homesteaders in the mid-late nineteenth century. Today, 150 years later, forests once again blanket 75% of New England and the Hudson Valley (formerly New Holland). Unbeknownst to the casual visitor, a large proportion of this seemingly pristine forest is the result of secondary regrowth on abandoned farmland. In many of these areas, traces of the original land use are still present in the form of the many stone walls that mark former field boundaries of the long-gone Yankee farmers. Although not all of this land is of the highest quality, much of it is potentially productive, and would be especially suitable for the cultivation of vegetable or orchard crops. Similar tracts of abandoned or set-aside land, totalling many tens of millions of hectares across the industrial world, could readily be brought back into profitable production, should the need arise.

South America

While set-aside and abandoned farmland might bring tens of millions of hectares back into cultivation, development of new farmland could open up hundreds of millions of hectares for food production. There are many areas of the world where new arable land can still be won, but by far the greatest potential lies in South America.[678] Expansion of arable farming in South America is a relatively recent phenomenon that promises to have a considerable impact on global food markets. According to recent surveys from USDA and other US agencies, as much as 100 million hectares may be available in just one area of the continent, namely the vast and relatively undeveloped *cerrado* (savannah) of central-western Brazil.[679] To put this figure into context, it is three times the total area of Indian arable crop production, and is 80% of the entire crop area of China; it is also equal to three quarters of the current arable land area of the entire continent of South America.[680] It had been thought previously that much of this land was unsuitable for agriculture, but this view has now changed, thanks in part to field trials undertaken by Brazilian scientists at EMBRAPA.[681] This work has demonstrated that much of the *cerrado* is suitable for sustainable, large-scale cultivation of major commodity crops such as soybeans, maize and rice.[682] The new land in the Brazilian *cerrado* region alone could add almost 10% to the global area of crop production.

A major impediment for the utilisation of much potentially cultivable land in developing countries like Brazil is the lack of infrastructure of all kinds. Investment in such areas also requires a stable polity, a reliable regulatory regime (especially for international trade), and favourable economic conditions. One of the most serious infrastructural problems bedevilling agricultural development in South America, and many other developing countries, is a dearth of transport links. Transport is a particularly acute concern in Brazil as farmers seek to export crops, such as soybeans, maize and rice, to growing global markets from Europe to China, in competition with other major exporters like the USA.[683] Inadequate road and port facilities in Northeastern Brazil can lead to 150-kilometre traffic jams, with unfortunate truckers sometimes queuing for weeks at a time to get to overcrowded wharves, so that they can load their soybean cargoes onto ships for export overseas. By contrast, farmers in the USA enjoy a vast network of publicly built and maintained highways, canals, irrigation systems and navigable rivers, which has benefited from over a century of government investment and public subsidy.[684] This expensive, taxpayer-maintained US network ensures the rapid, cheap and efficient passage of produce from Midwestern farmlands to Great Lakes ports for onward shipping to global export markets. As a result, farm-to-port transportation costs are as much as five-fold higher in Brazil than in the USA. Of course, this sort of situation makes a nonsense of notions like 'free trade', given that the starting positions (the so-called 'level playing field') of these two countries already enshrine such fundamental inequalities.

On a more upbeat note, and thanks to its improved economic performance, Brazil is now at last investing more heavily (often via public–private consortia) in crop-related infrastructure. A recent development with potentially profound consequences for Brazilian farmers, especially in the new *cerrado* land, is the epic transcontinental highway, the *Transoceanica*, scheduled for completion in the next few years. A consortium from Brazil, Peru and the Andean Development Corporation is planning to lay a new tarmac highway from Assis on the Brazil/Peru frontier to the Pacific at a cost of some $892 million. This will link up with existing roads so that a high quality, sealed tarmac highway will stretch from São Paolo on the Atlantic coast to Lima and beyond to Pacific Ocean ports.[685] The *Transoceanica* will open up the massively lucrative and ever growing Asian market, most notably in China, to Brazilian produce. In terms of its potential impact on agriculture, and the national economy in general, this enterprising development in Brazil may rank alongside the construction of the first transcontinental railroad in North America in the 1860s.[686]

While Brazil may be the major source of new arable land, it is far from the only example in South America. At the other end of the continent, in Argentina, the area of arable cropland has more than doubled, from 12 to 25 million hectares, between the 1980s and 2000. As with Brazil, much of the new Argentinian land is used for the increasingly profitable cultivation of soybeans and other cash crops, for export to Europe and China. However, there is still a large area of potential arable land that has yet to be utilised in this vast and still agriculturally underdeveloped country. For example, Argentina has 142 million hectares of prime pasture that is mainly used for beef cattle. Some of this land could be converted to arable use without loss of beef yields, by adopting livestock feedlot systems used in industrialised countries and also to an increasing extent in developing nations. Intensive livestock rearing is becoming more commonplace in response to ever increasing demands to supply global export markets, both in older industrial countries and in newly affluent developing economies, especially in Asia, as we will now see.

Increasing global demand for all kinds of meat has led to the adoption of intensive production systems by several developing countries, with a given country often specialising in a single type of livestock commodity. For example, Thailand currently produces over one billion chickens per year for export to supermarkets as far away as Europe, including much of the chicken meat on sale in the UK.[687] In 2004, the UK imported over 50 000 tonnes of chicken meat from Thailand. The Tesco supermarket chain alone purchased some $110 million worth of Thai chicken meat, mostly as de-boned portions for use in its ready meals. Intensive beef, dairy and hog feedlots in the USA, Canada, Denmark and the Netherlands have already proved to be much more economically efficient than conventional outdoor rearing.[688] Thanks to such practices, abundant supplies of inexpensive meat have for the first time been brought within reach of hundreds of millions of people, for whom any form of meat used to be a luxury.[689] In addition to supplying cheap meat to consumers, intensive livestock rearing practices liberate land from pasture and make it available for other uses, including arable farming. Obviously, intensive feedlot production also faces some significant challenges, such as animal welfare concerns and pollution from animal effluent. However, these concerns can and should be addressed by improved management systems that ensure compliance with existing international regulations.

Many other opportunities for cultivation of potentially productive land exist in countries throughout the enormous and diverse continent of South America, a region that is uniquely pregnant with agricultural potential. For example, over the past decade, Chile has hugely increased its agricultural production as the country emerged from a period of political and economic isolation that had inhibited its trade potential. Inexpensive, high-quality Chilean wines, and fruits of all kinds, are now available in most European supermarkets. Smaller countries have also shared in the new agricultural boom, with Peru increasing its area of prime arable farmland by over 50% (from 2.6 to 4.2 million hectares) in the period from 1992–2005.[690] Many of the improvements in Peruvian agriculture are due to a combination of improved security and better governance, plus the elimination of punitive tariffs on Andean imports by the USA and European Union. We will return to the thorny issue of tariffs and subsidies, and their role in suppressing food production, in the next chapter. Meanwhile, another good example of expanding crop cultivation can be found in the more humid north of the South American continent, where much desired tropical crops like oil palm are now being planted. During the two decades from 1980–2000, over 100 000 hectares of new oil palm plantations were established in Columbia alone. More recently, from 1994–2003, additional oil palm plantings in the northern region of South America produced 0.6 million tonnes of oil. This palm oil yield is equivalent to the oil yield from 5 million hectares of soybeans, or one sixth of the entire US soybean area.[691]

Ironically, the prospect of a South American agricultural renaissance (or, more correctly, neo-naissance) has now led to expressions of alarm in some of the traditional producer countries. There are obvious concerns about the cultivation of land in environmentally sensitive regions, but most of the Brazilian *cerrado* is savannah rather than rainforest.[692] Moreover, it is hardly appropriate for relatively wealthy Europeans and Americans to criticise others for doing what we have done to the land in our own countries, often for many millennia, in order to gain our own wealth.[693] Meanwhile, US farming interests are increasingly worried that newly emerging producer countries, like Brazil and Argentina, will soon become major competitors in the lucrative trade in commodities such as soybeans.[694] These fears are quite justified. It is quite likely that, if (or more likely, when) the South Americans can build up their transportation infrastructure to anything approaching what the North American taxpayer has already subsidised over the past century, the former will emerge as the major producers of some of the world's most valuable commodity crops over the next few decades. In doing so, of course, South American farmers could also go a long way towards meeting many of the additional food requirements of expanding populations elsewhere in the developing world, most particularly in Asia.

15

The roles of management, subsidies and breeding in crop improvement

Weak institutions and feeble contracts constitute the
most important challenge to the agricultural sector
Decio Zylbersztaijn (2005)
Economist[695]

Introduction

In the previous chapter, we saw that there will be significant challenges to global food production over the next half century, but that a combination of enhanced economic development and improved land use can go a long way to meeting future food demands. In this chapter, we will survey some of the other important strategies that can be used to enhance food production. Such options include improved management at all levels, from on-farm organisation of crops, through the entire transport and processing chains, and including any related physical and fiscal/legal infrastructure. Another important factor stifling food production in many developing countries is the rampant subsidy/tariff-led overproduction of agricultural commodities in many richer countries. We will examine ways in which the reform of this iniquitous system could greatly stimulate global agriculture. Finally, we will look at the prospects for improving crop production via plant breeding, using all available strategies, including transgenesis. Here, we will see that the outlook for continued increases in crop yields is more promising than many scientists and other pundits believe. Finally, it will be concluded that, while transgenesis may give breeders a few additional options, it is no panacea for the many challenges that confront twenty-first century agriculture. Indeed, transgenesis is neither necessary nor sufficient for the greatest forthcoming challenge to world agriculture, i.e. how to feed adequately an extra 2.6 billion people over the coming half century.

Improved management

Management issues are often overlooked when discussing potential crop yields. However, by implementing relatively simple aspects of good husbandry and the better organisation of supply chains, farmers and crop processors can often improve their yields to a far greater degree than can be achieved through many decades of conscientious plant breeding. Over the coming decades, some very significant increases in food yield should be possible by the better management of many existing crop systems, particularly in developing countries, where the yields of many crops are still only a fraction of those of industrial nations. This does not necessarily imply the use of intensive input regimes in developing countries, but it will require improved planning and management of existing crop systems to minimise inefficiency and unnecessary waste. As discussed previously, some of these management issues involve investment in improved infrastructure and communication networks, but many are very simple, readily achievable, inexpensive to realise, and capable of delivering huge productivity gains with minimal delay.

The most dramatic relative improvements in crop yields can frequently be realised by improved management of orphan crops by subsistence farmers. This is especially true in Africa, where it may be possible to double yields of some crops, simply by virtue of more knowledgeable and effective management regimes. To give a simple example: in the potentially productive country of Burkina Faso,only one third of the potential arable land is farmed and even this land is often mismanaged in a way that reduces its productivity. Mismanagement ranges from poor irrigation practices resulting in soil salinisation to allowing cattle to graze on fallow land during the dry season, which exposes the soil to erosion by wind and rain. These problems can often be tackled by simple and inexpensive ameliorative measures, but they require the cooperation of local tenant farmers who should be included as major beneficiaries of any reforms, i.e. not just the landowners. We will now examine two case studies that show the potential of improved management to increase crop productivity. Such gains are possible, not only with orphan crops in the poorest countries, but also with a relatively well-developed commercial crop like oil palm, which is a major export commodity in countries such as Malaysia and Indonesia.

Case study I: Malaysian oil palm

Large-scale commercial oil palm cultivation began in the early twentieth century in what was then British-ruled Malaya, using palms imported from Africa via the Bogor Botanical Gardens in Java. Oil palm is now grown on a massive scale in Malaysia, Indonesia and, increasingly, in Papua New Guinea and the tropical regions of Latin America. The following discussion relates to Malaysia where, despite a tradition of oil palm cultivation dating back over a century, there is still significant potential for large increases in yield via better management practices. However, similar considerations also apply to other oil palm growing countries where the potential for yield gains may be even higher because many of their plantations are less well developed than in Malaysia. In 2005, palm oil production finally outstripped that of soybean and this tropical crop became the number one source of edible vegetable oil in the world.[696] One of the major motors driving this increased production is the see-mingly insatiable demand to supply the expanding populations of India and China.[697] Already, the 2.4 billion people in these two countries make up 37% of the total world population. In addition to their growing populations, these two nations are becoming more affluent and, with their higher standards of living, there comes a steadily increasing demand for vegetable oil, as we will see below. The current average annual yield of useful oil from an oil palm plantation in Malaysia is about 3 tonnes per hectare. However, this yield can be doubled to over 6 tonnes per hectare by improving the way the crop is managed and harvested.[698] We know that such improvements are possible because some commercial oil palm plantations already achieve such high yields, simply by implementing more efficient management systems. Hence, the more efficient plantations often have the same types of oil palm, in similar soils, and in the same climate as other growers. The greater yield on well-managed plantations is due to such mundane measures as reducing crop losses from attacks by pests and diseases, poor harvesting methods, spoilage during transport and storage, and inefficient processing in mills.

By adopting some relatively straightforward measures, a good manager/husbandman can double the output. Large yield improvements can therefore be realised in existing plantations without recourse to expensive new technologies.[699] Palm oil yields could be increased even further, to as much as 8–10 tonnes per hectare, by using existing high-yield germplasm. But this will require a programme of clonal propagation and replanting of existing plantations that would take at least a decade. Finally, even higher yielding plants can almost certainly be selected using advanced screening methods like DNA marker-assisted selection and/or transgenic technology. It is not out of the question that the combination of advanced germ-plasm with improved management and processing methods could result in future yields of over 20 tonnes of oil per hectare. Indeed, I have personally seen some exceptionally high-yielding individual oil palm trees in commercial plantations that, if they were mass propagated without any yield loss, could theoretically produce as much as 60 tonnes per hectare.[700]

This impressive productivity of the oil palm tree is vastly superior to that of any of the annual temperate oilseed crops, such as soybeans, oilseed rape and sunflower.

These annual crops have a maximum yield of about 1 tonne per hectare, instead of the 3–10 tonnes or more that is already possible with oil palm. Oil palm has such a superior oil yield for two main reasons. First, it produces oil not only in its seed, but also (and rather unusually) in its fruit.[701] Second, the crop plant is a large perennial tree that is able to grow year-round thanks to its tropical habitat, whereas the temperate oilseeds only live for a single year and do not have the opportunity to become very large during their brief 4–6 month growing season. Given that the annually cultivated oilseed crops grown in western countries only produce up to 1 tonne per hectare, it is clear that oil palm has significant potential, not only to satisfy the increasingly demanding markets for edible oil in India and China, but also to act as a source of valuable non-food products for the oleochemicals industry.[702] The global production of palm oil in 2005 was about 32 million tonnes. If improved management resulted in only a 50% increase in yields (a very conservative figure), we could produce an extra 16 million tonnes of edible oil from the same area of land that is already in use for this crop, i.e. without the need to expand the area of cultivation. The same quantity of oil from one of the major temperate oilseed crops like soybean or oilseed rape would require the cultivation of 30–50 million hectares of land in Europe or the Americas.[703] This is just a single example whereby some basic improvements in management can unlock enormous additional food yield and also free up land elsewhere for other uses.

Over the next few decades, it will be especially important to foster the increased production of oil crops at a much higher rate than that of cereals. This is because the population of much of the developing world is not only growing, it is also getting wealthier. As we have already experienced in industrialised countries during the nineteenth and twentieth centuries, increasing affluence is generally correlated with an increase in the consumption of dietary fats and oils. In fact, one can make a case that per capita consumption of edible vegetable oils is a useful indicator of the economic wellbeing of a country. This may at first sight seem surprising, but there are excellent biological reasons underlying our craving for dietary fats. Dietary fats are so desirable because people on a starch-based subsistence diet are faced with relatively dull and tasteless foodstuffs, such as boiled rice, manioc or potatoes. Add some fats or oils, however, and the flavour, nutrient content (many unprocessed fats and oils contain lipophilic vitamins such as A, D and E), and the calorific value of the food are often greatly enhanced. There are other very practical reasons for the greater appeal of foods cooked with fats and oils, namely their enormous enhancement of taste and odour. Just think about the improved gustatory qualities of roast potatoes, French fries and fried rice when compared with their water-cooked counterparts. The reason for their enhanced appeal is that heated fats and oils produce a host of flavour compounds, many of them volatile, and also solubilise, and thereby enhance, otherwise cryptic flavours that may be present in non-fatty foodstuffs.[704]

People invariably find such flavours and odours attractive, probably because they are indicative of calorie-rich foods. Hence we have the perennial human craving for lipidic foodstuffs, which can be better satisfied once a population progresses beyond a basic subsistence diet. In line with this assumption, it was found that, following rising income levels across much of the developing world in the 1990s, vegetable oil consumption increased much faster than general food consumption. For example, during the 1990s, per capita vegetable oil consumption rose by 31% in Mexico, 35% in South Africa, 64% in China, 65% in Indonesia, and 94% in India.[705] This means that, as people became more affluent, they switched to a more satisfying diet containing much higher amounts of oil. Of course, if the theory is sound, the reverse should also be true. This means that when times are hard people should tend to cut back on 'luxuries' like fats and oils. Thankfully, there have been very few recent instances of widespread and sustained economic decline to test our vegetable oil theory. However, there has been one notable case, namely the economic collapse that followed the demise of the Soviet Union. During the worst years of this economic shock, edible oil consumption throughout the former Soviet Republics fell sharply. For example, between 1990 and 1994, consumption of food oil in Russia fell by no less than 35%, as people quite literally 'tightened their belts' and presumably reverted to a diet of boiled cabbage and potatoes.[706]

Just as oil palm yields can be greatly increased by improved management practices, similar principles also apply, albeit to varying extents, to many of the other major crops of developing countries. For example, in sub-Saharan Africa the yield of maize is only 1.7 tonnes per hectare versus a global average of 4 tonnes per hectare, while sweet potato yields are 6 tonnes per hectare versus a global average of 14 tonnes per hectare.[707] The reasons for this productivity gap are manifold, but improvements in management practices could contribute significantly to raising overall yields to levels that are closer to the global average and thereby increase food production in some of the countries that need it the most. We will now look at two more examples of the role, for either good or ill, that management in the broader sense (including economic) can play in crop production.

Case study II: Ukraine and Vietnam

As we saw in Chapter 10, the Ukraine was once known as the 'breadbasket' of Europe. It is a major area of cereal and oilseed production, as well as being an important centre of plant breeding research. In almost all of the former Soviet Republics, crop production declined, in many cases dramatically, following the dislocations of the post-1990 period. But probably the most catastrophic collapse of all occurred in the Ukraine. During the decade after independence, Ukrainian grain production fell by a half and the use of fertilisers by 85%. For some crops, the situation was even worse. For example, wheat production declined from 30 million tonnes in 1990 to a mere 10 million tonnes in 2000. Large areas of good, potentially productive farmland were abandoned, including as much as 40% of the area devoted to forage crops.[708] Much of the once world-leading plant breeding infrastructure also collapsed, with the result that some farmers were reduced to growing unimproved landrace varieties and to bartering amongst themselves for seed.

It was only in the years after 2001 that Ukrainian agriculture started a painful process of restructuring that has seen a slow but steady recovery in yields. Ironically, this restructuring largely involved the consolidation of the many small farms created by the privatisation of state collective farms after 1990. Therefore, in a sense, one could say that the large farm units of the Soviet era are now being recreated, although this time as privately owned and managed, and hopefully more efficient and profitable, business ventures. Provided improved management and breeding systems continue to be implemented, we can look forward to some very substantial increases in food production from this region over the next few decades. Despite some modest recent improvements, however, the Ukraine and the other ex-Soviet states still have a long way before they will be anywhere near to realising their full potential as crop producers.[709] In order to achieve this, it will be necessary to rebuild the plant breeding infrastructure, and also to inculcate a more open and entrepreneurial management culture into the agronomic practices of this entire region.

An instructive counter-example to the 1990s situation in the Ukraine is provided by Vietnam during the same period. Here, a series of reforms initiated in 1986, catalysed a huge improvement in agricultural productivity between 1991 and 2001.[710] The key to this success story was to unlock the entrepreneurial potential of Vietnamese farmers by returning control of their land, reducing taxation, and improving infrastructure, such as roads, waterways and market outlets. Farmers were stimulated to undertake their own initiatives to increase crop productivity, in the knowledge that they would be the direct beneficiaries. Increased food production then allowed farmers to diversify into cash crops like coffee, cocoa and pepper. This also reduced the levels of subsistence farming, with its associated high levels of poverty.[711] Increased crop production by newly empowered farmers enabled Vietnam to boost agricultural exports from $600 million to almost $2500 million in the eight years from 1991 to 1999. As the farmers increased their income, they invested in new crop varieties and other inputs, leading to further yield improvements and yet higher incomes. Within a few years, the burgeoning export of many types of produce from

Vietnamese farms was matched by an overall increase in trade throughout the national and regional economies.[712] Modern Vietnam does not just produce export crops, it has also become a valuable new import market for neighbouring crop-producing countries. Thanks to its rising level of affluence, Vietnam increased its imports of non-indigenous foodstuffs from $250 million to $1100 million between 1993 and 1995. As we saw above, imported foods like palm oil are highly desired by populations that are improving their economic wellbeing.

Therefore, these relatively basic managerial measures in Vietnam have led to a virtuous cycle of more general economic improvement that has been diffused to other countries in the region via the resulting increased volume of trade. As part of its continuing drive to improve crop productivity, Vietnamese breeders are now colla-borating more openly and equally with other organisations around the world. In one case, they are hoping to take advantage of crop genomic mapping data from the USDA, and to develop molecular markers to breed soybeans resistant to floods and diseases.[713] Adoption of new varieties might enable Vietnamese farmers to triple the yield of their soybean crop from the current level of 1.26 tonnes per hectare. Vietnam is also collaborating with India on projects involving use of new agbiotech approaches to breeding. Of particular interest is the development of methods for farming in drought-stricken land, as well as the cultivation of new high-yielding hybrid varieties of rice. As an indicator of recent progress in Vietnam, in 2006 a new biotechnology research centre was set up in Ho Chi Minh City to focus on vaccines and pharma-biological products related to agriculture, health and the environment. The message here is clear: farmers and breeders in many developing countries have the capacity to produce far higher crop yields and avail themselves of new technologies, but are often inhibited by poor management practices of all descriptions, many of which can be simply and promptly remedied.

To conclude this section, we can assert with some confidence that the improvement of management and infrastructure in the developing world could make a massive contribution to the improvement of crop yields and the hence to the economic status of local farmers. As stated in a recent review on global food security from the International Food Policy Research Institute, Washington DC: 'Although agroecological approaches offer some promise for improving yields, food security in developing countries could be substantially improved by increased investment and policy reforms.'[714] Taken together, the cultivation of new arable land and the improvement of agricultural management can probably meet a sub-stantial proportion of the additional demand for food over the next 40–50 years. Meanwhile, there are other economic and political factors that can potentially contribute as much as better management or breeding to agricultural improvement, as we will now discuss.

Agricultural overproduction and subsidies

We will now examine the stifling effects on developing country agriculture of the iniquitous subsidies and surpluses that are still tolerated by taxpayers in industrialised countries.[715] Removing such practices could greatly stimulate food production and related businesses in many needy nations, ensuring that farmers will still be able to feed the world fifty years from now. At present, we do not just produce enough food to feed the world comfortably; we are also overproducing food in many regions. Globally, we probably produce 20–50% more food than we really need simply to deliver enough calories to sustain the current world population at a basic level. Overproduction of food is reflected in lower prices, and its diversion towards other non-food uses, such as biofuels or animal feedstuffs.

Of course, a measure of deliberate overproduction of food crops is needed as a buffer against contingencies like adverse weather or crop disease. An example of such an eventuality was the massive drop in wheat production in Southeastern Europe in 2003, following an unusually cold winter and a prolonged spring drought.[716] The result was a huge shortfall in wheat production, amounting to about 16 million tonnes, which far exceeds the entire annual UK wheat harvest. The fact that hardly anybody was even aware of this relatively recent European agricultural disaster demonstrates the enormous resilience of our agro-economy. In bygone years, the result would have been mass hunger in the immediately affected region, and large increases in grain prices further afield. But after the 2003 wheat harvest failure, the lost grain was readily made up with surplus production from other wheat growing regions. There was no famine or food shortage, even in the most seriously affected areas of the Ukraine. Similar climate-related vagaries in crop output are regular occurrences throughout the world, although we cannot predict their location or timing. This is why an individual small farmer normally stores more grain than he really needs, and why many governments, especially in countries that are more vulnerable to fluctuations in crop output, maintain strategic grain reserves. However, this cannot justify the grossly excessive scale of artificially maintained overproduction of crops in many industrial countries.

Subsidies and tariffs stifle development

Some of the major brakes on agricultural production in developing countries are subsidies and tariffs imposed by industrial countries. This problem has been especially difficult to resolve because, despite their obvious drawbacks for global trade, such subsidies tend to be sensitive domestic political issues in many of the richer nations, especially in Japan, continental Europe and the United States. Industrial countries have been subsidising surplus production for many decades, to the tune of a massive $350 billion per year, to maintain their farmers at an artificially high level of prosperity.[717] To put this huge figure in a broader context, the Global Policy Forum estimates that it would cost a mere $13 billion to provide basic nutrition across the world and only $6 billion to provide universal education for all.[718] The amount spent by rich countries on agricultural subsidies is ten times the cost of global crop research and, even more shamefully, one thousand times the entire budget of CGIAR and its eighteen Research Centres.[719] It is now recognised that subsidies are inimical to the interests of developing country farmers and their already difficult quest for crop improvement. Subsidies are also against the letter and the spirit of free trade, as espoused by the WTO and supposedly practised by industrial countries. Accumulations of subsidised produce depress prices and inhibit international competition. Subsidies also reduce motivation to improve productivity in many developing countries, where farmers might otherwise grow more and better crops. And by stifling crop production in regions where higher yields may be possible (as with sugar crops[720]), subsidies and tariffs greatly restrict the overall global capacity for crop production.

These factors apply not just to food crops, but to all types of agricultural produce. An especially topical example is cotton, *Gossypium hirsutum*, which is a potentially valuable cash crop that can be efficiently grown in many parts of Africa and Asia. For example, cotton provides about 60% of merchandise exports in the impoverished Sahelian country of Burkina Faso. But efforts to improve cotton farming in this region have been held back by the dumping of cheap cotton from US growers, who collectively enjoy government subsidies of more than $4 billion per year for this crop alone.[721] In 2005 and 2006, the World Trade Organization stated that such subsidies distorted world trade and breached limits that had been agreed previously at the Uruguay Round of discussion completed in 1994.[722] The WTO also ruled that the grievances of countries like Burkina Faso should be addressed 'ambitiously, expeditiously and specifically'.[723] Indeed, no less a figure than World Bank President, Paul Wolfowitz, stated during a visit to the country in late 2005 that: 'The key to tackling the problem of cotton subsidies, which obviously hurts farmers here in Burkina Faso and in other poor countries ... is to tackle agricultural subsidies across the board'.[724] In July 2006, WTO Director General, Pascal Lamy stated:

One significant part of our Doha round negotiations involves reducing subsidies which distort trade while encouraging governments to use other forms of support which can facilitate development and environmental protection ... Shifting support in this way is politically difficult and requires determination and courage, but the evidence is clear that such reforms can level the playing field and provide real rewards across the board.[725]

Although many (but by no means all) governments, and the overwhelming majority of world public opinion, favours reform of agricultural subsidies, it has been pointed out that a rapid change in the system may turn out to be a mixed blessing for many developing countries. This is because some of the most penurious nations are also net importers of agricultural goods, often from the richer countries. If subsidies were to disappear suddenly, so it is argued, the cost of imported food would increase, to the detriment of these already disadvantaged poorer regions.[726] This argument may be true in the short term, but it misses the point that poorer countries often import food because their own agricultural produce has been rendered artificially uncompetitive by cheap subsidised imports. One of the best longer term stimuli for agriculture in developing countries will therefore be the removal of such subsidies. This will allow prices to rise, benefiting local farmers, and encouraging further investment in crop improvement.[727] One needs only to note the example of Vietnam (see above) to realise how the stimulation of local food production in one country can give a significant boost to an entire regional economy. The stifling effect of subsidy-driven low prices is one of the most serious structural problems facing attempts to improve global crop performance. By denying developing country farmers fair access to markets for their produce, subsidies and tariffs remove incentives for increased crop production and inhibit efforts to breed better varieties or update management practices.

Improved crop breeding

This brings us to our final set of tools for boosting global food production, namely good old plant breeding itself. Although improved management and politico-economic reforms are vital for agricultural advancement, we must not forget the raw materials – the crops themselves. There is a degree of disagreement and confusion about the future role of conventional crop breeding, especially as regards the debate about 'feeding the world'. The prognosis for the ability of conventional plant breeding to increase yields of food crops sufficiently to provide for the inevitable future population increases over the next forty years is still the subject of some controversy, even among many scientists.[728] There is a widely held view by many of the more research-based plant scientists (as opposed to breeders) that we have reached a yield plateau for most of the major crops. Interestingly, as with the Malthusian pessimisms of the 1970s, concerns about diminishing yield gains are nothing new. As long ago as 1982, respected US agricultural economist, Earl Heady, was expressing very similar qualms.[729] As we now know, these concerns were groundless and cereal yields made impressive gains for two more decades. So is there any substance in the current worries about a plateau in crop yield gains?

On the one hand, the spectacular increases in yields of the 'big three' global food crops, namely rice, wheat and maize, have unquestionably levelled off in recent years.[730] By 2003, the average wheat yield in developing countries was almost the same as that of industrialised countries, about 2.5 tonnes per hectare. However, many experts believe that genetically based yield gains in our major commercially traded crops could continue for several more decades.[731] It is perhaps telling that many (but by no means all) scientists in industrial countries believe that a yield plateau has been reached, but the opposite tends to be true in developing countries.[732] There also appears to be a tendency for agbiotech focused scientists, and especially those who are remote from practical breeding, to devalue the potential of conventional breeding strategies. Meanwhile, their more pragmatic colleagues, who are closer to the practical problems of breeding in developing countries, tend to be more confident about prospects for non-transgenic crop manipulation.

Quite apart from the debate on the yield prospects for wheat, rice and maize, there is little doubt that there are still massive gaps in yield that can be improved by breeding in many other 'orphan' crops in developing countries. One example is the tropical plantain, which is the fourth most important source of edible calories in the tropics. Tropical plantains are part of the genus *Musa*, and are close relatives of commercial bananas. Plantains are important staple crops grown throughout the tropics. Sometimes called cooking bananas, the edible part of the plantain is a hard starchy vegetable with little flavour, which is cooked and eaten like a potato.[733] The yield of this immensely important staple tropical crop has barely improved (up by a mere 3%) over the last thirty years.[734] As we saw in Chapter 6, agencies like CGIAR are applying modern molecular breeding techniques to plantain and other orphan crops in developing countries. These efforts are still at a relatively early stage and there is enormous untapped genetic potential waiting to be released in such crops. Hundreds of millions of small farmers cultivate these crops as subsistence foods, often gaining a modest income from sale of surplus produce in local markets. Development of higher yielding varieties will therefore stimulate local economies in some of the poorest and most food-insecure regions of the world. These measures will require concerted action from both public and private agencies in each country, plus international bodies like the FAO.

Advanced breeding methods like marker-assisted selection and clonal propagation may make a useful contribution by making it easier to produce improved cultivars of the mainly tropical and subtropical crops in regions of maximum population growth. But, as ever, such efforts could be seriously jeopardised in some areas by renewed wars or civil unrest (see Chapter 16). It is also important to remember that simply producing more food to keep pace with an increasing population is no guarantee against malnutrition or hunger. The world already produces enough food to feed us all, but 800 million people still suffer from food shortages. The key factor in insuring against hunger is to address issues such as poverty and access to a balanced food supply. Higher yielding staple crops may also contribute to the amelioration of rural poverty as farmers get more income from their land. As we saw in the recent example of Vietnam, rising incomes can then produce a virtuous cycle of increased investment in farming inputs, better seed stocks, improved knowledge/education and diversification into new crops, all of which is translated into higher incomes for farmers and others.

What is the role of transgenic crops?

In this chapter, we have looked at various ways in which agricultural production might be improved in the coming decades in order to feed the increasing global population. It should be apparent that this is a complex issue that goes far beyond the confines of plant breeding per se. We have seen that improved crop production will depend on many other factors, including the cultivation of more land, the better management of existing farming and processing systems, and the ending of discriminatory subsidies and tariffs in richer countries. Taken together, the issues that we have discussed above should make us justifiably optimistic that, providing these challenges are addressed, farmers will be able to feed the world quite adequately, without recourse to any simplistic 'magic bullet' solutions. As we seek to devise strategies to improve food availability around the world, it is important that we do not permit ourselves to be distracted by the allure of superficially attractive, but relatively unproven, plant breeding technologies. Feeding the world is possible, but not solely via a quick-fix, transgenic approach. Rather, the realisation of this eminently achievable goal will depend on much hard work in applying some very straightforward principles of plant biology, economics, and agricultural management, in an interlinked, coordinated and common-sense manner.

The issues around management and ownership of agbiotech resources also need to be resolved by developing real partnerships between the public and private sectors. This applies to all forms of molecular technology, including DNA markers, genomic resources, and the various proprietary transgenic technologies. Such partnerships will enable the development of viable public-good ventures in developing countries. We need genuine long-term partnerships over decades rather than short-term initiatives. Meanwhile, transgenic technology should be recognised for what it is, i.e. just another mechanism for generating variation for breeders. At present the scope of the technology is relatively restricted and it is most unlikely to contribute to a quantum leap in agronomic performance of developing country crops in the short-to-medium term. As with the other more recent tissue culture and DNA-based breeding technologies, transgenesis most assuredly has its place in the breeder's toolkit, but it does not constitute a toolkit in itself.

The way forward in confronting the very real agricultural challenges posed by population growth is to resuscitate our depleted public sector, reinvigorate and diversify the oligopolistic private sector, and develop new working practices for plant breeding. Crop scientists will also need to work more closely with farmers in developing countries, especially in regions that have received little or no benefit from the Green Revolution. Crop breeding strategies to confront the entire range of such agronomic problems should be devised using a mechanism of informed choice based on all of the available technologies, and using pragmatic criteria such as cost, timescale, feasibility etc. Sometimes, no doubt, newer technologies like transgenesis or marker-assisted selection will be appropriate and their long-term potential is still very exciting. But for the next decade or two, we should recognise that such relatively expensive tools will remain fairly restricted in the range of their uses for global crop improvement.

In Part V, we have seen that global food production faces many challenges to deliver adequate food to increasing populations, especially in areas like sub-Saharan Africa. However, the solutions are also many and varied, ranging from economic and political measures to improved plant breeding efforts, particularly for orphan crops. In Part VI, we will discuss the major challenges for private and public sector breeding in the twenty-first century.

Part VI

Plant breeding in the twenty-first century

Eaque est scientia, quae sint in quoque agro serenda ac
facienda, quo terra maximos perpetuo reddat fructus.
It [agronomy] is as well a science, which teaches us what crops
should be planted in each kind of soil, and what operations are to
be carried on, in order that the land may regularly produce the
largest crops.

Marcus Terentius Varro (116–27 BCE)
Rerum Rusticarum de Agri Cultura[735]

16

The future of international plant breeding

Give fools their gold, and knaves their power;
Let fortune's bubbles rise and fall;
Who sows a field, or trains a flower,
Or plants a tree, is more than all.
John Greenleaf Whittier (1807–1892)
A Song of Harvest

Introduction

In this chapter, we consider some of the challenges facing both public and private sector plant breeding, as we move into the post-privatisation era of the early twenty-first century. For example, given the increasingly parlous state of public sector breeding, should the private sector now be considered as a major future provider of breeding-related R&D? After all, this already happens in other key areas of the economy. Nearly all pharmaceuticals are researched, developed and marketed by private companies that often enjoy near-monopoly status in their supply to consumers. In richer countries, these expensive products are generally disseminated, cheaply or freely, via taxpayer funded healthcare systems. If the public is willing to subsidise private sector dominance of the provision of life-saving drugs, why should our food supply be any different? The answer is that the current pharmaceutical industry paradigm is proving deeply flawed, especially in supplying cost-effective drugs to poorer consumers.

A particular problem with the present organisation of the pharmaceutical industry is that cash-poor governments in many developing countries cannot subsidise sales of expensive drugs to the poor. This becomes a serious issue with a major disease like AIDS, where the cost of medication is beyond the reach of most sufferers and the state. The alternative use of cheaper generic drugs has been challenged by the pharmaceutical industry, resulting in bitter international rows. Meanwhile, and to our great shame, thousands of people die daily from the entirely preventable effects of AIDS. The existing industry paradigm has also signally failed to address what we

241

might call 'orphan diseases' (by analogy with orphan crops), such as malaria and tuberculosis, that still needlessly kill millions of people each year.[736] We can therefore conclude that the structure of the pharmaceutical industry is not a very useful model for international plant breeding.[737]

This is not to dismiss any role for the private sector in public-good crop improvement, either in industrialised or in developing countries. For example, it might well be possible to tap into the best of the entrepreneurial spirit of the private sector by allowing it to compete with, or even replace, public sector bodies for the provision of some aspects of public-good research. Governments or other organisations, including philanthropic trusts, might decide to contract out some or all of their research on a competitive basis, rather than set up their own research centres or institutes. Under such circumstances, a private company, or a consortium of enterprises, might end up winning the contract instead of the public body. Contracted work need not be limited to the kind of commercially orientated research that has already been largely privatised in most industrialised countries. There is no a priori reason why the private sector should not also be contracted by governments and/or donors to undertake public-good research, e.g. on orphan crops or unprofitable traits in developing countries, simply on the basis of open and competitive tenders for such services.

But is such a 'private sector service-provider' model realistic in the long term? The main problem with this model is the lack of continuity of funding and professional expertise that would almost certainly result. Whatever their faults and inefficiencies, the better public research centres are formidable repositories of knowledge and expertise of a kind that would be very difficult to recreate in a private sector contract research firm. Plant breeding is quite different to the majority of goods and services, from telecoms to tourism, that are eminently amenable to residing in the private sector. Moreover, as we saw in some of the previous chapter, the supposedly resurgent agricultural private sector has itself experienced some significant setbacks in recent years and is unlikely to be able to provide the full range of specialised services, never mind the extended attention span, required for pubic-good R&D. Perhaps these uncertainties should make us cautious in writing off the public sector just yet.

Enfeebled as it undoubtedly is in the early twenty-first century, the public sector still carries out many pre-competitive functions that cannot be adequately fulfilled by private enterprise. These include many aspects of research on commercially unattractive crops. Another key role that is strongly endorsed by companies is the collection, preservation and development of plant germplasm. Such collections include crop species and wild relatives, many of which carry potentially useful genetic traits. Public sector breeders also frequently play important, and relatively disinterested, roles in government advice and the determination of policy issues such as breeders' rights and intellectual property rights. This function is assuming greater importance as the general public becomes ever more engaged in debates relating to agriculture and food, ranging from transgenic crops to 'fair trade'. Informed statements by truly independent public sector plant scientists carry a great deal more weight with the public than even the most reasoned and articulate pronouncements from company spokespersons.

Most companies regard public researchers as a useful buffer against what might otherwise be a poorly regulated private sector, where two or three major players could end up controlling the major commercial crops and even dominating the regulatory process itself, to the eventual detriment of the industry as a whole.[738] The public sector will also continue to be the best provider of qualified scientists, as recognised by major private sector bodies such as the American Seed Trade Association, which recommends a continuing strong role for the public sector in the training of plant breeders, even in the case of crops like maize and sorghum where breeding is now dominated by the private sector.[739] The public sector is regarded as a key repository of up-to-date expertise in most of the cutting edge areas of new technologies such as genomics, bioinformatics and molecular markers. As with training of breeders, private companies recognise the key role of the public sector in applying new discoveries in plant biology and related disciplines to develop generic tools, such as genetic maps and mutant resources, for modern crop breeding.[740]

At the international level, there is still a relatively strong and effective group of public sector, crop development institutes, with CIMMYT and IRRI being especially noteworthy examples in Mexico and the Philippines respectively.[741] But even these international centres are struggling to maintain their focus on crop improvement, and on applied research and advisory services. As we saw in Part III, there has also been a marked trend in Europe and North America for public sector plant science to become more academic and less practically oriented. This situation can only be ameliorated by fundamental institutional reform in the countries concerned. Appropriate career structures and recognition systems should be put in place to sustain the work of good quality, well-motivated scientists. Alas, we are not all of the calibre of Norman Borlaug in being effective innovators, who are at times both iconoclastic and pragmatic (see Chapter 6). We need to nurture a new generation of researchers who can see beyond the next grant proposal or high-impact publication. The alternative is to hand over our future by default to the agbiotech companies. This would be bad enough for richer countries, but it would be far worse for the disenfranchised poor of the world who surely deserve to share the fruits of our hard-won, taxpayer funded scientific knowledge.

While public sector plant breeding has declined in most industrial countries, its importance in the developing world has continued to grow. Partially, this has been due to the slower pace and much reduced scope for privatisation in such countries.

The maintenance of mainstream plant breeding as a respected professional activity in many developing countries has also been helped by a relative dearth of the kind of research infrastructure needed to sustain the more glamorous high-tech alternatives, like molecular biology and genomics. But this situation is now changing, as funding for conventional breeding declines, while access to high-tech facilities across the world becomes cheaper and easier. As we will now see, even CGIAR is threatened by a form of 'mission creep', whereby the steady diversion of funding to other areas is diluting its effectiveness as the main engine of crop productivity improvement in the developing world.

Whither CGIAR?

As long ago as 2000, agricultural economists were commenting that: 'Expenditures on agricultural research in … the International Agricultural Research Centers (IARCs) have stagnated and in some cases, declined sharply in recent years.'[742] In the past few years, some of the largest and most effective centres, including CIMMYT, have been forced to cut back on research activities, even to the extent of cancelling field trials and curtailing the training of new plant breeders. Funding for germplasm research at CGIAR centres started to be reduced in the 1980s and early 1990s. This was not so much due to overall budget cutbacks, but rather because a large list of additional activities was imposed on the centres by their donors, with little or no commensurate increase in funding.[743] The additional activities included an entirely laudable series of ventures, from improving participation of women in economic activities to assisting natural resource management. However, since there was no corresponding increase in funding, core activities, such as breeding, were cut back. During the 1990s, spending on crop productivity at CGIAR declined at an annual rate of 6.5%, while research into environmental protection and biodiversity each increased their share of the budget. As a result of these cutbacks in germplasm research, CGIAR was not well placed to capitalise as effectively as it might have on some of the new opportunities that were emerging in agbiotech in the 1990s, including marker-assisted selection and aspects of transgenesis.[744] Total CGIAR research expenditure peaked at about $330 million in 1991 but by 2000 it had fallen back to $300 million.[745]

Today, CGIAR is facing a severe funding crisis. It may be instructive therefore to look at its value for money in international terms. One way of doing this is to compare the funding versus the impact of a typical CGIAR institute with a (far wealthier) US crop improvement research centre. To take one example of many such centres, Michigan State University Agricultural Experiment Station (AES) has an annual budget of $80 million. This sum is roughly twice the annual budget of the largest of the CGIAR centres (CIMMYT). However, it is in the human impact of the two institutions that the comparison is most telling. The Michigan State facility serves primarily a state clientele of 8000 full-time farmers, 16 000 part-time farmers, and 27 000 'hobby farmers' in the local region.[746] In contrast, CIMMYT works on two of the three most important food crops in the world (wheat and maize), as well as on their associated natural resource and economic problems. From 1966 to 1997, 85% of spring bread wheat varieties and spring durum wheat varieties released in developing countries were based on CIMMYT materials.[747] In other words, with half the resources of the Michigan AES, CIMMYT serves a global clientele that numbers in the billions in terms of consumers and hundreds of millions in terms of farmers that are directly using its seed.

Of course, this comparison is between two institutions that are not exactly the same; the mission of Michigan AES is obviously rather different to that of CIMMYT. However, it is interesting in view of the questioning by some of the value-for-money of international ventures like CIMMYT. When compared with most publicly funded crop improvement ventures in developed countries (there are many other centres that are very similar to Michigan AES in the magnitude of their costs versus the size of their client base), CIMMYT and the other CGIAR centres look like an unbeatably good investment.[748] This conclusion was reinforced with the publication in 2005 of a detailed assessment of the impact of CIMMYT activities during the 15-year period from 1988–2002.[749] Among the many impressive statistics in this report, perhaps the clearest indication of value for money is that the annual cost of wheat improvement research at CIMMYT is $9–11 million, whereas the annual benefits derived from its wheat varieties across the world is between $1–6 billion.[750] Few commercial investors would even dream about realising such a massive cost: benefit ratio.

The funding crisis at CGIAR prompted a review of its organisation by one of the major donors, the Rockefeller Foundation, in which several cost-saving and efficiency measures were identified. Following this review, in early 2005, CIMMYT and IRRI jointly announced a partial merger of the two centres. Under the terms of the new alliance, CIMMYT and IRRI will focus in future on coordinated research into the improvement of the three major crop staples, i.e. rice, wheat and maize. The main priorities identified were development of intensive crop production systems in Asia, more stress on informatics, training, education, and more research related to climate change. At present the long-term prospects for CGIAR are still far from certain. As noted in a recent management review, the CIMMYT/IRRI merger and the associated redundancies plus other restructuring activities, have had a serious effect on staff morale that might adversely affect the ability of the network to maintain its existing level of plant breeding research and germplasm conservation.[751] One interesting recommendation from the CIMMYT review relates to the use of transgene technology. The expert panel was critical of the balance between work on insect resistance in maize using inherent resistance traits, as opposed to transgenic Bt traits. Since 1990, the transgenesis programme at CIMMYT had cost $12 million, while the conventional breeding approach had only been allocated $2.9 million. This was despite the availability of non-transgenic maize varieties with good inherent insect resistance traits. It was felt that the latter type of inherent insect resistance would be particularly useful in the poorer areas of subsistence farming, the so-called marginal maize producing areas, or MMPAs. As the report stated:

Donor material with effective insect resistance has been identified but over the last decade it has not been used effectively in the most advanced breeding material. New modern breeding tools are now available for successful integration of such polygenic resistance during the breeding process for selecting better maize varieties. This level of resistance may be more sustainable [compared with that of transgenic varieties], much easier to handle in seed and trade systems, and can have enormous impact on stabilising yield at higher levels in MMPAs[752] [author's note].

This reinforces our conclusion that an over-fixation on a single technology, such as transgenesis, is not just scientifically unsound, it can actually impede progress towards crop improvement. This is especially true for the poorly resourced international breeding programmes. As we have seen, the network coordinated by CGIAR is under severe financial pressure and it must be prepared to make hard-headed choices about the most practical technologies to be employed in its breeding programmes. In spite of the attraction of high-tech methods, we should all bear in mind the sentiments of Norman Borlaug about 'academic butterflies', as quoted in Chapter 6. The primary role of breeders is to improve crops. In order to achieve this task, they need to be able to use the full range of technologies, but they should not be overly diverted by academic distractions. There are already more than enough 'academic butterflies' in the wider plant science community. What we need at CGIAR and national research centres are well-trained and objective breeders of a more pragmatic and practical mien.

In conclusion, we can say that, despite the many challenges that confront it, the CGIAR network is still excellent value for money, enriching the lives of hundreds of millions of poorer people around the world. Its annual running costs of $370 million pale into insignificance when compared with the tens of billions of dollars that are spent on comparable research in industrial countries. These richer countries have far lower populations and much less real need than the more than one billion people directly dependent on crop varieties developed by CGIAR centres. One can only hope that CGIAR and its staff will weather the storms of the coming years. This should be possible as long as the donor community maintains its relatively modest commitment, and the breeders are enabled to use the most practical, cost-effective technologies required for each crop improvement programme.

National research centres

Although CGIAR is probably the most important player in international plant breeding, there are other research providers and other producers of improved crop varieties. The most important of these are the national public sector networks now established in many developing countries, plus an emerging private sector. Over the past decade, in particular, there has been increasingly active private sector involvement in crop improvement in the improving economies of East Asia and South America. This has been most marked in regions that have embarked on the large-scale cultivation of major commercial export crops, especially in South America. To a great extent, the recent activity of companies like Monsanto in Argentina and Brazil was prompted by the prospects of expanding their area of proprietary transgenic commodity crops, such as soybeans and maize, in these increasingly important producer countries. This strategy has been at least partially successful, as farmers have in general welcomed and often greatly profited from the new crops. Before such ventures can be considered secure, however, there must be stable and reliable market conditions in host countries. As we will see in the next chapter, failure to agree on a suitable licence-fee structure for transgenic soybean cultivation in Argentina led to the (temporary) withdrawal of Monsanto from that country in 2004, to the detriment of both the company and the indigenous soybean farmers.

One way in which developing countries can better help themselves is to enter global markets as active technology developers alongside private companies. Hitherto, such a strategy would have been prohibitively expensive, even if the countries possessed the necessary scientific and marketplace expertise. Another formidable barrier has been the maze of breeders' and patent rights that have surrounded varietal improvement of crops since the 1960s. However, there are signs that things are beginning to change; for example, some developing countries, such as China, India, Brazil, Malaysia and South Africa now have expertise to compete with some of the best modern breeding systems of the richer nations. At the same time, economic and legislative reforms are making many of these markets more secure and reliable places for both internal and external investment. This leaves the third major hurdle to be overcome, namely the thorny issue of breeders' and patent rights, especially at the high-tech end of plant breeding. In Chapter 19, we will consider such IPR issues in more detail and will examine how these barriers to innovation might be surmounted, to the ultimate benefit of the global marketplace.

Participatory plant breeding

International plant breeding has made tremendous advances since the 1960s. However, there remains one singularly recalcitrant category of farmers where all the best efforts of the public sector alike have had only limited success. This refers to farmers in those parts of the world that lack much of the knowledge or basic infrastructure to profit from conventional programmes of crop amelioration, or who for a variety of cultural and related factors have been unable/unwilling to engage in crop improvement initiatives. Despite well-intentioned programmes of crop improvement by professional breeders and NGOs, many people in these regions continue to languish in a condition of poverty and malnutrition. Their plight is exacerbated by low crop yields, inadequate or unbalanced nutrition, and a high incidence of preventable disease. While there are many regions like this throughout the world, they are mostly concentrated in sub-Saharan Africa. A particular challenge is the multiplicity of small farms, even within a small area, that are often highly variegated as regards their crops, soil types, and climatic conditions. In the words of Gordon Conway, President of the Rockefeller Foundation:

... despite recent policy changes recognizing the importance of farming and despite a renewed interest in investing in the potential of small farmers, African agriculture and the livelihoods of its rural people continue to deteriorate. ... Those trends must be changed, and one way to do it is to invest in small-farm agriculture now.[753]

Over the past thirty years, this challenge has largely confounded the efforts of those with a stake in agricultural progress and the fight against poverty, including scientists, development workers, governments, and aid donors. It seems most unlikely that the private sector will penetrate this unrewarding market, and even the public sector seems to have largely failed so far. More recently, this dire situation has stimulated some innovative thinking by breeders and others who work with communities of small farmers, who often have no choice but to cultivate crops on poor soils of marginal quality. Several initiatives have addressed these challenges, but the most promising new approach is undoubtedly participatory plant breeding. This strategy of crop improvement does not require a great input of external funding or capital, and relies instead on informed self-help by the farmers themselves, as we will now see.

A decade of progress

Participatory crop breeding has been developed over the past decade as an alternative and complementary breeding approach to the kinds of formal, expert-led crop improvement that we have considered hitherto.[754] This form of applied research is novel because it includes non-experts, i.e. untrained farmers, as direct participants in the planning, execution, and evaluation of the research process itself. Recent external evaluations of the effectiveness of participatory breeding have been encouraging, especially if combined with an awareness of wider social dimensions of agriculture.[755] The strategy seems to be most effective when carried out in collaboration with the sorts of small farmers, often growing orphan crops, who live and work in marginal areas that may be unsuitable for cultivation of mainstream crop varieties. Such farmers frequently have to cope with unstable and difficult growing conditions and lack knowledge of and/or access to many of the fruits of modern plant breeding. The adoption of new plant varieties by this group has, therefore, been extremely low, most particularly in sub-Saharan Africa, but also in several other regions.[756] One of the reasons for the failure of conventional breeding in this context is that it has tended to focus heavily on 'broad adaptability', i.e. the capacity of a plant to produce a high average yield over a wide range of physical environments and growing seasons. In conventional breeding programmes, crop varieties that produce very good yields in one particular growing zone, but poor yields in another, tend to be eliminated from selection trials. But this sort of locally adapted variety is often exactly what small farmers might need. Another downside to broadly adapted varieties is that they are often deliberately selected to require substantial amounts of inputs, such as fertiliser, which is something that a cash-poor small farmer most assuredly cannot afford, no matter how good the prospective yield might be.

Socio-economic research has revealed additional and hitherto unsuspected cultural and other non-yield-related factors that can strongly influence the small farmer, but which are often invisible to the professional breeder working in relative isolation from such farmers. Breeders tend to focus strongly on the 'holy trinity' of yield, disease resistance and pest resistance. Although these traits will always be important, other characteristics that might be more relevant to a small farmer include the following: ease of harvest and storage, taste and cooking qualities, speed of crop maturation, and the suitability of crop residues as livestock feed. This wish list and its order of priorities can vary significantly from one region to another. The bottom line here is that, unlike maize and soybean in the US Midwest, the 'one size fits all' approach to modern commercial breeding most definitely does not apply to the diverse agronomic micro-environments that exist in many developing countries. Often, the only way to get such location-specific and culture-specific information to breeders is to co-opt small farmers into joining with the latter as genuine partners in the breeding process itself.

Partnerships with farmers

In the past, professional breeders may not have recognised the specific issues relating to the needs of small farmers.[757] Even if these needs were appreciated to some degree, it has not always been possible for breeders to address them, because of some of the limitations relating to the structural context of international plant breeding. For example breeders working on crops from developing countries are placed in a context that explicitly cuts them off from their farmer-customers. Hence, breeding is often done as a result of research by scientific experts from industrial countries; the breeding centres may be located in a different country or even on a different continent from the recipient farmer-customers; the whole effort is funded by international donors, also from industrial countries, who may have their own agendas and impressions of the needs of these farmers, which may not be congruent with the actual requirements of all the growers in a particular region. Meanwhile, the farmer-customers of the breeders tend to be poor and often ill educated (in the formal sense) people of low social status and little influence in their own country, who probably do not even share a common language with their would-be benefactors. In fact the gulf between the farmer-client and the scientist-provider is often even wider than this, encompassing profound differences in culture, knowledge, religion, race, gender, language and world-view.

This means that, unlike their engagement with the articulate and often vociferous farmers in industrialised countries, it is much more difficult for researchers to pick up market signals from poorer farmers in developing countries. Previously, farmer participation in conventional breeding programmes has been limited to evaluating and commenting on a few advanced experimental varieties just prior to their official release. This kind of rather token participation has meant that few farmers felt a sense of ownership of the research or its products. Neither were they able to contribute their own, often extremely apposite, informal technical expertise, e.g. about local soil conditions and crop diseases. Many varieties that progressed all the way to on-farm trials in such regions might have been eliminated from testing years earlier if the local farmers had been given the opportunity and confidence to assess them critically on their own terms. This form of farmer participation is not just more democratic and likely to result in an acceptable outcome; it can also be more efficient for the breeding programmes as a whole by screening out undesirable varieties well before the expensive field-trial stage.

Another key issue that was uncovered surprisingly recently is gender. Farmers have been the chief engineers of crop and variety development for thousands of years, and in many developing countries women have been the principal growers and harvesters. Research from South America and Africa demonstrates the overwhelming importance of gender issues in many aspects of plant breeding. For example, many traits involved in harvesting, threshing, milling and cooking of grains can be more or less invisible, even to the men in the local community, and may be overlooked by scientist breeders. However, these processing-related traits may be of paramount concern to the women who actually carry out such manipulations as they prepare food from the crops on a daily basis. Historically, these women would have been the people who carried out most of the empirical breeding. And in many cases, the women might possess useful local knowledge that they could impart to the modern-day breeders, providing they are given such an opportunity.

The importance of women in the success or failure of breeding projects has been shown in several case studies in Côte d'Ivoire. The studies revealed that selection of inappropriate traits by poorly informed scientific breeders can lead to the rejection of new varieties by women farmers, and a consequent negation of the entire breeding programme.[758] The process of farmer participation in more-or-less empirical breeding activities continues today in many parts of the world. In particular, millions of local farmers, most of them women, still actively select and breed many landraces of the so-called orphan or neglected crops, that are frequently the essential staples for the sustenance of their families. One of the strengths of participatory plant breeding is that it combines this local expertise with modern scientific breeding in a novel and so-far promising approach that may help the 'lost generation' of poorer farmers. Following the recognition of this need to bridge the divide between breeder and farmer, growing numbers of participatory plant breeding projects are now being trialled throughout the developing world.

Projects in India

Although participatory plant breeding has mainly focused hitherto on the simpler and more accessible technologies of crop improvement, it should not be regarded as a peripheral venture that is somehow divorced from 'mainstream' plant breeding. On the contrary, it should be thought of as a more sophisticated, customer-orientated approach to the overall strategy of crop improvement. As such, there is no reason why participatory plant breeding should not benefit from the full panoply of modern plant breeding technologies, providing they are appropriate in the context of a particular programme. As long as breeders are able to circumvent some of the more restrictive IPR challenges that limit the use of some modern breeding technologies, there are many possibilities for the use of the latest breeding approaches within the context of a participatory plant breeding programme. Indeed, there are already several projects employing advanced molecular strategies, such as DNA marker-assisted selection, as part of a participatory plant breeding approach.

This type of combined strategy is currently being pioneered by a team of UK and Indian breeders in the States of Jharkhand, Madhaya Pradesh and Karnataka for the selection of stress tolerance and other traits in several types of rice,[759] and in Gujarat for better yielding maize.[760] Most stress tolerance traits are genetically complex and can have a low heritability. This makes the task of the breeder very difficult, but marker-assisted selection can appreciably lighten this load and facilitate the development of new improved varieties. The same research group (from Bangor, UK) that is using marker-assisted selection in rice is also studying the potential for various innovative transgenic approaches to improving crops such as bananas and potatoes. Once these plants are available, their desirable new traits can be intro-gressed into locally adapted crop varieties with assistance from small farmers. Such molecular-based technologies are still very much long-term ventures, but they might be a useful component of participatory plant breeding in the future. As yet, this new public sector led approach to breeding is in its infancy, but initial results are encouraging. Given that participatory plant breeding may be the most effective way to reach those 700 million people who have missed out on the Green Revolution, it is certainly well worth persevering with similar programmes over the coming years.

Seed banks and germplasm conservation

National and regional seed banks

Seed banks and other centres of germplasm conservation are invariably 'public' goods held in national or international centres. I have qualified the word 'public' here because, as we will see below, certain national seed banks (most notoriously in the USA) are increasingly managed as national assets rather than genuine, inter-nationally accessible, public-good facilities. Hence, plant breeders based in certain proscribed nations are now excluded from access to US germplasm resources, solely for political reasons (we will return to this topic when we consider the future of seed banks later in the chapter). According to the United Nations FAO, there are almost 1500 gene banks worldwide maintaining about 5.5 million samples. These constitute virtually the entire *ex situ* stock of crop genetic resources on the planet.[761] The most effective agency for germplasm conservation and development in developing coun-tries is CGIAR, which holds an estimated 666 080 accessions in its 15 research centres.[762] Since CGIAR focuses on developing countries and their crops, there is no prospect of any significant private sector involvement in such an operation. On the other hand, CGIAR is vitally important because it acts on behalf of three quarters of the global population, and serves all the regions of projected high population growth and food demand over the coming decades.[763]

There is no realistic substitute to the work of CGIAR and any minor contribution to their crop portfolio that might be made by new transgenic varieties, whether from public or private sources, will almost certainly need to be channelled through existing non-transgenic CGIAR varieties anyway. For example, the transgenic 'golden rice' varieties developed in Europe are currently being bred into local Asian varieties for field trials by IRRI in the Philippines. Any new transgenic crop will normally be developed initially in a transformation-friendly variety for initial glasshouse and small-scale field studies. But that variety will then require subsequent crossing with a range of existing local cultivars that are better adapted for conditions in the various climates and soil types around the developing world. Such a process involves extensive field trials and can take five years or more. Without the involvement of IRRI, there would be no realistic prospect that this sort of transgenic laboratory variety of a crop such as rice would ever see the light of day on Asian farms.

The largest national germplasm resource centres are in the USA, China and Russia, but are themselves largely based on collections from sites around the world. Hence, while these collections are 'national' in the sense that they are funded and housed in specific countries, they are all completely 'international' because their collections are from many other countries in addition to the host nation. The US National Germplasm System has over 450 000 accessions; China has 300 000 in its Institute of Crop Germplasm; and Russia has almost 180 000 in the N. I. Vavilov Institute. Other important national collections are held in Germany (160 000), Japan (146 000), India (144 000), South Korea (116 000) and Canada (100 000).[764] In addition, there is an increasing effort to collect and catalogue flora from regions of wider biological interest and/or those deemed to be endangered (e.g. rainforests). Examples include the establishment of UK national and pan-European seed collections at the Millennium Seed Bank, managed by Kew Gardens, and the new Norwegian seed bank on the Arctic island of Spitsbergen.[765] The Kew initiative was started in 1997, thanks to a $90 million grant from the UK National Lottery. Like all such one-off disbursements, however, the real challenge for the future will be to fund the long-term running costs in the coming decades.[766]

It is precisely this sort of relatively unglamorous recurrent expenditure on existing fixed assets that is most under threat for plant breeding and germplasm centres in today's ultra short-termist world. Politicians are only too keen to fund construction of new research centres, especially if they involve exciting and photogenic new technologies guaranteed to capture the public imagination. But it is much more difficult to persuade politicians, and scientific policymakers, to use their funds instead to sustain rather less glamorous and non-photogenic ongoing commitments. Alas, compared with the nationally televised opening of a shiny new high-tech genomics centre, there is not a lot of headline-grabbing material in the mere maintenance of funding commitments to a dusty, thirty year old seed bank just off the main Damascus highway near Aleppo in the arid North of Syria (i.e. ICARDA). Neither is there much headline-grabbing mileage in the funding of seed-saving farmer networks in Afghanistan or Congo. Therefore, while we may be fascinated by exciting new discoveries in comparative genomics and rainforest botany, much of our basic infrastructure for the conservation and exploitation of existing plant germplasm is under threat from a gradual but steady erosion of funding and the benign neglect of a disinterested public.

Vulnerability of international seed banks

The continuing importance of seed banks across the world, and their ever present vulnerability, can be illustrated by three relatively recent examples of manmade disasters. The two Middle Eastern examples are from Iraq and Iran, and concern priceless seed banks containing thousands of crop accessions that were nonetheless threatened with senseless, wanton destruction by bombing and looting during various wars. The third example from Africa is of a potentially life-saving rice breeding programme that was gravely endangered by similar threats in the context of a chronic civil conflict.

UC Davis and the Iranian National Seed Bank

Iran is one of the centres of greatest diversity for wheat and its many wild relatives. The Zagros Mountains in the west of Iran were one of the sites of the earliest experiments in wheat cultivation and a crucible of our present civilisation. Most of the Iranian landraces of wheat, and their nearby wild relatives, have yet to be investigated for valuable traits such as disease resistance or stress tolerance. In the mid-1980s, over 11 000 wheat varieties that had been painstakingly collected and assembled at the University of Tehran over many decades, came under imminent threat from bombing by Iraqi forces as part of the ongoing Iran–Iraq war.[767] In order to guarantee their safety, it was reluctantly decided to move many of the wheat samples from the University of Tehran to a safer location in the USA.

From those wheat seeds, University of California breeders were later able to identify about 8700 new genotypes with resistance to serious pests and diseases, including wheat aphids, various rusts and viruses, plus several new varieties that were highly salt tolerant.[768] These Iranian varieties will now be used in public and private wheat breeding programmes worldwide, including in the USA itself. But the entire collection might well have been lost for good had it not been for the disinterested actions of a few far-sighted Iranian and American scientists. The discovery of the salt and disease tolerance traits in the Iranian wheat varieties was made by a group from the Genetic Resources Conservation Program at UC Davis, led by Calvin Qualset. In 2005, Qualset appeared at a US congressional briefing to promote greater US investment in the CGIAR-led Global Crop Diversity Trust that needs $260 million for an endowment to support struggling food crop gene banks around the world.[769] This is a modest investment given the near certainty of discovering new commercially exploitable varieties that could, like the Iranian wheats, directly benefit the US economy.

ICARDA and the Iraqi National Seed Bank

Ironically, while the previous example related to Iran, in the next example we will look at the seed bank in Iraq, the country that was the prospective perpetrator of the attack on the Iranian seed centre in the 1980s. Iraq was the cradle of much of early cereal agriculture, and of course is where the first farming-based urban cultures developed in Mesopotamian cities such as Uruk, Eridu and Lagash. Like Iran, Iraq has a rich endemic population of cereals, but it also grows a host of other crops that are specially adapted to the hot, dry, irrigated conditions that have sustained crop cultivation along the Tigris and Euphrates Valleys for over ten millennia. This region includes the centre of temperate cereal domestication and is home to a host of wild relatives with potentially useful agronomic traits.[770] Most of the key germplasm resources that had been collected and used by Iraqi breeders, including samples of all the regional crop varieties, were kept at the National Seed Bank, in the (now notorious) Baghdad suburb of Abu Ghraib.[771] While a great deal of overseas attention was given at the time to the pillaging of Iraqi museums and other national treasures in the chaos that followed the US-led invasion of 2003, the fate of the National Seed Bank went largely unremarked. Regrettably, once the Iraqi authorities had ceased to function in April 2003, no provision had been made by the occupying forces for the protection of the National Seed Bank or for any of the other botanical research facilities at Abu Ghraib.

In an unprecedented outbreak of looting and vandalism, virtually all the carefully collected plant material at the National Seed Bank, and the entire growth and storage facility, were destroyed. Luckily, some of the staff had previously managed to remove a duplicate set of the more valuable seeds to a place of safety across the border, at ICARDA in Aleppo, Syria. These seeds were saved and propagated by ICARDA breeders. The Syria-based breeders at ICARDA are now gradually returning this precious germplasm to Iraq for planting by local farmers, who otherwise faced the loss of many varieties specifically adapted for their growing conditions.[772] Once again, only the initiative and bravery of a few breeders saved an irreplaceable resource for the future use of the people of the region. A further twist to this story occurred in mid-2005, when it was reported that hostility of the US government to the Syrian government was in danger of jeopardising efforts to ensure the future of ICARDA itself as an international plant germplasm repository.[773] Such talk serves only to undermine confidence in, and the possible willingness of donors to support, vital resources like the germplasm conservation and research centre at ICARDA. After all, it is not the fault either of the breeders at ICARDA, or of their cereal collections, that they happen to be located in a region of political instability.

WARDA in Côte d'Ivoire and Liberia

The third example of the risks sometimes faced by the international plant breeding community concerns events in November 2004 at the CGIAR-coordinated, West African Rice Centre in Côte d'Ivoire. This centre (also known as WARDA) has spent the past few decades developing special rice varieties suitable for African farmers: the so-called 'nerica' (New Rice for Africa). Farmers using nerica varieties of African rice can achieve between 25% and 250% higher yields with no inputs, apart from small amounts of fertiliser. The crop is especially suited to the poorest regions where farmers are plagued by a combination of insect pests, weed infestation and poor soils. This work was recognised internationally when the 2004 World Food Prize was awarded to Sierra Leonean plant breeder Monty Jones, for his research that led to development of nerica at WARDA.[774] From this we can get some idea of the importance of WARDA for the future of farmers throughout Africa. The work of the WARDA breeders was dramatically interrupted in November 2004 by a bombing raid on the city of Bouaké, where the WARDA headquarters were located. The raid was carried out by government forces in breach of a ceasefire that had been negotiated in the ongoing civil war in Côte d'Ivoire. One of the WARDA senior scientists, Robert Carsky, was killed in the raid and most of the scientific staff were promptly evacuated to Cotonou in neighbouring Benin.[775]

As of mid-2006, only a small contingent of dedicated staff still remained behind at Bouaké to guard and maintain the precious germplasm collection that included all the nerica lines.[776] As a result of the violence, research at WARDA was significantly disrupted and development of nerica and other African crops was further delayed. The events of 2004 were only the latest in a series of setbacks to WARDA due to civil strife. An even more serious disaster had occurred in Monrovia, Liberia during the fighting that erupted in the city from 1987–1988. At that time there was an extensive WARDA germplasm collection held in Monrovia. In the wake of the fighting, the main seed stocks were pillaged and all the rice breeding lines were destroyed.[777] In some cases, looters actually threw away the priceless seeds, simply in order to make off with the cheap plastic bottles in which they had been stored. Luckily, in this case, much of the collection was also held in duplicate elsewhere in Western Africa. This has allowed WARDA to restore 80% of the rice collection and to redistribute 1075 sample varieties of rice to six countries in West Africa. These accounts of the tribulations of WARDA, and of the Iraqi and Iranian seed banks, highlight the need for duplicate seed collections to be held in facilities that are readily accessible (to all) across the world. This is especially important in regions that are prone to serious human conflict or environmental disasters.

Seed banks and reconstruction

The tsunami of 2004

It is not only manmade disasters that have affected seed conservation and exploitation efforts around the world. There have been several high-profile natural disasters that have had similarly grave effects. We now recognise that seed banks are a vital part of the reconstruction efforts following either manmade or environmental disasters. For example, the tsunami of 2004 caused widespread damage to agriculture in coastal regions around the Indian Ocean. In addition to its immediate impact on existing crops, the tsunami also caused severe salt damage to extensive areas of fertile rice growing farmland. Much of this farmland is located in heavily populated coastal areas that were inundated, often for several weeks, by salt water. This left saline deposits in the soil that in some areas will take decades to wash out. Soon after the disaster struck, the International Rice Research Institute (IRRI) received urgent requests from Malaysia and Sri Lanka for salt tolerant varieties of rice. Thanks to its seed bank of rice accessions from around the world, IRRI was able to identify 40 different salt tolerant varieties. Six varieties were immediately sent to Sri Lanka and Malaysia. Meanwhile, breeders at IRRI also began working with local farmers in the affected regions on a longer term project to transfer salt tolerance traits to local varieties. In the aftermath of the tsunami, some observant Indian scientists showed that there may yet be a silver lining from the catastrophe. Breeders in the Southern State of Tamil Nadu found that several tall, relatively low-yielding, local varieties of rice had survived immersion in salt water, while all other crops had died. The breeders now hope to cross these varieties with higher yielding lines to produce new salt tolerant rice varieties for the benefit of coastal farmers, and those who cultivate rice on degraded saline soils.[778]

Rebuilding agriculture across the world

IRRI has also provided new seed to countries such as Afghanistan, Rwanda and Cambodia, where seed banks had been destroyed during recent civil conflicts, and to Honduras, Nicaragua and Cuba where crops had been devastated by hurricanes. However, such efforts are merely the tip of a very large iceberg, as was shown in a 2005 report from CGIAR entitled '*Healing Wounds; How the International Research Centers of the CGIAR Help Rebuild Agriculture in Countries Affected by Conflicts and*

Natural Disasters'.[779] This report documents dozens of examples, from about 50 countries around the world, where the CGIAR network has provided seed and breeding advice following a succession of environmental and manmade disasters. Examples range from famine in North Korea to post-hurricane assistance across the Caribbean. The report stresses that the restoration of agriculture is a key initial step in enabling developing countries to recover from such disasters. In the medium to long term, such assistance is also likely to facilitate greater self-sufficiency and economic progress than more newsworthy palliative measures like food aid. Interestingly, less well organised and relatively informal local seed banks are almost always much more resilient during short-term conflicts or disasters than more elaborate, centralised germplasm repositories. Hence, the dispersed seed network in Rwanda soon recovered from the disruption of the genocide/civil war of 1994, whereas the centralised Iraqi seed bank was completely destroyed during the relatively brief invasion/looting episode of 2003.

Seed banks and duplicate germplasm collections are also important for assistance after the kind of prolonged conflicts, such as in Cambodia or Afghanistan, which often result in destruction of most or all seed stocks in the affected country or region.[780] Destruction in such prolonged conflicts will occur whether national seed collections are dispersed or centralised, and will probably also lead to local erosion of agrobiodiversity.[781] These centres are critical for future agricultural growth, so their vulnerability requires special attention, for example the establishment of more dispersed and redundant international systems for germplasm storage. In the past few decades, regional and international networks of expertise and gene banks have proved themselves, time after time, as priceless safety nets that have provided the knowledge and materials needed to restore agrobiodiversity and re-establish seed and food production systems after these sorts of human and environmental disasters.

The future of gene banks

It seems inconceivable that international germplasm collections, or gene banks, will ever be located in the private sector. It may well be possible for private contractors to assist in the operation of germplasm centres, and it may also be useful for the private sector to assist in their funding (perhaps in return for access to plant varieties), but there is little doubt that this kind of long-term international activity can only be sustained by public funding from governments around the world. As we have seen, there is no such thing as a truly 'national' agriculture. We all grow crops that have been brought in from elsewhere and we are continually searching for new crops and new varieties throughout the world.[782] This means that even a major national seed bank, e.g. in the USA, is a centre of international interest and (arguably) there should be an element of common ownership, or at least common access to its resources. After all, most of the crops grown in the USA (and Europe) originate from elsewhere, and overseas countries, from Iran to Russia, continue to provide US farmers with invaluable germplasm to sustain the ever improving performance and profitability of their crops. As we will now see, the ownership of national germplasm collection, especially in the USA, remains an especially contentious and unresolved issue.

International patrimony or restricted property?

There is a certain irony in the increasing centralisation of plant germplasm resources in industrial countries, nearly all of which are almost entirely bereft of indigenous crop species, and where genetic erosion in commercial crop cultivars has been most extensive. This process has been continuing since the 1970s and was commented upon in 1983 by the economic botanist, Garrison Wilkes, as follows: 'The centers of diversity are moving from natural systems and primitive agriculture to gene banks and breeders' working collections with the liabilities that a concentration of resource (power) implies.'[783] Germplasm accessions for most major crops are increasingly concentrated in a few industrialised countries with their better funded and more secure (as discussed above) storage facilities. These countries therefore control much of the future genetic potential, as well as the technologies for unlocking and exploiting such potential, in such crops. At the same time, a side-effect of international crop improvement is that many landraces of traditional crops in those developing countries in which most of the world's crops originated, are rapidly disappearing as farmers focus on higher yielding, but more genetically uniform, improved varieties. Hence the developing countries that are the centres of origin of many crops are gradually losing their remaining heritage of genetic variation, as traditional landraces and *in situ* germplasm resources continue to disappear. One of the most serious problems with this process is that once a country of origin loses its genetic resources to a more powerful industrialised country, it may never get them back.[784]

This highlights the importance of the ready and free availability of germplasm resources as a common heritage of humankind. In the case of the Iranian wheat accessions moved to California for security reasons, and then found to have valuable agronomic properties, it would seem perverse in the extreme if Iranian breeders were to be denied free access to such material. However, there is also a counter-argument that the agronomic potential of these accessions was not discovered until they were tested by the US breeders. Interestingly, this argument was rejected by US Secretary of Agriculture, Warren Christopher who in 1994 urged the Senate to ratify the Convention on Biological Diversity, with its guarantee of the rights of centres of origin nations to the fruits of their own germplasm.[785] The reasons for Secretary Christopher's exhortation to the Senate were entirely pragmatic. He recognised that the US would continue to require access to overseas germplasm for its own agricultural industry, and the Convention provided a single international framework for the management of such acquisitions.

A challenge from the USA

The acquisition of overseas agriculturally related germplasm as an explicit aspect of US government policy was made clear in the following statement in 1986 by Henry Shands, Head of the US National Plant Germplasm System:

The U.S., moreso than most nations of the world, is deficient in having centers of origin of important food and fiber crops. In fact, the U.S. has no economic crop native to its lands. One goal of this nation is to become self-sufficient in germplasm. Attainment of the goal presents a major task in the acquisition and characterization of germplasm.[786]

Access to such germplasm is worth a great deal to the national agricultural industry, as exemplified by just two of the major US grown crops, soybeans and maize. In 1994, the US government estimated that access to exotic germplasm added a value of $3.2 billion to soybean production, and about $7.0 billion to the maize crop.[787] This means that there is a great deal of pressure from agribusiness for the US government to adopt the most restrictive policies towards access to what is a demonstrably lucrative economic resource. Superimposed on such economic considerations are political factors that restrict further the access of some countries to international plant collections that are held in US facilities. For example, any material held in the main US germplasm repository, regardless of its origin, becomes the property of the US government and is then unavailable to certain countries that are deemed 'unfriendly'.

The policy of using plant germplasm resources as a political weapon in modern times goes back at least as far as the 1970s, as shown in a letter from a senior USDA/ARS administrator to the chairman of the International Board for Plant Genetic Resources, or IBPGR (now called the International Plant Genetics Institute, and a CGIAR centre). In 1977, the IBPGR chairman invited the main US germplasm storage centre at Fort Collins, Colorado to participate in a newly established international network of collections of key types of globally important plant germplasm. The uncompromising reply from the US agency was as follows:[788]

We are willing to accept selected collections for long-term maintenance at Fort Collins. They would become the property of the U.S. government, would be incorporated with our regular collections, and made available upon request on the same basis as the rest of the collection ... As you know it has been our policy for many years to freely exchange germplasm with most countries of the world. *Political considerations have at times dictated exclusion of a few countries*[789] (my stress).

No other national germplasm centre that participated in this UN-led initiative had imposed such onerous conditions on the storage of plant material. The selective restriction of access to germplasm by scientists from, or based in, countries deemed to have unfriendly governments is virtually unique to the USA. Among countries denied access to germplasm by USDA/ARS were Afghanistan, Albania, Cuba, Iran, Libya, Nicaragua and the former USSR.[790] This policy is still very much in evidence, as shown by recent threats to Syria, which is itself host to one of the most important international germplasm collections at ICARDA. The ICARDA collection houses over 100 000 plant accessions, including more than 30 000 wheat and 24 000 barley varieties. As we saw above, it was scientists at ICARDA who helped Iraqi breeders to save their own seed collections following the wanton and avoidable destruction of the National Seed Bank at Abu Ghraib in 2003.

Following the crucial role of ICARDA in saving the Iraqi cereal collections, it was especially poignant to read in an editorial in the journal *Nature* in June 2005 that US threats could even lead to the evacuation of the ICARDA collection itself – all 122 000 accessions. In the words of the editorial: 'One response (to the threats from the US government) would be to pack the seeds into storage boxes and airlift them out of Syria...'[791] Hopefully matters will not come to such dire straits. And it would be perverse indeed if US sanctions and other threats to Syria were to result in ICARDA collections being taken 'for safety' to the USA, thereby to become US government property. In the present political climate, this could have the effect of making the seeds unavailable to the very Syrian farmers whose enterprising ancestors domesticated the first cereals at the village of Abu Hureyra on the River Euphrates, over twelve millennia ago.[792] This dismal prospect may be something for US politicians to reflect upon as they take their morning toast, lunchtime sandwiches, or evening pasta, and chew contentedly on the progeny of crops originally developed by the wheat farmers of ancient Syria.

Most plant breeders are rightly concerned about the lack of clarity in the current situation regarding ownership rights of biological resources and especially the case of exploitable plants.[793] There is a framework for an international agreement on this issue in the Convention on Biological Diversity, but this initiative has little meaning unless it is endorsed by all of the major countries that are either centres of diversity or users of germplasm. The failure of the US government to sign this agreement, despite the urging of eminent national politicians, including former Secretary of

Agriculture, Warren Christopher, means that the status of plant germplasm remains in limbo, to the detriment of breeders and others who seek to collect, catalogue, maintain and ultimately exploit these valuable resources that have hitherto been regarded as part of the patrimony of all humankind. In the meantime, there are several initiatives that seek to establish international gene banks on a more secure footing. According to the executive summary of a UK-based report published in 2002:

The data points us to one major conclusion: genebanks can no longer rely on uncertain annual sources of funding – as most now do – to fulfil their perpetual responsibility for maintaining the diversity of plants that underpin our food security. They need a major new endowment – a fund generated by public and private sources – that can support, in perpetuity, this essential work.[794]

While the funding crisis for germplasm storage and conservation is most acute in the national research centres, it has also affected CGIAR, which between 1994 and 2002 lost half its core funding.[795] The situation is even worse in many national centres in some of the most gene-rich regions of the world. For example, in the very region that stands to benefit the most from improved crops, namely sub-Saharan Africa, almost half of the gene bank centres have had their budgets cut. In some cases, the refrigeration bill alone consumes so much of the budget that staff numbers are being reduced at a time when the demand for additional storage is still rising. A more positive development is the establishment of the Global Crop Diversity Trust. This is an emerging curation initiative designed to help seed banks, which was initially set up jointly by the UN FAO and CGIAR. The first batch of funding has come from government bodies in Australia, USA, Switzerland, Samoa, Egypt and Colombia, and the United Nations and Gatsby Foundation, but more is needed.[796]

A way forward?

It is estimated that about $260 million will be required to set up a permanent endowment to generate secure long-term funding for future maintenance of international germplasm centres.[797] Given the multi-billion dollar value of seed banks for private sector companies and governments in industrial countries, it may be appropriate if, in future, more funding is forthcoming from such companies and governments, rather than from poorer donor countries like Samoa, Egypt and Colombia. In the longer term, it is also important to consider the maintenance of plant genetic resources, not only in seed banks (*ex situ*) but also in their native environments (*in situ*). There are obvious challenges in the practice of *in situ* conservation in regions with poor access and insecure conditions (e.g. Afghanistan), but these should not detract from more attention being paid to this important aspect of agrobiodiversity.[798]

In this chapter, we have looked at some of the challenges currently facing international plant breeding and seed conservation. Even within the public sector, there are often tensions between the traditional transnational, free-access, public-good stance of most plant research organisations and the more narrow nationalistic, protectionist stances of the wayward few. Unfortunately, the latter tendency seems to be gaining ground, as powerful nations like the USA restrict access to their public germplasm collections. At the same time, some developing countries are increasingly claiming ownership of all biological resources that happen to be found within their boundaries. This sometimes has the unfortunate side effect of blocking access to potentially useful plant resources, sometimes even to indigenous researchers who are seeking knowledge that may benefit their host country. These unresolved disputes about the ownership of plant resources have led to an atmosphere of uncertainty that is in danger of inhibiting progress by restricting access to much needed research tools for crop improvement. The general air of uncertainty extends to the future direction of international plant breeding. As we saw previously (in Chapter 10) in the industrialised countries, we also have a creeping process of academisation and an over-fascination with molecular breeding technologies in the international arena that is also infecting some CGIAR centres. But there are also many solid success stories as breeders engage more with local farmers, as in the participatory approach to breeding. In the next two chapters, we will explore ways in which public sector breeding can regain its confidence, rebalance itself, and engage in more equal partnerships with a changing private sector that is also facing its own series of challenges.

17

Rebalancing our approach
to crop improvement

[Imagination] reveals itself in the balance or reconciliation of
opposite or discordant qualities: of sameness, with difference; of the
general, with the concrete; the idea, with the image ...

Samuel Taylor Coleridge (1817)
Biographia Literaria

Introduction

In preceding chapters, we have followed some of the seismic shifts in the balance
between public and private sector activities in plant breeding over the past few decades.
While some developments have been positive, others have fuelled concerns about the
overall direction of plant breeding research, particularly its future capacity to deliver on
the primary mission of crop improvement. Two key priorities should be the revitali-
sation of the public sector and the re-empowerment of plant breeding as a valued and
socially necessary scientific discipline. To achieve this, we must seek to re-establish
those structural balances in plant research that have gone so seriously awry over the
past few decades. For example, we should restore the balance between the following:
the public and private sectors as they relate to agriculture; transgenic methods and non-
transgenic variation enhancement plant breeding strategies; academic research and
applied R&D in the plant sciences; and between pragmatic, public-good crop
improvement, especially for developing countries, and the more inward-looking topical
issues (such as the debate on GM crops) that currently preoccupy many richer industrialised nations.[799] In this chapter, we will begin by discussing how to revitalise our
much depleted public sector, and re-establish some of the balances within plant science
in general. We will then look at how we might restore some much needed perspective to
the polarised public discourse on agriculture and agbiotech.

264

Revitalising the public sector

In earlier chapters, we saw how public sector research in plant science has been significantly reduced and diverted towards more basic areas of study. This has been a global phenomenon with a multiplicity of causes, although many of the key developments, such as privatisation and academisation, were largely pioneered in the UK. Underlying this complex series of events are many political, economic and technological factors not necessarily directly related to plant breeding. Hence, privatisation and rationalisation of institutes was arguably pursued for mainly political (ideological) and bureaucratic (efficiency) reasons. Economic factors encouraged increasing use of short-term funding of research. Target-orientated managers and bureaucrats also found it expedient to fund research on a piecemeal project-by-project basis, with a typical project lifetime of two or three years, rather than at a more strategic level involving several linked projects over several decades, as in the past. Meanwhile a beguiling new set of DNA-based technologies promised a new and possibly revolutionary shift in the crop improvement paradigm. Many public sector scientists, policymakers and politicians believed that these innovations would be best driven forward by the expanding, confident, and seemingly well-resourced private sector, rather than what was perceived to be a more academically focused public sector.[800] This combination of factors led the public sector to retreat back into its cloistered laboratories, where it focused on using its wonderful new genomic and biochemical tools to investigate the intriguing world of fundamental plant biology.

The seductive allure of basic research

And what a wonderful activity such basic research is! As a practitioner, I can quite understand the fascination of this area of science. We are now in one of the most exciting periods, many of us would say the most exciting period, for research into the fundamental mechanisms of plant growth and development. Thanks to newly developed biological tools (wetware), physical equipment (hardware), and informatics (data/software), we can achieve in a few days what would have taken a lifetime of research when I was a Ph.D. student in the 1970s. I have been embedded in this world for most of my scientific career and, although I moved into more applied research during the heady years of the early 1990s agbiotech boom, I was eventually attracted back to the intellectual ferment and excitement of basic research, international conferences, and publication in high-impact journals. I believe that I am far from alone in this regard, a feeling that is borne out by similar expressions of concern from colleagues in the plant science community. Indeed, several plant scientists have referred to what they call the 'black hole' of basic research, which sucks in much of the best scientific talent, never to release it. It seems clear therefore that the balance is now dangerously skewed away from applied research (see Chapter 10).

One consequence of our currently unbalanced research focus is that we are creating such huge amounts of new raw data that we are now unable to digest and assimilate them into our existing corpus of information. These data proliferate and accumulate to create bottlenecks that in many cases will take years to be digested into comprehensible information (many of the new genomics programmes are a case in point, as we will discuss in the following section). Even more remote is the prospect of applying the information that is yet to be derived from the raw data and transforming it into useful knowledge, never mind true wisdom.[801] This concept was elaborated in the research context by Milan Zeleny and Russell Ackoff in the 1980s as a hierarchy of the assimilation and utilisation of facts that progresses from data, to information, to knowledge, to understanding, and finally to wisdom.[802] Often referred to as the DIKW hierarchy, the concept is nowadays much used in the fields of information science and knowledge management. The enunciation of such concerns in relation to plant science research may seem rather novel, unfamiliar, and even perhaps alarming, to some researchers. However, as we will now see, the question of balancing investment in basic research with our ability to exploit its fruits has been a perennial concern that is by no means unique to today's world.

Rebalancing plant science research

Concern about the balance between acquisition of scientific data for its own sake and the utilisation of knowledge, i.e. basic versus applied research, has been a perennial theme of public discourse over the centuries. Back in the nineteenth century, Russian writer Peter Kropotkin[803] noted that:

At the present moment, we no longer need to accumulate scientific truths and discoveries. The most important thing is to spread the truths already acquired, to practice them in daily life, to make of them a commonplace inheritance. We have to order things in such wise that all humanity may be capable of assimilating and acquiring them; so that science, ceasing to be a luxury, becomes the basis of everyday life.[804]

In this quotation, Kropotkin argues that mechanisms to assimilate and apply knowledge from basic research are just as important as the creation of such knowledge in the first place. One of the first challenges to plant breeders, after the explosion of scientific discovery during the Enlightenment, was to find a mechanism whereby new knowledge could be distilled, transformed and eventually utilised for practical crop improvement. Thinkers from Descartes to Darwin struggled with this challenge. In the end, the most enduring and productive solution emerged at around the turn of the twentieth century, with the largely US-led establishment of a cadre of scientifically trained, public sector plant breeders with expertise to link developments in plant biology through to practical breeding. Without these professional breeders, we would not have generated the improved crops and extra food that sustained the massive population rises of the twentieth century.[805] Several decades after Kropotkin's article, US scientists used similar arguments about plant research. The 1908 report of the Commission for Agricultural Research looked forward to 'an era of the diffusion rather than the acquisition of knowledge'.[806] Also in the same 1908 publication is the following statement from the plant breeders, Hambridge and Bressman, which is interesting in the light of many of the post-1985 developments in plant breeding that we have already considered:

the field of breeding and genetics has become so large, it is so dependent on progress in basic research, and it requires such continuous effort on projects running over many years or even more than one generation* that it obviously becomes a function of government institutions capable of devoting the necessary money and time to the work and doing it with a sufficiently disinterested attitude. This is especially true because the results are for the benefit of all people rather than one group.[807] (*Presumably this refers to human, rather than plant, generations.)

The point made here by Hambridge and Bressman is that, not only do we need basic research per se, we also need it to be a continuous, sustained, long-term effort, as best pursued in the more disinterested domain of the public sector. Linking these Russian and American viewpoints, we can say that the most powerful mechanism for advancing plant breeding, and hence agriculture, is a well-founded, secure, and confident cadre of professional, public sector basic researchers who are closely linked with more applied breeders, plus the associated infrastructure for dissemination of knowledge and improved crops all the way from laboratory and field plot to the farmer. Unfortunately, it is precisely this linked basic and applied plant breeding infrastructure that has been decimated across the industrialised world over the past two decades. And many of the 'brightest and best' plant scientists have succumbed to the allure of a more isolated (from direct utility) form of academic research.

The world of basic plant science in the twenty-first century is characterised by immensely data-rich systems, such as databases of genomic, proteomic and metabolic information, huge 'libraries' of hundreds of thousands of individually mutagenised plants, microarray data showing thousands of gene expression profiles, and much more. The mining of these data and their conversion into useful knowledge will take thousands of research lifetimes; and in the meantime we are adding to the databases at a near-exponential rate. It is only too easy to become metaphorically lost in this cornucopia of potential knowledge. Eventually, no doubt, we will be able to describe how plants function in the most exquisite biochemical detail, and how their genomes and the external environment interact to give rise to the many physiological processes that underlie plant development. Such an understanding will then provide a rational basis for many radical types of crop manipulation that we can only dream of at the present time. But this will take many decades, and in the meantime there are other more urgent and immediate challenges demanding the attention of the practical plant scientist, such as crop improvement programmes focused on yield, quality and stress tolerance traits.

One of the strongest criticisms of UK biological research made in a BBSRC report from 2004 was the: 'lack of impact of basic plant science research on applied crop research'.[808] This sentiment was subsequently echoed in the final report that was published by AEBC, shortly before its abolition by the UK government in April 2005.[809] Recommendation 4 of the AEBC report states: 'We endorse the recommendation of the BBSRC sustainable agriculture group for a review of the capacity for more systems-based, longer-term sustainable agriculture studies.' As I have stated previously, some of the best plant science research in the world is done in the UK, and this should be allowed to continue. But that same excellent work is also in danger of losing some of its direction and becoming impoverished, in the absence of mechanisms to link it to the improvement of agriculture, as used to exist in the not-so-distant past. Recommendations 5 and 6 of the same AEBC report on UK plant science research state respectively that: 'The public good should be a more explicit objective within research agendas.' 'Public funding for near-market research should not be ruled out where it contributes to the sustainability of farming.'

A lack of balance between basic and applied public-good research is not just a UK problem. We should also be clear that the solution to what is a worldwide deficit in plant breeding research is most definitely *not* a series of highly publicised, short-term, 'quick-fix' initiatives by national research councils or other funding bodies. What we have here is a fundamental structural problem of both UK and international science, requiring well thought out, long-term measures to address it properly. In the UK context, such organisational measures might include the reinstatement of plant breeding research as a core activity, both in the few remaining public sector research institutes, and in those universities that can support such work. It may also be appropriate to shift some of the more basic research work currently done in public institutes back to the universities, where it would be more appropriately located within the broader culture of knowledge discovery and transmission, coupled with the synergistic combination of teaching and research, that prevails in a good university environment.

It is a matter of acute concern that, over the last two or three decades, many public sector scientists and research institutes have tended to focus more intensively on high-profile, prestigious 'discovery phase' research and less on its application as a public good. We are now at a point where the prospects for using new knowledge about plant biology, and especially genetics, are truly awe inspiring, but we are still decades away from realising many of these aspirations in terms of real-life crop improvement, especially for developing countries. We must move away from the culture of valuing taxpayer funded basic research simply as an activity in itself, with little recognition of its wider context in the public interest or for public-good utility.[810] This aspiration will require more focus on applied aspects of science at all levels, from the undergraduate curriculum onwards. We should also ensure that plant scientists are trained in all aspects of plant growth, development and manipulation so that we do not continue to produce graduates who know all about the intricacies of molecular genetics, but who have never heard of quantitative variation, heterosis or wide crossing. The purpose and consequences of research assessment exercises also need to be thought through more carefully, lest they distort and divert the focus of both scientists and their science.[811]

One of the most difficult challenges facing plant breeding in the twenty-first century is to attract and retain good scientists into the discipline. At present, the odds are so highly stacked against a researcher seeking to forge a successful career in applied plant science that, notwithstanding an urgent wish to strengthen the area, it is difficult, even for a strong advocate such as me, to recommend such a career to my students as a convincing alternative to more basic research studies. We should look forward to the time when our students are able to recognise the value of plant breeding research and practical fieldwork as being at least the equal of laboratory-bound basic studies of plant biology. We still have a long way to go to achieve this aim. Perhaps one way in which we might progress is to broaden out the somewhat navel-gazing and ill-informed public debate on GM crops, and to discuss instead the global challenges to agriculture and the potential solutions that could be offered by practical and truly disinterested plant breeders. We will now look at the public debate on agriculture, and its influence on the perception and conduct of crop-related research in the UK during recent years.

Rebalancing the public debate on agriculture

Why is the UK so anti-GM?

Several aspects of agricultural practice have emerged as major topics of public debate around the world over the past decade, although the emphasis varies in different countries. In the UK, as with most of Europe, public debate on agriculture has tended to be informed by the traumatic effects of recent food scares, such as the BSE outbreak, and food poisoning episodes caused by the *E. coli* and *Salmonella* pathogens.[812] General concern also exists about the consequences of agricultural intensification, for example its effects on the physical environment, on wildlife, and on livestock welfare. Finally, there is unease about the increasing corporate domination of the food chain, from global seed companies to industrialised mega-farms to the multinational food producers, and finally on to the supermarket oligopoly. The phenomenon of the supermarket oligopoly is especially marked in the UK where four vast supermarket chains control over 80% of the grocery retailing market.[813] Into this seething cauldron of concern, transgenic crops were thrust in the mid-1990s, in a guise that virtually guaranteed that they would raise the hackles of an already sensitised and suspicious population.

Without knowing anything about the complexity and intrusiveness of existing forms of scientific plant breeding, as already employed for almost a century, consumers were informed that transgenic technology constituted a mould-breaking revolution of almost unlimited potential (see Chapter 15). Puzzlingly for the consumer, this 'revolutionary' technology was then used for the rather mundane purpose of deterring insects and weeds in the field. The crop itself was unchanged and the consumer derived no benefit from the resulting foods, which were neither cheaper, better tasting, nor more nutritious than non-transgenic counterparts. In the meantime, however, transgenesis had acquired the aura in the public mind as an experimental, and yet-to-be-perfected technology. After having had their expectations raised so highly about the forthcoming 'gene revolution', consumers were then told that the new genetically engineered food products that appeared after the mid-1990s were actually more or less identical to existing 'conventional' foods. Indeed, the GM foods were said to be so similar to existing equivalents that they would need neither labelling nor segregation. The public was rightly confused.

At this juncture, many pundits, including a few scientists, began to voice (mostly unfounded) concerns about possible environmental or food safety effects inherent in the technology. These concerns were duly amplified by the media, further raising the level of public suspicion and distrust. This was all bad enough but, in 1996, unlabelled soybean shipments, including a small amount of GM seed were exported for the first time to Europe.[814] Because about 60% of processed foods in supermarkets contain material from soybeans, this meant that a large proportion of the European population would involuntarily consume products of transgenic technology – whether they wished to or not. At this stage, UK consumers were still relatively accepting of GM food products, and an openly labelled (as GM) tomato paste (made by Zeneca) was a modest commercial success.[815] Hence, most of the earliest organised opposition to

GM crop imports came, not from the UK, but from environmentalist pressure groups on the European Continent.[816] However, the presence of unlabelled GM products in the UK soon heightened public unease there as well. This unease was then skilfully exploited by professional anti-GM campaigners.[817]

The real meltdown in public acceptance of GM foods in Europe came in 1998–1999, with release of data from the incomplete and much overhyped study of Pusztai *et al.* on the effect of GM potatoes on rats.[818] A degree of mishandling of this affair on all sides, especially the apparent victimisation of Pusztai by his (public sector) employer and by parts of the UK scientific establishment, played into the hands of anti-GM campaigners and worried many independent scientists.[819] In the UK, picketing of a few supermarkets in early 1999 led to the precipitate decision by retailers to ban GM-derived foodstuffs from their shelves.[820] To the general public, this drastic action by supermarkets appeared to confirm many of their worst fears. The result was an extensive backlash against transgenic crops in general. It also distorted scientific discourse for years to come, leading to redirection of agriculture-related research funding in the UK to address issues of public and political concern, rather than more scientific priorities. Meanwhile, in the public arena, the subsequent GM debate has conflated a whole series of issues including globalisation, countryside management, corporate control of the food supply, scientific accountability, transgressions of the 'natural order', to name but a few. In these earnest, but ultimately unenlightening and disputatious diatribes, the actual scientific questions were often lost in a morass of unproductive verbiage that has yet to lead to any kind of remotely satisfactory resolution of the 'GM issue'.

The UK farm-scale evaluations

As an example of the prevailing reality-disconnect in the broader debate about agriculture in developed countries, many people in Europe and North America have argued at great length, and at vast expense to the taxpayer, about how many weeds are permissible, nay even desirable, in a field crop. From much of the resulting media coverage one would have supposed that the mission of the UK cereal farmer was more concerned with the cultivation of weeds such as cleavers, *Gallium arapine*, wild oats, *Avena* spp., and mayweed, *Matricaria* spp., instead of his crop of winter wheat or barley.[821] Perhaps the most egregious instance of this occurred with the farm-scale evaluations (FSEs) of transgenic crops, the results of which were published in 2003–2004.[822] This $9 million, four-year research study was meant to inform government policy on regulation of transgenic crops, and to address public concerns about the cultivation of such crops in the UK. The principal aim was to observe the impact on weed and invertebrate populations of growing transgenic herbicide tolerant varieties of three crops, compared to non-tolerant varieties of the same crops. In retrospect, one can point out three fundamental problems with the rationale behind this costly and questionable exercise.

First, the FSEs did not address the issue of transgenic crops at all. The herbicide tolerant varieties of the three selected crops, maize, sugar beet and oilseed rape, were indeed transgenic, while the control non-tolerant varieties were not transgenic. However, as any breeder knows, there are many crop varieties that are herbicide tolerant without being transgenic. There are also numerous transgenic crop varieties that are non-tolerant to herbicides. Therefore, all this study measured was the effect of growing a herbicide tolerant crop compared with non-herbicide tolerant varieties of the same crop. Because herbicide tolerant plants are insensitive to some of the most effective broad-spectrum herbicides, farmers can apply these herbicides later in the season in order to control weeds growing alongside the crop. One would expect therefore that the population of weeds, and their associated invertebrates, would be somewhat different in fields growing tolerant, as opposed to non-tolerant, crops. To nobody's surprise, the $9 million study duly demonstrated that this was indeed the case.[823]

The FSEs found that two of the three tested herbicide tolerant crops had significantly lower weed and invertebrate populations than the controls. The UK government proceeded to use these results to withhold approval of these (GM) varieties of sugar beet and oilseed rape. The same data were widely quoted around the world as 'proof' of the adverse environmental consequences of transgenic crops in general. But, as we have just seen, the results from the FSE study had absolutely nothing to do with transgenesis per se. The opposite conclusion could have been generated simply by comparing non-transgenic herbicide tolerant and transgenic non-tolerant varieties of the same crop. In this case, cultivation of the (non-transgenic) herbicide tolerant crop would have led to lower weed and invertebrate populations in the field. One could then have proceeded to argue that the use of such non-transgenic crops (that happened to be herbicide tolerant) was bad for the environment because they tend to reduce on-farm biodiversity. Quite apart from the fact that herbicide tolerance is only one of thousands of traits that could potentially be altered by transgenesis, the entire design of the study was so flawed as to invite such misinterpretations and misuse of the data. As we have seen above, one could have devised an exactly parallel study that would have shown non-transgenic crops as the environmental culprits, and banned the non-GM varieties instead.

The second serious flaw with the FSEs was the implication from their design, and from the surrounding public rhetoric, that the conservation of weeds on cropland should be one of the key goals of farmers, and that GM crops threatened this goal.[824] Hopefully, the intelligent reader will appreciate the inherent absurdity of this concept. One of the major reasons for the decline in rural plant and animal species, which peaked in the UK during the 1980s but is still ongoing to a much more limited extent, is the intensification of farming. This process has been under way since the end of World War II and is a completely separate issue to transgenesis. Indeed, as we saw in Chapter 6, it was conventional plant breeding, not transgenesis, that opened the way to most agricultural intensification regimes. During conventional breeding programmes in the 1960s and 1970s, new varieties were developed that were specifically designed to grow well in the presence of fertilisers, fungicides and the like. We could therefore argue, and with much better logical justification than from the FSEs, that all crops produced by conventional, non-transgenic, breeding over the past fifty years should be banned forthwith. After all, are not such crops the real culprits that have enabled the process of crop intensification that has had such profound effects on the agro-environment?

One of the justifications used for the FSEs and the subsequent ban on some transgenic crop varieties was that they compounded the effects of agro-intensification and reduced on-farm biodiversity. Agro-intensification is a particularly important political issue in the UK because it relates to one of the parameters used by the government to compile its 'quality of life index'. Over the past fifty years, crop intensification has not only reduced weed and invertebrate populations, it has also affected larger and therefore more visible and 'popular' species such as farmland birds. By the late 1980s, it was realised that several well-known farmland bird species including the corn bunting, *Miliaria calandra*, and skylark, *Alauda arvensis*, were in steep decline.[825] These birds are two of the key indicator species that make up the Farmland Bird Index for England, which is currently used by the UK government as one of its biodiversity indicators of quality of life in the country.[826] There is a clear danger here of conflating a high-profile, politically defined measurement (quality of life index) with a single, relatively remote causative factor (herbicide management) in what is in reality a highly complex, multi-factorial phenomenon (on-farm biodiversity). In fact, the Farmland Bird Index declined steeply from 1978–1988, but then levelled off, and has been stable since 1998. Unfortunately this led to the argument that by encouraging more on-farm weeds, countryside biodiversity would increase and the skylarks would return (hence boosting the UK quality of life index and pleasing the politicians).

However, if the goal were really to encourage greater diversity of plant and animal life in the countryside as a whole, any increase in cropland weed populations would have precisely the opposite effect. This is because weeds compete directly with crops for space, so more weeds mean fewer crop plants. The result would be that farmers would need to expand the cultivated area across the country, so as to maintain the overall volume of crop production. Hence, a 5% yield loss due to the encouragement of weeds would require the cultivation of an additional 230 000 hectares of the countryside. This area would be conveniently supplied by ploughing up the New Forest, Dartmoor, and Exmoor National Parks.[827] Perhaps the loss of a few National Parks in the UK might be considered a worthwhile price for an increase in cropland weeds, but one doubts that such ideas would carry a great deal of resonance with the general public.

The third problem with the farm-scale evaluations is quite simply the cost of the whole process, which was in excess of $9 million. Added to this are the many indirect costs, including researcher time, plus the hundreds of hours of committee time, public debate, tonnes of paperwork etc. The end result of this inordinate expenditure of funds, that could have been used elsewhere, e.g. for crop improvement research, was as follows: two transgenic herbicide tolerant varieties of sugar beet and oilseed rape were not approved for cultivation in the UK, while one maize variety was approved. The process has now created a dangerous precedent whereby any new transgenic crop variety might have to be subject to a similar battery of tests, at vast expense to the UK taxpayer, without even addressing the central issue of whether, and if so how, transgenic plants might differ in any biologically meaningful respect from non-transgenic plants expressing similar traits. In my view, this entire exercise has been a diversion of resources and attention towards a single and relatively trivial problem of our overly introspective society. In the meantime, we are ignoring far more serious and urgent challenges facing agriculture in the wider world.

Developing a sense of perspective

Another example of the diversion of public attention towards ill-informed and ultimately trivial debates is the issue of whether the transfer of genes between species is an unwarranted transgression of the 'natural order'. This debate fails to acknowledge that humans have been hybridising different species for millennia and have been using wide crosses to effect radical transformations in crop plants for over fifty years (see Chapters 2 and 3). Over recent years, these and similar debates have consumed hundreds of millions of dollars of public funds, as well as an untold amount of time and energy from the participants. Meanwhile, back in the real world of poor crop yields, poverty and hunger that are still the norm in many developing countries, public-good institutions like CGIAR are becoming progressively more and more starved of the funds that are necessary to confront these urgent challenges. If the pro- and anti-GM campaigners of the more affluent world had devoted just part of their formidable energies to the fight against poverty, lack of opportunity, and poor crop yields that still plague many developing countries, the prospects for CGIAR and other international agricultural improvement agencies would be much brighter.

In this respect, and the three other types of balance mentioned in the Introduction to this chapter, we seem to have lost our sense of proportion. Metaphorically speaking, we have become obsessively fixated on a few rather paltry trees to the exclusion of the surrounding bountiful forest of opportunities. Rather than focusing so much on the doings of a few agbiotech companies and the existence of farmland weeds, we should concentrate on the reform and empowerment of plant breeding as a whole. Instead of the current focus on transgenesis and input traits, we should lift our gaze to the whole array of breeding technologies, whatever their vintage or provenance. But we should do this with a severely pragmatic eye, not to be dazzled by specious novelty or fashion. And, most difficult of all for many active plant scientists, we should try to value and foster more effectively those more applied and practical areas of research. These research areas may not necessarily be regarded as the most intellectually challenging fields of botanical endeavour, but without them a lot of plant science would be effectively relegated to the realms of esoteric arcana.

18

Where do we go from here?

Coming together is a beginning.
Keeping together is progress.
Working together is success.
Henry Ford (1863–1947)
attributed

Introduction

In this chapter, we will examine how we might progress beyond the present unsa-
tisfactory state of plant breeding. In doing so, we should free up breeders to use the
best of modern technology and scientific knowledge, while using such tools to address
the key important challenges confronting twenty-first century agriculture. From our
previous analysis, we can identify three serious issues relating to the evolution of plant
breeding R&D over the past few decades. First, there is the withdrawal of the public
sector from most aspects of practical research in many major countries, and the
consequent academisation of much of its work. The dire straits of practical plant
research and breeding are seen most acutely in Europe and Australasia. Although the
problem is less marked in the USA, where a somewhat reduced, but still relatively
vigorous and effective, practically orientated public sector continues to function, things
could be greatly improved here as well to harness the full potential of researchers.
Indeed, across the world, public sector bodies need to adopt a much more practical
and outward-looking attitude towards plant breeding and crop improvement.

The second issue is the gap between an increasingly academically inclined public
sector and a rather uncertain commercial private sector that appears to be in transition from its current seed-based, input trait dominated business models. These two
potential partners in global crop improvement need to be brought together on a
more equal basis for their mutual benefit. One example would be to the harnessing of
some of the entrepreneurial skills of the private sector (perhaps on a contract basis)

276

to drive improvements in practical breeding and crop management in developing countries. Third, the private sector itself needs to start moving in a more innovative, diverse, and genuinely entrepreneurial direction in order to realise its true potential. It should move away from today's oligopolistic agbiotech business model that is focused on low-value input traits and begin to diversify towards higher value, more consumer-friendly output traits. All parts of the private sector should pay more attention to relevant market signals, whether from customers demanding product labelling or from major food crop producers concerned about molecular pharming. In this chapter, we will review some ways in which the private and public sectors can exploit each other's strengths, for their mutual benefit and to move forward the agenda of international, public-good plant breeding.

Empowering and recruiting the private sector

As we saw in previous chapters, there have been several useful developments in the private sector in recent years, with a new generation of small agbiotech companies experimenting with non-food, biopharmed crops and improved methods of trans-gene expression. But many small companies still find it difficult to compete, or find sufficient freedom to operate in today's over-patented and over-regulated agbiotech market, particularly in Europe. One way to empower small companies, public sector researchers, and developing country breeders will be the development of alternative technologies, both transgenic and non-transgenic. We will now review the challenges posed by restrictive rights, before going on to discuss some novel solutions such as 'open access' technologies or the formation of public sector consortia to develop their own proprietary technologies. Other emerging opportunities include methods for the introduction and domestication of dozens of new plants to augment our very restricted crop repertoire. By investing in new crops, we open up the prospect of massively extending the genetic diversity and environmental range of the plants that we exploit both for food and for non-edible uses.

Private sector plant breeding has experienced numerous changes over the past two decades, many of which have been as tumultuous and unsettling as anything happening in the public sector. From a very small research base in the 1970s, private sector groups grew rapidly in size and scope, until by the late 1990s they were significant contributors to a wide range of plant research activities, from genomics to biochemistry. By the 1990s, the larger companies could afford to carry out the kind of basic and strategic research that had hitherto been the exclusive domain of university scientists. Many such projects resulted in high-quality papers published in the better peer-reviewed journals. The prospect of doing cutting-edge science, with the added bonus of lucrative commercial applications, attracted many good graduates into the commercial sector at this time.[828] But the agbiotech sector of 2007 is very different from that of 1992. Many companies, both small and large, have gone out of business or been acquired by competitors. Downsizing and consolidation of the private sector research base has adversely affected the scope for collaboration with the public sector. For example, most of the multinationals have largely abandoned Europe and moved their major biotech operations to North America. We have also seen how commercial agbiotech has come to be dominated by a tiny number of large corporations with an arguably limited capacity or desire for radical innovation, and an overly narrow technology and product base.[829]

Diversifying and outsourcing

For more than a century, it has been repeatedly demonstrated that the most efficient mechanism for the advancement of modern crop breeding is a mixed economy, comprising active and vigorous elements of both the public and private sectors. While basic scientific knowledge emerges primarily from university research groups, the private sector can be a powerful and dynamic source of the technical innovations required to bring more basic discoveries to fruition as useful products. The rapid expansion of private sector involvement in crop R&D during the 1990s caught many people by surprise at the time. Although it initially promised great things for agricultural improvement, agbiotech is now in danger of turning into a retrograde development. One of the best ways for the private sector to address the challenges facing contemporary plant breeding would be to diversify its own operations, and to work more closely with the public sector. Both partners in the global enterprise of crop improvement are ailing and they each merit prompt ameliorative measures.

   In some respects the private and public sectors in plant breeding are already becoming closer, with each acquiring characteristics of the other. Hence, some companies have made their genomics databases freely available to the research community, while others sponsor development projects in places like Africa.[830] By the same token, some public sector breeders now receive much of their funding from industry and in a sense act as breeding departments for such companies.[831] But we should not allow the distinctive missions of the public and private sectors to overlap too much, particularly in the international arena. For example, we should not expect biotech companies to 'feed the world' – that is not their mission and shareholders in Bayer, Monsanto and Syngenta should be informed of the rationale for ventures that are not in the ultimate interest of the profitability of their companies.[832] After all, we do not expect Glaxo, Dunlop or Microsoft to solve respectively the medical, transportation or IT problems of the third world; and neither should such charitable ventures be a primary concern of a shareholder owned agbiotech enterprise.[833] Companies normally exist to maximise shareholder value and, while loss-leading ventures like community projects have their place, they are peripheral activities that should not be regarded as part of the core mission.

Most private sector seed companies are fully aware that their long-term future will be best served by the continued existence of a strong public sector commitment to plant breeding. Companies will always require access to new types of innovative experimental germplasm and new sources of genetic diversity to combine with their elite materials for the creation of new commercial varieties. The drawback is that development of such resources is generally regarded as being too high risk, and the return on investment is too long term, for the average investor.[834] For these reasons, private companies regularly contribute to the support of public plant breeding, either directly or indirectly. This is especially true in the USA.[835] Arguably, one of the main reasons for the better performance and dynamism of plant breeding in the USA is that the privatisation agenda was never carried to the extremes that occurred in many other industrial countries. This means that, in the USA, germplasm from public breeding programmes is still the principal source of genetic diversity in most commercial crop varieties.[836] Later in this chapter, we will examine other ways in which structural aspects of US R&D funding regimes can and do assist public/private plant breeding and agbiotech partnerships, to the great benefit of overall technological innovation.

As industries go, commercial plant breeding is still a relatively immature enterprise, relying mainly on two public sector inspired technologies (inbred-hybrids and trans-genesis). It also operates in a highly volatile and unpredictable environment with respect to government regulation and consumer acceptance. The main agbiotech business model still involves a vertically integrated, technology-driven (rather than market- or consumer-led) product development, of a sort regarded as outdated in many other sectors of industry. However, things are gradually changing in agbiotech, with a newly emerging group of small startup companies. Some of these companies focus on small market niches, such as a single type of non-food, biopharmed crop, while others specialise as providers of expert high-tech services, e.g. in genomics or metabolomics. None of the new companies seeks to become another Monsanto or Bayer, and they often have innovatory capacities that seem lacking in larger cor-porations. The more canny multinational corporate CEOs might exploit the advan-tages offered by such small companies by outsourcing services like technology development to them. This is a familiar paradigm in unrelated sectors of industry and commerce, and is increasingly encountered in the biomedical area of biotechnology. In what may be a sign of the times, in April 2006, agbiotech giant Syngenta announced the formation of a $100 million venture capital fund, called LSP BioVentures.[837] The fund will invest in growth companies and new technology startups in areas such as crop biotechnology, crop protection, biomaterials and biofuels.

The outsourcing paradigm might also be usefully explored by public sector organisations to create a new type of innovatory private–public partnership for crop improvement. This might be especially applicable in developing countries, where the cost of establishing in-house expertise in the full set of expensive and rapidly evolving new molecular technologies may well be prohibitive. One solution is to outsource such work to state-of-the-art private sector service providers. As an example of what this might entail, we can consider the mineral oil industry. Many oil-rich developing countries were historically dominated by Western oligopolies, as famously exemplified by the 'Seven Sisters'.[838] Most of the oil industries in such countries were nationalised as the latter became independent in the years after World War II. Nationalisation had distinctly mixed results with poor management, corruption, and lack of expertise contributing to largely negative outcomes for the population and economies of many newly independent nations, such as Nigeria and Venezuela.[839] As some of these countries matured, however, they endeavoured to revitalise their inefficient and ill-managed national oil companies by outsourcing key areas of new and emerging technologies. For example, specialised Western oil-service companies with access to the latest technologies for prospecting and exploitation of reserves were hired by national oil companies in some developing countries.[840] Today, the much reformed national oil companies of nations like Mexico and China often compete successfully with major multinational oil giants in world markets.[841]

In terms of plant breeding, the analogy would be for a developing country to hire a commercial service provider with the latest agbiotech expertise to help it to develop its own crop resources. Some of the more progressive and outward-looking developing countries, such as Malaysia, are now beginning to move in this direction, in order to advance their crop breeding programmes.[842] The provision of such expert services, to both public and private sector clients, may open up lucrative business opportunities for a new generation of agbiotech entrepreneurs. In the future, small service companies could play a key role in facilitating the work of a wide range of organisations, including the large corporations, international public research centres, and government organisations across the whole crop improvement sector, both public and private. Of course, this will only come about if governments, other public sector players, and the private multinationals can relinquish their peculiar aversion to outsourcing in this area. In the end, there seems little doubt that we need a mixed economy of strong partners from the public and private sectors in order to obtain the optimal value from the application of science to crop improvement.

A new market-based public sector paradigm

Efficient markets need some regulation, e.g. patent legislation to protect innovators, but too much regulation will just as surely stifle future innovation. The current agbiotech regulatory regime is too lax in some areas, and too restrictive in others, as regards the encouragement of both innovation and diversity in the marketplace. Meanwhile, the public sector has been allowed to decline far too much for the health of the industry as a whole. We need a more diverse and entrepreneurial private sector to bring products to the market, but this will only be healthy in the long term if it is complemented by a more vigorous public sector.[843] I have suggested above a 'service-provider' model as one possibility of enhancing the agbiotech private sector and perhaps building novel and mutually beneficial relationships with public sector organisations. But this is only one of several new options that are available for renewed partnerships in plant breeding. In the next section, we will tackle the thorny issue of technology ownership (the IPR problem), and then look at a new paradigm for a more market-based public sector that could both complement, and compete with, the private sector. The IPR problem in relation to developing countries was highlighted in a 2005 study sponsored by the American Association for the Advancement of Science (AAAS) and by Science and Intellectual Property in the Public Interest (SIPPI). As stated in the executive summary:

Our goal is to identify intellectual property approaches that can promote access to and use of health and agricultural product innovations by poor and disadvantaged groups, particularly in developing countries. The paper encourages more public sector IP managers to understand and employ strategies that will accomplish these goals'.[844]

Solving the IPR problem

During the past few decades, exploitation of improved crop varieties by the public and private sectors alike has been greatly complicated by a maze of protective legislation, such as patenting, intellectual property rights (IPR) and plant breeders' rights (PBR).[845] Of course, the original reason for patent and PBR legislation was to stimulate innovation by rewarding the inventive entrepreneur. Yet herein lies a paradox, whereby well-meaning attempts by governments to stimulate the marketplace via such regulatory mechanisms can sometimes end up restricting healthy competitive forces. Hence, an overgenerous patent-granting regime can give an inappropriately large advantage to early market entrants and/or those who acquire the IPR held by such companies. As we have seen in the case of agbiotech, such a situation may sometimes lead to the development of oligopolies or quasi-monopolies that, via their dominant IPR positions, can effectively control the market, sometimes stifling its further development. It is not only private companies that are affected by the perceived IPR stranglehold of the major agbiotech giants. One of the reasons for a lack of any significant public sector presence in the development of transgenic crops (in contrast to the initial 'discovery phase' research) is the perception that there are serious IPR problems due to broad-spectrum patents owned by biotech companies. To quote from an article in the journal, *Science*:

'Just about everybody in public institutes has been incredibly naïve about IP rights,' says Gary Toenniessen, director of Rockefeller's rice biotechnology program. 'It's been a shock to us to realize that you cannot use the results of research you funded because almost everybody's product is tied up in IP [disagreements].'[846]

This problem was highlighted for academic researchers in 2000 by the 'golden rice' episode. Here, a Swiss-based public sector research project, largely funded by the Rockefeller Foundation, had produced a transgenic vitamin A enriched rice variety for the use of poorer farmers in developing countries. Much to their surprise, the researchers discovered that, during the development of their transgenic 'golden rice' plants, they had inadvertently used proprietary techniques protected by no fewer than 70 patents, which were originally held by 31 different organisations.[847] Most of these patents are held by private companies, although a few are owned by universities.[848] This is a good example of the immense complexity of the IPR situation in agbiotech, a complexity that can itself act as a serious disincentive to innovation, especially by smaller and less well-resourced organisations, both public and private. In this particular case, the companies and other institutions were eventually persuaded to waive their IP rights for the use of the vitamin A rice (albeit only in the poorer developing countries where there would be no commercial market anyway), but who knows if these companies will, or indeed should, be so charitable in the future?

In a recent initiative to address this issue, the Rockefeller Foundation has agreed with four major agbiotech multinationals, Dow Chemical, DuPont, Monsanto and Syngenta, to share freely patented technologies with the African Agricultural Technology Foundation (AATF).[849] Such initiatives are, of course, most welcome in the short term. But it is also the case that they are targeted at a region where there is little or no prospect for these companies to extract significant profits anyway, so they have nothing to lose by waiving their rights on this occasion. Moreover, as with other examples of corporate largesse that were discussed in Chapter 12, these initiatives rely on the whims and caprices of individual companies, rather than addressing the more fundamental and serious long-term issues in African agriculture. For example, it would be surprising if agbiotech companies would remain quite so charitable with the AATF if one of its more active members, such as Kenya, suddenly developed competing varieties of transgenic maize or cotton that threatened their global markets.

Fortunately for the future of innovation in agbiotech, there are three emerging developments that should ensure that the current situation of overly burdensome IPR will not be with us for much longer. The first development is that many early broad-spectrum patents will expire during the next decade or so. For example, in March 2005 the Hoffman La Roche patent on polymerase chain reaction (PCR) technology expired. This came as a great relief to thousands of molecular biology researchers across the world who had endured years of what many considered exorbitant prices for proprietary enzymes, such as Taq DNA polymerase. The end of this patent protection will save greatly in research costs at universities and research institutes. It should also stimulate the wider growth of PCR-based techniques, ranging from DNA fingerprinting in medicine to marker-assisted crop improvement in developing countries. Finally, cheaper PCR technology will open up a host of new commercial opportunities in fields from forensics to human genetic profiling.

The second development concerns the historic over-willingness of patent agencies to grant broad claims, especially in agbiotech (see Chapter 7). As the various patent authorities gain more expertise, they should not be lured into the granting of some of the sweeping claims that we have seen over the past twenty years. This is already happening to some extent. For example, in 2001, the US Patent Office implemented a series of guidelines designed to tighten up scrutiny of new patents.[850] Unfortunately, the process is not retrospective so many inappropriate patents from the last twenty years remain in force. These patents can, of course, be challenged but such time consuming and expensive procedures are beyond the resources of small companies or public sector research groups. Meanwhile, the US Court of Appeals ruling in 2005 on the non-patentability of ESTs, in a case involving Monsanto, is an encouraging sign of a more realistic approach to the enforcement of utility criteria in the granting of patents.[851] Within the next decade, one hopes that the scope of bio-patents will belatedly be brought more into line with those currently granted in other technology sectors, such as chemistry and engineering. The third useful development concerning agbiotech IPR is the emergence of so-called 'open access' technologies for plant transformation and other core techniques for modern breeding. Free access to such technologies may in future allow smaller companies and public sector groups to develop novel transgenic varieties that are unencumbered by exclusive, and sometimes inaccessible, proprietorial IPR. We will now discuss this important and relatively new development in more detail.

Open access technologies in plant breeding

To date, so-called 'open access' technologies have been largely limited to market sectors like computing (cf. the Linux operating system). Although prospects of

Linux-type 'open source' technology in agbiotech seem rather remote, there are moves to develop similar monopoly-breaking opportunities here too. These efforts received a distinct boost in 2005 with publication of a method of transferring genes into plant cells that bypassed patented methods owned by major agbiotech companies. The most commonly used technology for gene insertion is *Agrobacterium tumefaciens*-mediated transformation, which is owned by Monsanto and therefore subject to a fee if used for commercial purposes. The new 'open access' method uses several different bacterial families, including *Rhizobium* species NGR234, *Sinorhizobium meliloti* and *Mesorhizobium loti* to transfer genes to plants.[852] Because the use of bacteria or other vectors to deliver transgenes into plant cells is a core technology (rather like a saw for a carpenter) for transgenesis, its ownership by a single major company can potentially act as a disincentive to others to innovate in the field.[853] In the case of the new 'open access' method, however, the Australia-based group CAMBIA, who developed the technology, has stated that it would be freely available.[854] CAMBIA described the new methods as constituting what they call a: 'protected technology commons ... having no commercial restrictions other than covenants for sharing of improvements, relevant safety information and regulatory data and for preserving the opportunity for others to freely improve and use the technology.'[855]

Publication of the paper outlining this new technology elicited widespread media interest and invited comparisons with open source technologies in other commercial areas. Examples of the press coverage included articles in the Economist and Wall Street Journal entitled respectively: 'The triumph of the commons – Can open source revolutionise biotech?' and 'Sharing your innovations is potentially profitable'.[856] The CAMBIA group has also launched an initiative called BIOS, or Biological Innovation for Open Society.[857] Partially funded by the Rockefeller Foundation, this initiative is specifically aimed at developing countries and encompasses all forms of biological innovation, including crop breeding, genetic and natural resource conservation, and crop husbandry. In late 2005, the CAMBIA initiative was further boosted when the Norwegian government agreed to provide $2.5 million to catalyse a joint venture between IRRI and CAMBIA aimed at developing country crops.[858] The aim is to operate within the current IPR and PBR systems but to utilise these for the greater public good, instead of the private gain of the few. In some interesting passages from the BIOS prospectus, it is stated that:

The public sector science community is complicit by neglect, as virtually all practices of academic scientists promote the belief that 'good science' can, almost by magic, transform itself into public or private goods.

... data, the genetic materials and the published science (are) routinely captured and hijacked – enclosed – by those entities, usually large multinational corporations that have access to the means of converting that information into economic value through goods or services. This enclosure rarely ensures a sustainable competitive advantage, and is sometimes an inadvertent and very unfortunate side effect of a strategy for industry survival.

And finally:

The clearly visible manifestation is the dramatic increase in the use of intellectual property protection by both public and private sector, the concomitant low standard but broad scope of such IP grants, and the trend towards exclusive licensing and exclusionary use of IP portfolios.[859]

One of the refreshing differences between initiatives like CAMBIA/BIOS, and older paradigms of public good research, is a willingness to engage more closely with the commercial marketplace. Such initiatives will, where appropriate, employ market tools for innovation and product development. A second initiative in this area is the US-based Public Sector Intellectual Property Resource for Agriculture (PSIPRA), as developed by the Rockefeller and McKnight Foundations, in collaboration with the ten major US Land Grant Universities. As with CAMBIA, the US initiative aims to support plant biotechnology research in developing countries. University members of PSIPRA have undertaken that, when they license patented agricultural technologies to private companies, they will in future retain rights to the exploitation of these technologies for humanitarian purposes, e.g. in developing countries. In some cases, the new technologies could also be applied to small specialty crops, such as peanuts, broccoli, lettuce and tomatoes, in which the seed and agbiotech industry does not have strong commercial interests. Another example of an open access initiative from the USA is known as MASwheat. This initiative is coordinated from UC Davis and involves a consortium of researchers at twelve public institutions in the USA. MASwheat members will use DNA markers to select for 23 new traits in wheat, conferring resistance to fungi, viruses and insect pests, plus improved grain quality for bread and pasta making. In true open access tradition the consortium will make all marker sequences and research protocols freely available to all via its website.[860]

These types of initiative can serve as templates for public sector researchers around in the world. In particular, a selective open access strategy (i.e. aimed at some developing countries) could be made a condition for the award of grant funding by government agencies and charitable foundations. Some examples of such bodies include the Framework Funding initiatives in the European Union, national Research Councils, such as BBSRC in the UK, and charitable trusts, such as Leverhulme or Gatsby. In addition to the proactive initiatives from PSIPRA and MASwheat discussed above, US universities such as Stanford and Yale have started to retain rights to the non-profit dissemination of their licensed technologies.[861] It is noteworthy, however, that European agencies are lagging behind many public sector and philanthropic organisations in the USA in the implementation of such an agenda. As discussed in the 2005 AAAS/SIPPI report mentioned above, and as part of a new AAAS Humanitarian IP Management Initiative, a key challenge in the better use of publicly generated IP pro bono publico will be to highlight 'the importance of managing public sector IP to facilitate humanitarian use and applications.'[862]

The new vision conjured by such initiatives, which many researchers would share, is of a reformed public sector that no longer behaves as a quasi-state enterprise acting apart from, and sometimes in conflict with, the private sector.[863] Rather, we should now look forward to a mixed public/private economy in which all players would participate, and perhaps seek to reform it from within if/when they deem appropriate. In the case of BIOS, members of the public community join together in order to enjoy open and free access to technologies that they develop themselves. The technologies are fully patented and the community undertakes to act collectively in their defence in the event of infringement. This means that a small university laboratory in Wales or a plant breeding centre in Senegal can share equally in the benefits of the latest technologies of crop improvement by acting together as a kind of collective multinational enterprise. The main difference between such an enterprise and firms like Monsanto or Syngenta is that the former is predicated on non-profit, public-good crop improvement rather than the maximisation of profit and shareholder value.

Re-entering the marketplace

There is room in the new marketplace for both of these models of agricultural enterprise (and, indeed, many more), especially as many of the interests and goals of the different organisations will be complementary rather than competitive. The public sector should appreciate this changing paradigm and enter the marketplace more vigorously, if it wishes to pursue the realistic long-term development of public-good transgenic crops, e.g. for the benefit of poorer countries.[864] A few ventures along these lines are already under way, and it is noteworthy that a particularly important recent initiative has come from a developing country, namely Brazil, which has developed a new monopoly-breaking herbicide tolerance technology. We have already seen that herbicide tolerance is now the most widely used trait in transgenic crops. The best-known examples are the various Roundup Ready® varieties owned by Monsanto. This herbicide tolerance trait alone accounted for some three quarters of the entire global area of transgenic crops in 2005.[865] In 2004, the Brazilian crop research agency, EMBRAPA, announced development of new transgenic varieties of herbicide tolerant soybean with resistance to imidazolinone, rather than Roundup. Commercialisation of the new varieties could end the current near-monopoly enjoyed by Monsanto's Roundup Ready® soybeans.[866] The new EMBRAPA varieties are now being field tested for regulatory appraisal and once available to farmers they could further boost prospects for soybean farmers in Brazil.

As we saw in Chapter 14, Brazil is rapidly becoming a major global supplier of soybeans. Transgenic Roundup Ready® varieties have proved to be very productive. Indeed, they were so popular with farmers that seeds were illegally smuggled into the country on a massive scale, before the Brazilian government legalised cultivation of transgenic soybeans in 2004. However, as soon as the transgenic seeds from Monsanto and others became legally available, Brazilian farmers discovered that they were expected to pay the unpopular and contentious licence fees charged by the company. Avoidance of licence fee payments by farmers in Argentina had already led Monsanto to suspend sales of soybeans in that country in 2004, amidst great acrimony on all sides.[867] The new EMBRAPA developed transgenic varieties will enable Brazilian farmers to grow herbicide tolerant varieties of soybeans that are owned by their own public sector research institutes, rather than by an overseas company. Hence, one assumes, they would be much less expensive than imported seeds. The lower cost of the new soybeans would hopefully stimulate the agro-economy of the country and justify the public investment in EMBRAPA that was required to develop the technology in the first place. By 2005, EMBRAPA was earning over \$25 million per year from the private sector, 90% of which was from seed, both conventional and transgenic. It is estimated that the agency will earn about \$21 million on transgenic soybean royalties alone by 2006.[868] In addition to soybeans, EMBRAPA breeders are now developing transgenic varieties of papayas, potatoes, and maize.

These Brazilian developments are to be welcomed. Not only do they show what the public sector can achieve, they should also result in increased choice and marketplace competitiveness for farmers. Such a strategy is also likely to be the only way that low or zero-profit crops, e.g. transgenic salt tolerant millet for subsistence farmers in Africa, would ever realistically be made available. (That is apart from the odd bit of PR/philanthropy from biotech companies, which as we have noted, is neither desirable nor sustainable in the long term.) Ironically, it is probably only a reinvigorated public sector that will have the interest or sustained capacity to develop large scale, public-good applications of transgenic technology, especially in the developing world.[869] Hence, the truly radical, global potential of this technology will probably only be recognised once it is recaptured by the commons and put to more general public good, including in a commercial context.

Such demonstrably public-good applications of transgenesis would also be more difficult for NGOs, environmental groups, and other anti-GM organisations to oppose. This would be especially true if technologies are developed in, and are owned by, indigenous peoples in poorer countries, such as the Philippines, Paraguay, China, India and Romania. This process is already under way; in 2004, farmers in these five countries collectively planted about 6 million hectares of transgenic crops, many of which were locally developed varieties.[870] The increasing use of transgenesis in developing countries, providing it is genuinely the best option for crop improvement, will be a useful mechanism for disseminating the technology, recruiting it to public-good applications, and removing it from its current unfortunate connotations as an aberrant and dangerous tool, wielded exclusively by what is all too frequently caricatured as a malign global agribusiness sector.[871]

Domesticating new crops – an alternative to transgenesis

Transgenesis has considerable future potential as an extra tool for breeders to enhance genetic variation in our rather narrow portfolio of existing crops. However, the new genomic technologies may also enable us to domesticate some of the tens of thousands of additional useful species, which constitute the astonishing cornucopia that is the botanical kingdom. Between 13 000 and 11 000 years ago, our earliest staple crops were domesticated, by a fortuitous mixture of genetic happenstance and human intervention. Over the next five thousand years, virtually all of our remaining food crops were brought into cultivation. Since that time, farmers have achieved much in the selection of improved varieties of these ancient crops. This process was greatly accelerated after the eighteenth century, with the introduction of new breeding techniques, based on research into plant science. But it is still a remarkable fact that, notwithstanding all of our advances in crop breeding, not a single new species of our major crop group, the cereals, has been domesticated in the past two thousand years. Today, the vast majority of our global agricultural production relies on a very limited number of widely cultivated staple crops, such as wheat, rice, maize and soybean. Although there are at least 10 000–50 000 edible plant species, we only actually grow about 150 of them. And our major crop staples can be counted on the fingers of just two hands.

Our current dependence on such a tiny number of major crop species has both advantages and drawbacks. On the plus side, it simplifies the process of breeding and cultivation if we only need to focus on a limited number of plant species. Since the major crops were domesticated so long ago, we also have had several millennia of breeding and selection for the optimal cultivation traits of these species. Wider introduction of transgenic varieties of these major crops is likely to increase their dominance even further. However, concentration on a very small number of major crops with ever narrower gene pools may be undesirable for ecological reasons, leading to extensive monocultures that can be particularly susceptible to pests or diseases. These ecological risks are exacerbated by the ongoing genetic impoverishment and varietal erosion of all the major crops, and most particularly those species in which the private sector is the dominant player (see Chapter 8). A further factor is the likelihood of an increasingly unpredictable global climate over the coming centuries. Since the last great climatic shift to affect most of the globe (the Younger Dryas event of 12 800–11 600 years ago[872]), global climates have been uncharacteristically stable. All of our current major crop species are adapted to such relatively benign conditions. But there are increasing indications that the climate may be re-entering a more unpredictable period, due to anthropogenic or non-anthropogenic factors (or, more likely a combination of the two). Whatever the causes of climate change, it would seem prudent to begin research on the domestication of entirely new crop species as a possible hedge against future failure of our current major crops, such as wheat and rice.

From observing indigenous cultures, many of which previously cultivated a wide variety of plants, we know that at least 1650 tropical forest species could be grown as vegetable crops. Many of these plants are specifically adapted to areas where conventional commodity crops do not grow very well. There are also tens of thousands of under-utilised or as-yet unused plants that could furnish us with a vast range of additional food and non-food products.[873] Thanks to recent advances in our understanding of the genetic basis of domestication, and the availability of advanced breeding techniques, we now have an historic opportunity to recruit many of these species into our portfolio of useful crops. To my mind, one of the most compelling reasons to be interested in new crops is that we are now in the process of developing the kinds of technology that will make it feasible to domesticate some of these species within as little as a few decades. As we continue to uncover the mechanisms that regulate the key elements in the domestication syndrome, it will become increasingly straightforward to cultivate many of the tens of thousands of potentially useful plants that have hitherto been beyond our control.[874]

To date, much of the pioneering research on new crop development has been done by public sector groups in the USA.[875] A few European initiatives were launched in the 1990s, but these were never given the long-term funding needed for plant breeding research. Should we ever decide to rebuild our sorely depleted public plant breeding infrastructure, the development of new crops would be an admirable addition to its research portfolio. Indeed, it is precisely this kind of novel, challenging and socially worthwhile research area that could potentially attract some of the biology graduates who currently end up in basic research or in industry. New crop research will involve many cutting-edge molecular technologies and will probably also create exciting commercial opportunities on the way, but within an overall public-good context. Such a mixture of vision and risk might particularly appeal to the more adventurous younger plant researcher. And it would be a fitting tribute to continue and extend the efforts of our Palaeolithic and Neolithic ancestors who stumbled into crop cultivation more than ten millennia ago. Since then, we have successfully domesticated less than 1% of the known potentially useful plant species. So, why not use our new knowledge and technologies to tackle the remaining 99%? After all, it is not every scientist who could claim to have helped domesticate one of the first new crops for humanity in several thousand years.

Innovative applied R&D – the USA leads (again)

Over the next decade and beyond, international agencies (most notably CGIAR) and rapidly developing countries, such as China, Brazil and India, will hopefully work together to develop more sophisticated capacities for crop improvement. Meanwhile, there is little doubt about the location of the most important centre of innovation for crop improvement technologies. It is, of course the USA. One of the most significant boosts to innovatory capacity in the USA is the multiplicity of funding sources, both public and private, that can be found at national/federal, State, and local levels across the country. While European governments are frequently (and sometimes accurately) caricatured as over-subsidising locally based businesses, the US government is itself a massive funder of private sector R&D, especially in newer sectors like biotechnology. In the area of agbiotech alone, there are large funding programmes available to companies from numerous federal agencies including the National Institute of Standards and Technology, Department of Agriculture, Department of Defense, National Institutes of Health, Department of Energy, and National Science Foundation. These funding agencies support early and mid-stage commercially orientated R&D in the most exciting areas of modern agbiotech. Projects include joint ventures between small startup companies and agbiotech giants, single-company ventures, and collaborations between companies and academic laboratories in universities. We will now look at a few examples of such projects and their funding mechanisms.

In Chapter 12, we met Sangamo Inc., a California-based firm developing technologies using engineered zinc finger repressor proteins (ZFP TFs) to regulate genes in plants. Much of the current Sangamo R&D work in plant agriculture is supported by an Advanced Technology Program (ATP) grant awarded by the National Institute of Standards and Technology (NIST).[876] This federal programme 'co-funds high-risk, high payoff ventures with industry', and has funded projects valued at over $4.1 billion since 1990. Sangamo has had three ATP grants totalling $6 million, including its ongoing $2 million venture on zinc finger gene regulation in plants.[877]

Another five-year, $4 million, ATP-funded project, this time to a different California-based agbiotech company called Mendel Biotechnology, is for development of a genetic engineering toolkit for plant breeders. The joint grant holder on the project is Semenis Vegetable Seeds, a seed company acquired by Monsanto in 2005. This federal agency (NIST) also provides multi-million dollar grant support to several R&D ventures by the larger agbiotech companies, such as Monsanto, DuPont and Dow Chemical.

Even more generous than the impressive NIST–ATP initiative is the Small Business Technology Transfer Program (STTR), which 'encourages small business to explore their technological potential and provides the incentive to profit from its commercialization.'[878] This federal initiative disbursed over $2 billion for research projects in 2004 alone. Over $500 million/year of this money is channelled via the National Institutes of Health (NIH) and several agbiotech areas are funded. One of the beneficiaries has been Arcadia Biosciences, of Davis, California, which in 2005 received a grant from the STTR/NIH to develop wheat lines with reduced levels of celiac disease-causing proteins.[879] The grant will be split equally between Arcadia and an academic collaborator at Washington State University. Another NIH grant to Arcadia was to develop soybeans with enhanced levels of specific isoflavones.[880] In 2005, Arcadia Biosciences also benefited to the tune of $2.9 million from a Department of Defense (US Army Natick Soldier Center) contract to develop longer lasting fresh vegetable produce.[881] In this project, the company will use its proprietary TILLING technology (see Chapter 3), with an initial focus on tomatoes and lettuce.

Moving to the State and local levels, US companies and university researchers benefit from a plethora of public funding schemes providing multi-million dollar support for high-risk projects. Among the dozens of examples of such ventures in the agbiotech area is the collaboration between Large Scale Biology Corporation of Vacaville, California and scientists from the J. G. Brown Cancer Center at the University of Kentucky, Louisville to develop a biopharming system for production of anti-cancer therapeutics in tobacco plants.[882] The work was funded by the Statewide Kentucky Economic Development Finance Authority. Large Scale Biology Corporation was also funded for separate projects by US federal NIST–ATP and NIH programmes, similar to those described above.[883] It is also increasingly common for State agencies to attract agbiotech companies to relocate by offering financial and other support. In 2005, biopharming company, Ventria Bioscience[884] was set to relocate to Missouri following an expected $10 million pledge from the Missouri Development Finance Board to underwrite a new Center of Excellence for Plant Biologics at Northwest Missouri State University.[885]

A further lucrative source of research funds in many US States is the large settlements from lawsuits against tobacco companies, totalling about $250 billion nationwide.[886] For example, in Arkansas, the Tobacco Settlement Proceeds Act of 2000 provides $62 million/year for 25 years 'to improve and optimize the health of Arkansans'. Some of this money has been used to establish the Arkansas Biosciences Institute (ABI) at Arkansas State University in Jonesboro. The Executive Director of ABI is Carole Cramer, who co-founded two biopharming companies, CropTech Corporation and Biodefense Technologies, which are focused on development of tobacco-based manufacturing of human therapeutic proteins and vaccines. The Cramer group had been lured from their previous location, at Virginia Technical University, by the prospect of enhanced funding, improved facilities, and other benefits in the package offered by Arkansas. To quote the *Memphis Business Journal*: 'Another attraction to Jonesboro was the blatant attitude toward encouraging technology transfer to the private sector, and the opportunity to create an environment where entrepreneurs can thrive.'[887] The Arkansas Biosciences Institute is mandated by the State of Arkansas to maintain at least 25% of its faculty in a so-called 'new entrepreneur' status, whereby they are encouraged to set up their own biotech companies with assistance from the College of Business at the State University. Similar ventures are being set up across the USA and the overall climate, both for scientific research and for commercial entrepreneurship, contrasts sharply with the more depressed picture in Europe and Australasia.

The scale of funding involved in agbiotech and other types of public/private collaborative crop improvement R&D in the USA is in the many hundreds of millions of dollars per year. Moreover, many funding programmes have operated continuously for several decades, providing long-term continuity and allowing projects to be refunded several times over if they show promise. This level of commitment completely dwarfs anything on offer either by national governments in Europe or by the EU Commission. Several large multinational projects on crop development have been funded by the EU over the past decade but these tend to involve unwieldy consortia that are often selected as much on the basis of their mixture of the 'appropriate' participating countries as on their scientific credentials.[888] Also, these EU projects rarely last for more than a single three-year funding round and are therefore unable to provide any of the long-term continuity of effort that characterises many US initiatives.[889] In general, national funding of public/private crop-related research in Europe is at a lower level, is more sporadic, and is manifestly less productive than in the USA.[890] Indeed, several recent government initiatives in Europe attracted proposals from companies and academic researchers that were generally so mediocre in quality that a sizeable proportion of the funding was not even disbursed.

This is symptomatic of a wider malaise in Europe, which is characterised by a lack of connectivity between the plant science research base and its application, both in the public-good domain and in the commercial sector. While there are a few islands of good applied research into crop improvement, these are relatively isolated, poorly resourced, and lacking in practical outlets. European R&D is in danger of regressing to a point where it will eventually lose the critical mass required to sustain a substantial contribution to international crop improvement. Meanwhile the USA continues to power a new generation of high-tech innovations, while largely keeping intact conventional public crop improvement programmes, including farmer education and extension activities. At the same time, the more advanced developing countries such as China, Brazil and Malaysia are rapidly catching up with the USA as agricultural producers and may collectively threaten its status as the breadbasket of the world within the next decade or so. The message to Europeans is clear. The necessary resources for a better managed and more effective crop improvement strategy are already available in Europe, but they need to be channelled towards longer term and more strategic research programmes. Political factors, such as the overreaction to GM crops, have been allowed to distort the debate on the future of agriculture and to distract attention (and funding) from the real priorities for both European and international crop improvement.

Europeans would benefit from following the US lead in nurturing a longer term, more sustained approach to plant science research and its application in the private and public sectors. In this chapter, we have discussed how these two sectors might work together more closely, while retaining those distinctive qualities that give them their respective strengths. As IPR problems gradually recede and open access technologies become available, there are increasing grounds for optimism about the future of high-tech developments for public-good and commercial plant breeding. Among the exciting challenges that would face a reinvigorated public/private sector partnership is the beguiling prospect of beginning to domesticate some of the thousands of new potential crops that could serve as sources of more nutritious foodstuffs and of a new generation of renewable, biodegradable, 'green' chemicals that will eventually replace our rapidly depleting petrochemical feedstocks.[891]

19

Conclusions and recommendations

No, a thousand times no; there does not exist a category of science
to which one can give the name applied science. There are science
and the applications of science, bound together as the fruit to the
tree which bears it.

Louis Pasteur (1871)
Revue Scientifique

Introduction

In this book, we have surveyed the evolution of modern plant breeding, and its
application in crop improvement, over the past two centuries. We have examined
crop improvement from a variety of scientific and socio-economic perspectives and
have seen the sometimes surprising ways in which these interact to affect the trajectory of agriculture. Overall, there is no doubt that scientific plant breeding has
been an outstanding success, enabling farmers to feed the more than seven-fold
increase in the global human population that has occurred since 1800.[892] But we
have also seen that all is not well with plant breeding, especially in the public sector.
Over recent years, researchers and some science policymakers have started to realise
what breeders had been aware of since the early 1990s, namely that plant breeding is
sliding ever more deeply into crisis. One of the key aims of this book has been to
highlight such concerns, to explain their provenance, and to lay out some of the
options that might allow us to surmount these challenges in the years to come.

In this final chapter, we will first consider some emerging ideas from the USA that
were published after much of the present book had been written. These ideas are
interesting in their similarity to some of the arguments presented here, albeit from a
transatlantic perspective. In particular, there is an increasing stress on the importance of a greater engagement from plant scientists with the application of their
work, especially in developing countries. It is interesting that so many eminent US
commentators are voicing similar concerns, but also disappointing that so little has

294

been heard on these topics from European scientists, economists and public policymakers. We will then look at the possible future of agriculture as the twenty-first century progresses, with its attendant challenges of balancing the pressures of population increase and economic growth, together with the wild card of possible climate change of an as-yet unknown magnitude.

We will finish both chapter and book by drawing together some of the main points to emerge from our discussions as a series of recommendations. These points are intended to stimulate further debate about the future organisation, mission and mechanisms of plant breeding and crop improvement. An extended and in-depth debate on the wider contexts of plant breeding and agriculture will be an important mechanism to improve public perception and knowledge of the many diverse and interrelated issues that confront decision makers. It will also be a useful mechanism for moving towards a consensus on divisive topics like GM crops, patenting of plants, and public-good R&D. Participants in such a debate should include all potential stakeholders in breeding and agriculture from industrialised and developing countries. Examples of stakeholders might include farmers, public and private sector breeders, companies, plant science researchers, the general public, governments, politicians, economists, and NGOs. And, of course, the debate will be of little value unless it remains non-partisan, evidence based and rational.

Perspectives from the USA

In Chapter 10 (section entitled *The penny drops*), we saw that a few official bodies in the UK have belatedly started to recognise the existence of a crisis in applied plant research, and especially the widening gap between such research and the more academic studies that are still such a great strength of UK science. Although the USA has not cut back its applied research nearly as severely as the UK, American observers have also recently commented on the gap between the knowledge created by basic researchers and its application in real-world breeding, in both the commercial and public-good domains. In late 2005, Deborah Delmer of the Rockefeller Foundation cogently summarised some of the challenges confronting plant breeding in her inaugural article in *Proceedings of the National Academy of Sciences*.[893] Talking about what I have already termed the academisation of research (Chapter 10), and a culture that neither values nor promotes applied plant science, Delmer says:

Despite all of the complaints, scientists of the 'North' do have strong and relatively stable sources of funding for basic plant research, in particular for plant genomics. On the other hand, donors who support work on global agriculture are largely constrained to fund downstream applications relevant to the developing world. What seem to be lacking are systems that promote and reward efforts to create a strong interface between fundamental and applied research in support of global agriculture. .... for all approaches to crop improvement, we clearly need more efforts that promote meaningful dialogs between bench and field scientists, better systems of reward within academia for such collaborative efforts and, perhaps most critical of all, substantial new sources of funding for serious projects that aim to apply the exciting innovations in plant science to problems faced by poor farmers throughout the world.

Delmer goes on to discuss the case for applying some of the newer technologies for public-good projects in developing countries:

What has been sadly lacking in the public sector is an understanding of how to make strategic assessments of which projects can have the highest impact; ... there must be a serious resolve on the part of public-sector scientists to move beyond just proof of concept with a few transgenic events, and they will also need adequate resources and infrastructure for such efforts. I would also argue that, because the approaches for many different crops are similar, it would make sense for the CGIAR centers to come together to form one serious biotechnology unit where a team of skilled and interactive scientists can work together in an environment that provides the kinds of high-throughput capabilities and ability to do easy field testing that are needed for these types of efforts.

Delmer's call for greater focus on realistic goals by researchers in industrialised countries, especially in relation to the challenges of developing country crops, was complemented a few months later in a report sponsored by the American Association for the Advancement of Science. One of the major aims of this report was to raise awareness among public sector researchers, research managers and policy-makers about the need to make the fruits of their science more readily available for public-good applications. As the report points out, this should not be at the expense of commercial applications of research; with some judicious IPR management, research can be simultaneously profitable in the commercial sphere and available as a free public good for humanitarian applications.[894] Broadly similar sentiments were expressed at a meeting of a joint US–European Commission Task Force assembled in 2005 to identify challenges for plant science in the next few decades. A recurring theme was the need to support the application of new discoveries to downstream practical work in crop improvement, including establishing programmes to fund imaginative uses of plant biotechnology.[895]

I concur with these views, with the important caveat that we must not become overly fixated on transgenic technologies, especially in the short term. As noted in a recent review of CIMMYT research, for most applications in developing countries, conventional breeding approaches will be cheaper and faster for many years to come.[896] Meanwhile, in order to promote more meaningful interactions between basic researchers and developing countries, the scientific community must incentivise such collaborations by according them equivalent recognition and status to what is currently extended to cutting-edge academic studies. At the same 2005 US–EC Task Force meeting, sociologist Lawrence Busch (who chaired the investigation into the UC Berkeley/Novartis affair[897]) commented that in agbiotech, the 'choice of research trajectories is too narrowly defined', while 'Private sector goals, based largely on profitability, limit too greatly what products can be delivered by either the public or private sector'.[898]

Busch also commented on the anomalous regulatory environment that is 'simultaneously too cumbersome and inadequate', with R&D blocked by an overcrowded patent market, while lax regulation in areas like biopharming jeopardise the future prospects of the whole industry. Many of us would echo Busch's plea for researchers to end their relative isolation and engage more with the rest of a society that is understandably both excited and apprehensive about our work: 'Plant biotechnologists can either dismiss these problems as outside the field and watch support for research wither, or they can directly address them. If they do the latter, then plant biotechnology will have to become a truly social science.' Busch goes on to say:

The best approach to identifying key constraints in 'translational biology' is to establish a much better dialog between 'bench scientists' and those key breeders in the developing world who actually talk to farmers and understand local agriculture. Better systems of reward are needed within academia for collaborative efforts with breeders, and 'perhaps most critical,' substantial new sources of funding are needed for serious projects aimed at applying exciting innovations to problems faced by poor farmers.[899]

Finally, it is apposite to quote from the concluding part of a recent review by agricultural economists Brian Wright and Philip Pardey, respectively from UC Berkeley and University of Minnesota:

... it is important to recognise that the dominant constraint on agricultural innovation for the majority of developing countries is not at present IPR. It is a lack of sufficient and sustained funding in the face of well-documented high social returns on research *using non-transgenic technologies*, a problem also observed in developed economies with strong IPR, for all but a handful of crops[900] (my stress).

With a few notable exceptions, hardly any significant voices from Europe have been heard on these important global issues relating to the uses of plant breeding technologies, especially for public-good applications. It sometimes feels as if Europe is instead consumed by an unproductive, inward-looking, and often semi-hysterical discourse on minor aspects of GM technology, while the rest of the world gets on with the business of tackling the broader issues of how best to use all available and appropriate technologies for crop improvement.[901] This has had the unfortunate effect of frightening most reputable European scientists away from the public arena, thus leaving the stage to be filled with all manner of often dubious commentators and their shrill and disputatious, but ultimately sterile, discourse. One hopes that this situation will change and that scientists around the world, and most especially in Europe, will re-engage in the kind of public dialogue that Huxley (on evolution) and Einstein (on nuclear weapons) did not shrink from joining in their day.

Late twenty-first century agriculture

It is always risky to extrapolate from present trends in an attempt to predict the future, but such exercises can often be both provocative and instructive in informing our current actions and plans. Hence this speculation on how global agriculture might look like in the latter years of the twenty-first century. Firstly, it seems likely that transgenic crops will be cultivated more widely than today, although the extent of their eventual uptake will depend on the incorporation of more farmer- and consumer-relevant traits than the current restricted portfolio of simple input traits. In turn, this will depend on improvements in the innovation record of private companies, plus a greater willingness of public sector researchers to develop and disseminate transgenic traits as public goods, especially to developing countries. If real consumer-friendly traits can be produced, and appropriately marketed, in industrial countries, previous experience of technology uptake suggests that even sceptical Europeans will eventually accept transgenic crops. Once the current over-hyped expectations of transgenesis have subsided, the technology will become a routine option for the enhancement in many, but by no means all, agronomic traits. But transgenesis will always remain just one component of the breeder's toolkit. Transgenic strategies will not be appropriate for most developing country challenges in the short to medium term, and may never be able to replace non-transgenic approaches for some of the more complex genetic traits, such as aspects of stress tolerance.

Secondly, it seems fair to assume that the taxpayers and consumers of industrial countries will eventually tire of subsidising the overpriced production of many of their major agricultural commodities, such as sugar beet, cotton, rice and soybeans. As we saw in Chapter 15, this process has already started with pressure to reform the tariff and subsidy regimes of the US and EU trading blocs. This will free up the market and enable more cost-effective crop production in developing countries, such as Sahelian Africa and India (cotton), Vietnam and Brazil (rice), Central America and the Caribbean (sugar cane), and temperate South America and Southern Africa (vegetables). As already seen in Vietnam, Malaysia and parts of South America, greater productivity can lead to a virtuous cycle of additional investment in farming, higher yields, and improved regional prosperity. Providing its infrastructural deficiencies can be resolved, it is likely that the South American continent will emerge as a major supplier of many global commodity crops. This is already beginning to happen with soybeans, and with its diverse climatic range there is no reason why the continent cannot produce other major crops such as maize, rice, oil palm and cotton.

The majority of farmers in North America, with their high labour and equipment costs, will eventually be obliged to move from the cheaper mass-commodity crops to higher value, identity-preserved niche crops. In this respect, they should be well served by the relatively dynamic and innovative research infrastructure that still exists in the USA. Here, a somewhat weakened, but still reasonably strong, public sector is already well engaged in working alongside an increasingly diverse and innovative private sector. With strong and consistent government support, coupled with a favourable business environment, the formidable innovatory capacity of US plant researchers is likely to have been translated into a host of new crops and products that better suit the country's high-wage economy. It is possible that some low-value commodity farming will continue in places where cheap (possibly immigrant) labour is available, but this will not sit comfortably alongside the rest of the more high-tech economy.[902] Meanwhile, the 1990s agbiotech paradigm of input trait modification in a few major commodity crops dominated by a large-company oligopoly will have been replaced by a more productive and diverse commercial ecosystem populated by scores of specialist companies, many of them niche service providers to each other, to larger companies, and to the public sector.

The prognosis for European farming is somewhat less sanguine. Europeans could continue to try to produce food at a much higher cost than the rest of the world, using expensive and at times environmentally questionable intensive input systems. But this is unsustainable in the long term. The alternative is to follow the USA and move into higher value niche markets, rather than low-value, high-volume commodities like wheat, sugar and basic vegetable oils. For such a strategy to succeed, both private and public sector R&D in Europe should focus on facilitating the development of these new crops. Unfortunately, it is in Europe that the greatest damage to the once effective plant breeding infrastructure has been wrought by several decades of privatisation, rationalisation and sheer neglect. Also, as discussed in Chapter 15, the innovatory capacity of European researchers has been hindered by a combination of erratic and short-term public funding initiatives, a lack of connectivity between researchers and breeders, and a relatively weak private sector that is further hampered by regulatory constraints. It is probably not yet too late for a dynamic and successful plant breeding sector to re-emerge from Europe, but the clock is ticking and midnight is at hand.

Recommendations

The following brief list of recommendations is an attempt to distil some of the major points from the discussions in the previous eighteen chapters of this book. These points might serve as part of a framework for a debate on the future of plant breeding in its wider sense, and especially on how it might best be exploited for both commercial profit and the public good.

Public sector

Education and career structures

Make the application of plant science in breeding and other aspects of agricultural improvement an integral part of school and university curricula. Career structures, staff reward/promotion systems, and research assessments should reflect the value of breeding-related careers as being at least the equal of the more academic professions.

Applied research

Accord more recognition, prestige and funding to applied research. Public research institutes should return to their former (pre-1980s) mission of focusing on applied plant research, with academic research being based mainly in universities. Institutes and universities should work more closely with each other, with the private sector, and with developing countries, to ensure the efficient uptake of research, both for public good and for private profit.

Outsourcing

Public institutions around the world should consider outsourcing more crop improvement technologies and expertise to external providers, from both private and public sectors. This applies particularly to developing countries without the critical mass or lead-time to develop the full range of constantly evolving new technologies in-house.

Seed banks

Governments, the United Nations, charities, and private companies, should collectively set aside $150–250 million of new money for a public/private CGIAR-coordinated consortium to secure future funding in perpetuity for international seed banks. National seed banks should be treated as true public goods, available to *bona fide* public breeders of all countries, irrespective of their official ideologies.

Participatory plant breeding

A sustained (ten-year plus), coordinated international R&D programme should be set up to establish methods of best practice in PPB and recruit more breeders from public sector institutions in industrial countries.

New crops

Establish an international EU/US/etc. programme to develop a toolkit and working methods for crop domestication by initiating case studies on at least four new crops over the next two decades. This is a hitherto unrecognised opportunity to use the latest tools of genomics and conventional breeding to domesticate a wide range of new crops for both food and non-food use. Some of the chosen new crops should be selected on the basis of their performance under the altered climatic conditions (e.g. salinity, mild winters and aridity) that are already becoming evident in some regions of the world.

Open access technologies

Funding agencies should foster and encourage these ventures and public sector institutions should be strongly encouraged, and perhaps compelled, to exploit open access technologies wherever possible. Researchers, research managers, and funding agencies in industrial countries should also acquire and use the expertise that will enable them to make their own technologies available to developing countries without jeopardising their potential for commercial exploitation closer to home.

Private sector

Innovation

Establish more effective mechanisms to translate public sector research innovation into commercial utility. Some parts of the private sector, especially many larger companies, already spend a great deal on R&D that does not produce commensurate innovation. In too many cases, innovation still comes as an accidental by-product of public sector academic research.

Diversification

The agbiotech sector as a whole needs to become more diverse in its use of technology, in its business models, and in its crop/geographical range. Diversification in some areas will require a tightening of over-lax patent regimes that currently limit opportunities for new market entrants (see below).

Outsourcing

Larger agbiotech companies should follow the precedent of other industrial sectors in outsourcing more of their cutting-edge R&D to smaller high-tech service-provider businesses.

Governments and judiciaries

Patents and breeders' rights

Revise UPOV, PBR and IPR systems to recognise that transgenesis is not a special case in technology development. In the past, over-generous patent protection has been given to transgenesis, while other innovative technologies have much less legal protection, and hence suffer a competitive disadvantage.

Deregulation of crop production

Reform crop regulation systems to include a single internationally recognised protocol for traded crops. Crop production is over-regulated in many respects and the case for treating transgenic crops differently from other crops is scientifically unsound.

Subsidies/tariffs

Reduce or eliminate subsidies and tariffs wherever possible. They stifle output by potentially more efficient producers, especially in developing countries, and increase taxation burdens in the subsidising countries. Rural support programmes in richer countries should not be at the expense of poorer farmers overseas.

My final recommendation is that we bear in mind the words of Anton Chekhov as we seek to strike a balance between basic plant science and its applications, both commercially and for public good, in crop breeding:

'Knowledge is of no value unless you put it into practice.'

Notes

1 Archaeologists divide the prehistorical development of humans into a number of chronological stages. The so-called Palaeolithic period, or Old Stone Age, lasted from about 150 000 BP until 13 000 BP and was characterised by extensive use of chipped stone tools. Around the period 50 000–70 000 BP, at the onset of the final phase of the Palaeolithic (called the Upper Palaeolithic), the types of tools became much more diverse and people created increasingly elaborate art forms. The Mesolithic, or Middle Stone Age, lasted from about 13 000–11 000 BP and was characterised by several dramatic climatic changes, especially in the northern hemisphere. Finally, the Neolithic, or New Stone Age, began about 11 000 BP with the introduction of superior grinding processes for the manufacture of stone tools and the increasing adoption of sedentary/agricultural lifestyles. Of course, all these dates are approximate and overlap with each other to a great extent, with some cultures developing or acquiring such innovations while their contemporaries in different parts of the world did not. For example, as late as the early twentieth century, isolated cultures in South America and Asia were still maintaining an essentially Palaeolithic lifestyle.
2 Weiss *et al.* (2004); Piperno *et al.* (2004).
3 The Palaeolithic and Neolithic origins of agriculture, and the key role played by plant genome organisation, are examined in detail in my forthcoming book entitled *People, Plants, and Genes* (Murphy, 2007).
4 Darwin (1859, 1883).
5 Although Gregor Mendel had sent Darwin a copy of his treatise on heredity in 1866, it appears that Darwin failed to read it and he died without knowing how inherited variations were transmitted from parents to offspring. Late in his life, he admitted that 'the laws governing inheritance are for the most part unknown'. The word 'genetics' was not even invented until 1906, when William Bateson decided that a new term was needed to describe the emerging scientific study of heredity. 'I suggest ... the term Genetics, which sufficiently indicates that our labours are devoted to the elucidation of the phenomena of heredity and variation.' (quotation from William Bateson in his address to the Third Conference on Hybridisation, Cambridge, 1906).
6 Zohary (2004).
7 Zohary and Hopf (2000).
8 Murphy (2007).
9 The term 'variation' is used here in the sense of genetic, i.e. inherited, variation. There is another equally important source of variation that is caused by environmental effects. Farmers seek to minimise those environmental effects that are under their control, for example by trying to ensure that all parts of a field are equally watered and fertilised. But

some environmental effects, such as the weather, are more difficult to control. Even here, however, breeding can help by providing crop varieties that are genetically more resistant to environmental insults, be they biotic (e.g. pests and diseases) or abiotic (e.g. salinity and climate). For a more technical discussion of variation and crop selection, see Chapter 9 of Forbes and Watson (1992).

10 In this respect, breeding could be said to be just a special case of the process of evolution, as indeed it is. Darwin (1859) stated in his book *Origin of Species* that: 'The preservation of favourable variations and the rejection of injurious variations, I call Natural Selection ... ' The early events in crop domestication by humans were really an example of co-evolution between the two partners, to their mutual benefit. It is only in the past few centuries that people have extended the possibilities by creating new forms of variation and using more rapid and effective tools for selection, but all we are doing is speeding up a biological process that has been happening anyway for many millennia.

11 Weston (1645).

12 Ambrosoli (1997), pp. 317–318.

13 Examples of late seventeenth century 'improvers' are legion, but they include Thomas Locke in his *Treatise* (Locke, 1690), who implicitly advocated enclosure, John Worlidge's *Systema Agriculturae* (Worlidge, 1669), John Houghton's *Collection of Letters for the Improvement of Husbandry and Trade* (Houghton 1681–1683, 1692–1703), which is considered to be the first agricultural periodical, Richard Blome's *The Gentleman's Recreation* (Blome, 1686), Leonard Meager's *The Mystery of Husbandry* (Meager, 1697), and Timothy Nourse's *Campania Felix, or a Discourse of the Benefits and Improvements of Husbandry* (Nourse, 1700), which set out the new technologies and argued the need for enclosure if agricultural yields were to increase to the level perceived necessary for general economic growth (examples taken from Clark, 2004).

14 Rudolph Jacob Camerarius (1665–1721) was director of the Botanic Garden at Tübingen in Germany. In the late seventeenth century he showed that sexual reproduction occurred in plants and proposed the role of pollen in fertilisation (Camerarius, 1694).

15 Kölreuter (1761).

16 Early in the nineteenth century, John Lorain realised the possibility of growing maize as a hybrid crop, as quoted in his posthumous memoir (Fussel, 1992), although the same source notes that a US farmer called Cotton Mather had described the effects of maize hybridisation as early as 1716.

17 So-called 'terminator technology' is actually just one example of a wide range of the 'genetic use restriction technologies' (GURTs). For more detail of the science of GURTs, see Various (2004); for an assessment of the potential agronomic impact of GURTs, see Goeschl and Swanson (2003).

18 In 1828, Wiegmann described hybrid forms of pea, which were similar to the maternal or paternal plants, or combined the traits of the two (Wiegmann, 1828).

19 Gärtner (1849).

20 Naudin (1863).

21 Benzene is most frequently encountered as a volatile gas in diesel and gasoline fuels. In some European countries, gasoline formulations often contain 5% or more benzene. Mustard gas is a relatively simple but extremely toxic compound, di(chloroethyl) sulphide, that was used widely as a chemical weapon in World War I.

22 One especially powerful group of mutagens produced by normal cellular activity is the reactive oxygen species (ROS) that generate destructive hydroxy, peroxy and oxy radicals. These chemicals are an inevitable by-product of life in an oxygen-rich environment and can cause great damage to DNA and proteins in living cells, leading to mutations and certain cancers. The normal protective mechanism of plants and animals is to produce antioxidants

in order to neutralise the ROS before they can damage the cell. Hence the desirability of including antioxidant-rich foods, such as fruits and vegetables, in the diet. In both plants and animals, the ageing process is associated with a gradual decline in the ability to defend against ROS attack which in turn leads to a corresponding increase in the rate of somatic mutations and in the incidence of cancers of various kinds.

23 The recent depletion of the stratospheric ozone layer and the corresponding increase in ultraviolet radiation, especially in the Southern Hemisphere, has led to concerns that the mutation rate of crops may increase and possibly adversely affect their yields.

24 De Vries (1901).

25 Nowadays we would also add in the increasingly important group of epigenetic factors that are known to regulate key aspects of plant development and response to the environment, including many agriculturally relevant processes such as pathogen resistance, organ formation and polyploidy. For up-to-date accounts of this rapidly emerging area, see the brief review article by Bender (2002), the more detailed reviews by Grant-Downton and Dickinson (2005, 2006), or the comprehensive book *Plant Epigenetics* by Meyer (2005).

26 Walsh (2001).

27 Fisher (1918).

28 The two other people who contributed to the so-called 'modern synthesis' of evolutionary genetics were Sewall Wright in Chicago and J. B. S. Haldane in Oxford, Cambridge and London, but Fisher was the only one of the three who focused on plant genetics and breeding.

29 For a modern account of the practical application of quantitative genetics to plant breeding see Kang (2002).

30 This argument is made by Walsh (2001).

31 Cook (1937).

32 Shull's research unit at the Station for Experimental Evolution at Cold Spring Harbor, New York was supported by the philanthropy of the Carnegie Corporation. East was at the publicly funded Connecticut State Agricultural Experiment Station in New Haven.

33 Cook (1937).

34 Wallace published several well-regarded books on plant breeding, including Wallace and Bressman (1949) and Wallace and Brown (1956).

35 Crabb (1947).

36 Pioneer Hi-Bred remained as an independent company until 1999, when it was purchased by DuPont in a deal that valued the business at $9.6 billion.

37 Duvick (1999).

38 These arguments have been discussed by Simmonds (1979), Lewontin (1982) and Kloppenburg (1988), and are also presented here in Chapter 8.

39 Virmani (1994).

40 In the three decades following the initial 1974 breakthrough, planting of the new hybrid rice crop has spread so widely that now almost half of China's rice production area is planted in hybrid varieties that average a 20% higher yield than previous varieties. This translates into food to feed an additional 60 million people per year in China alone. Yuan shared his knowledge and technology with overseas scientists, providing them with crucial breeding materials for the commercial production of hybrid rice in their respective countries. By 2005, farmers in more than ten other countries besides China, including the USA, had benefited from Yuan's work. (See full citation for Yuan Longping online at: http://www.worldfoodprize.org/Laureates/04laureates/yuan.htm.)

41 Doggett (1988).

42 Duvick (1999).

43 The article by Pickett (1993) is a very comprehensive review of both early and more recent research on wheat hybridisation. A more up-to-date account of developments in hybrid wheat technology is in Cisar and Cooper (2004).

44 During the late 1990s, there was a great deal of controversy about the possible introduction of various sterile-seed (usually called 'terminator') technologies in transgenic crops. According to anti-GM campaigners this would deny poor farmers the chance of saving their seed and force them to repurchase seed from the biotech company year after year. In the end, no such GM crops were produced. However, as we have seen here, non-GM hybrid crops, which have exactly the same effect on seed saving by farmers, are now spreading across the world with little or no fuss.

45 The first wheat/rye cross is considered to have occurred in Scotland in 1875. Several publications outline the historical progress of triticale in the twentieth century: Stoskopf (1985); Vietmeyer (1989); Villareal *et al.* (1990).

46 Muntzing (1979).

47 Karpenstein-Machan and Heyn (1992).

48 According to Fedoroff and Brown (2004), the following information was present on the label for triticale flour from Bob's Red Mill Natural Foods in Milwaukie, Oregon:

'Triticale is a hybrid grain, a cross between wheat and rye. It averages 28 higher protein than wheat and contains all the essential amino acids, thus making it a more complete protein than the parent grains. It has an interesting nutty flavor and is high in fiber. ... Since Triticale Flour has inherited the best qualities of its parental grains, wheat and rye, and comes with a delicious flavor all its own, it needs to be discovered. Right now ... by you ... '

49 For a useful overview of the main techniques used in induced mutation, see Maluszynski *et al.* (2003).

50 Gager and Blakeslee (1927); Stadler (1928). Earlier workers, such as Gager in the USA and Pirovano had also experimented with various forms of radiation mutagenesis in plants, as detailed in the comprehensive text on mutation breeding by van Harten (1997).

51 In the 1940s, and despite the war then raging across Europe and beyond, Laibach's group in Germany developed *Arabidopsis* as a genetic model that could potentially serve as the *Drosophila* of the plant world (Laibach, 1943). A flavour of the challenges confronting these workers is given by the story of one of Laibach's students, called Erna Reinholz. Her project involved developing X-ray mutagenesis to create additional variation in *Arabidopsis*. The manuscript of her work was seized by the occupying forces of the US military in 1945. They were looking for anything that could possibly relate to the attempted development of nuclear weapons in Germany and noticed that her work involved the use of radiation. Luckily for plant science, the Reinholz Ph.D. thesis was published in 1947 by the Joint Intelligence Objectives Agency of the Occupying Powers, who cryptically described it as 'an unclassified captured document', doubtless not realising the biological significance of her work (Reinholz, 1947).

52 The FAO/IAEA Agriculture and Biotechnology Laboratory at Seibersdorf in Austria offers a free-of-charge service for the acute irradiation of plant materials (see the FAO/IAEA website at: http://www-infocris.iaea.org).

53 As of mid-2005, the FAO/IAEA mutant varieties database lists over 3400 crop varieties that include rice, wheat, maize, millet, barley, beans, brassicas, lettuce, soybean, sorghum, pepper, chickpea, oats, linseed and potato. The two countries with the largest mutation breeding programmes are China and India, but dozens of other developing countries are also using the technology. For more information, including the up-to-date list of crop varieties, see the FAO/IAEA mutation-breeding database (Maluszynski *et al.*, 2000), available online at: http://www-infocris.iaea.org/MVD/.

54 For recent reviews of mutation-assisted plant breeding, see Mohan Jain *et al.* (1998) and Mohan Jain (2005).

55 For an up-to-date account of the uses of physical and chemical mutagenesis in both research and breeding, see Kodym and Afza (2003).

56 Durrant (1962).

57 Sala and Labra (2003) provide a useful account of the practical use of somaclonal mutagenesis in crop breeding.

58 Spontaneous abortion is especially important in the larger mammals, such as humans, where relatively few offspring are produced and there is a high level of maternal investment *post partum*. It is estimated that, in humans, 35–50% of successful fertilisations are aborted for various reasons, either at an embryonic stage or an early foetal stage (Wang *et al.* 2003b).

59 In recent years, it has become possible to culture fertilised ovules in vitro and hence to prevent the normal spontaneous abortion of a hybrid embryo resulting from a cross between very dissimilar parental species (Chi, 2003).

60 Barclay (2004).

61 This threat to potentially useful wild relatives of crops has already been mentioned in Chapter 2 in relation to the wheat family, which has dozens of related species and genetically distinct subpopulations spread across Central and Western Asia. The situation of the major Asian crops may be even more serious, with losses of irreplaceable germplasm accelerating under pressure of population growth and industrial development. Gurdev Khush, former principal breeder at IRRI, and a developer of wide crosses of rice and 1996 World Food Prize laureate, has described the wild relatives as 'truly priceless seeds' (Barclay, 2004). Hence the International Rice Genebank at IRRI can be regarded as a public treasure chest of inestimable value, especially for future generations.

62 For an overview of the many new technologies related to plant tissue culture, a useful resource is Volume III of the *Encyclopedia of Applied Plant Sciences* (Thomas *et al.*, 2003) in which pages 1341–1431 are devoted to the techniques of modern tissue culture and their application in plant breeding programmes.

63 Phillips (1993); Forster *et al.* (2000).

64 Examples of such agents include oryzalin, trifluralin, amiprophosmethyl and N_2O gas as discussed in Bouvier *et al.* (1994), van Tuyl *et al.* (1992), and Taylor *et al.* (1976).

65 The use here of emotive terms like 'toxic' and 'natural' is of course deliberate. For a discussion of the use of language in the discourse on agbiotech, see Cook (2004).

66 Some popular nut crops come in varied forms that can be derived from several different species, while others come from just one species, as follows. Walnut: there are several edible nut-bearing members of the walnut genus, *Juglans*, of which the Persian walnut, *J. regia*, is the major commercial species. Black walnuts, such as *J. nigra* and *J. californica* are popular in the USA, as is the butternut, *J. cineria*. Macadamia: there are two edible species of macadamia (*Macadamia integrifolia* and *Macadamia tetraphylla*). Almond: *Prunus dulcis*. Pecan: this tree is closely related to the walnuts and was formerly called *Juglans illinoensis*, although it is now known as *Carya illinoensis*. Hazel: *Corylus avellana*. Brazil nut: *Bertholletia excelsa*. Pistachio: *Pistacia vera*. Note that, although only a few dozen nut tree crops are grown commercially, there are thousands of wild tree species that bear edible nuts, some of which many well be nutritionally superior to the popular commercial crops.

67 Miscanthus is a rapidly growing bushy crop harvested biennially for its wood. Automated micropropagation systems have been developed for rapid production of millions of new plants (Lewandowski, 1997). Miscanthus is one of the new, so-called 'biomass crops' being grown as alternative, renewable sources of fuel for power stations (O'Connor, 2003). While the idea may be good in principle, the development of such biomass crops requires massive public subsidy and there is scepticism about their economic viability (Murphy, 2003b).

68 Hayashi *et al.* (1992).
69 Corley (2000).
70 Corley and Tinker (2003).
71 Wong *et al.* (1997).
72 Blakesley and Marks (2003).
73 Arcioni and Pupilli (2004).
74 Carlson *et al.* (1972).
75 The culture and regeneration of plants in vitro required their exposure to a variety of hormones and synthetic growth regulators. The optimal growth and regeneration conditions may vary a great deal between species and even between individual varieties of a crop. To a great extent, plant tissue culture remains something of a 'black art' where the individual scientist may need to adapt conditions and recipes to suit local circumstances.
76 In 1978, cells of *Vicia faba* were fused with *Petunia hybrida* to create a hybrid plant (Binding and Nehls, 1978), and in 1979, the *Arabidobrassisa* was created by fusing Arabidopsis with Brassica cells (Gleba and Hoffman, 1979).
77 In 1976, an inter-kingdom fusion of human and tobacco cells was reported, although this was done more out of curiosity than for any practical purpose (Jones *et al.*, 1976).
78 For example, Mendes-da-Glória *et al.* (2000) for limes; Latado *et al.* (2002) for lemons; and Scarano *et al.* (2002) for a range of citrus species.
79 The quotation is from the July 2005 version of the company website: http://www.greentec-gmbh.com/e03soma.html.
80 Guha and Maheshwari (1964, 1966).
81 Kasha *et al.* (2001).
82 Ahloowalia *et al.* (2004).
83 Singh (1995).
84 The book by Burdick (2005) is a comprehensive account of invasive species and their consequences.
85 DEFRA (2003, pp. 13 and 85).
86 The topic of exotic and invasive species is explored at length in the books by Cox (2004) and Pimentel (2002). *Rhododendron ponticum* is an especially persistent plant in the UK where its high andromedotoxin production makes it unpalatable to most herbivores. Giant hogweed exudes painful blistering agents that can cause skin damage leading to infection, especially in the presence of sunlight.
87 Lee (1988).
88 A recent review of biolistic transformation is Kikkert *et al.* (2005).
89 See Chilton (1988) for an early account of *Agrobacterium* technology and Chilton (2001) for a more recent retrospective view.
90 This genetically modified variety, also called the 'lipstick strawberry' is not officially classified as transgenic, or 'GM', because even though it contains genes from a different species, it was produced by so-called 'conventional' breeding.
91 The Europeans were from the University of Ghent in Belgium and the Max Planck Institute in Köln, Germany (Caplan *et al.*, 1983); the US groups were based at Washington University (Barton *et al.*, 1983) and at the Monsanto Company, both in St Louis, Missouri.
92 The tomatoes were launched using some poorly performing breeding lines, and there was also reportedly significant commercial mismanagement that together blighted the prospects of this novel crop. The somewhat sorry, and at times farcical, saga of the Flavr Savr® tomato is well told in Belinda Martineau's book *First Fruit* (Martineau, 2001).
93 Many of the vegetable varieties, contain a more desirable range of glucosinolates that cause the slightly sharp taste that is common to all *Brassicas* from broccoli to Brussels sprouts.

This characteristic *Brassica* flavour is not exactly to everybody's taste, but is responsible for some of the most important positive nutritional qualities of these often underappreciated vegetables. See Mithen (2001) for a general review of this topic and Mithen *et al.* (2003) for a more technical account of progress in breeding anti-carcinogenic varieties of broccoli. Although definitive proof has yet to be published, the National Cancer Institute in the USA currently advises that broccoli, along with its cruciferous family members, may be important in the prevention of some types of cancer (http://www.nci.nih.gov/).

94 In the 1980s, the European Union ruled that all rape seeds grown for edible use could not exceed a maximum erucic acid content of 2% total seed fatty acids and that seed used for meal could not exceed a glucosinolate content of 25 μmol/g seed. The legislation also stipulated that batches of seeds could only be tested by one of the approved analytical methods, such as X-ray fluorescence spectroscopy (XRF).

95 Oleic acid is named after the olive, *Olea europaea*, and this monounsaturated fatty acid has been shown to reduce serum cholesterol levels in humans. Olive oil is one of the key components of the so-called Mediterranean diet that is associated with lower levels of cardiovascular disease than typical dairy- and meat-rich Western diets that contain more saturated fat and fewer vegetables. Ironically, although modern versions of rape oil have as much oleic acid as olive oil, rape oil tends to be used as a cheap commodity oil and is rarely promoted as the highly nutritional food that it undoubtedly is.

96 Oilseed rape is an allotetraploid species with a genome about six times larger than that of *Arabidopsis thaliana*, which is known to have about 25 000 genes (Murphy, 1998). Therefore there are about 150 000 potentially mutable genes in rape of which only a handful are involved in regulating erucic acid accumulation in the seed oil. Spontaneous mutations are only produced at the rate of one mutation per million cell divisions. Given these enormous odds, Downey's search for a zero-erucic plant (Downey and Craig, 1964) would have been virtually impossible without the dramatic advances in analytical technology, such as gas–liquid chromatography, that improved his chances by almost nine orders of magnitude.

97 All of the edible oilseed rape grown in Europe is of the high-oleic type. In the UK, rapeseed is the third most important national crop, comprising about 10% of total arable cultivation. Rapeseed oil is the major vegetable oil and margarine component and is present in about 50% of all processed food products in a typical UK supermarket.

98 Total sales from the agbiotech industry (seed sales plus technology fees) are estimated at $24 billion from 1996–2004. In 2004, the annual sales of agbiotech products were estimated at $5 billion (James, 2006). However, all of these sales are simply substitutions for non-GM seeds so no new crops have been produced. Hence, GM soy or maize varieties simply replace other non-GM varieties of the same crops; no new crops or crop products are created. In contrast, the canola varieties of oilseed rape, with an annual value of $6 billion, represent a completely new crop with new markets that did not exist before the 1960s.

99 Note, however, that as a sop to public sensibilities, the word 'nuclear' (with its connotations of 'radiation') is usually omitted and the technique is referred to as magnetic resonance imaging (MRI) when it is used in a medical context.

100 Gill *et al.* (1985).

101 For example, in a recent review of the use of DNA markers in oilseed rape breeding, the following agronomic traits have been identified from such markers: resistance to five major diseases, seed fatty acid content, glucosinolate content, cold tolerance, flowering time, and plant height (Snowdon and Friedt, 2004).

102 For a recent account of the use of marker-assisted selection in the breeding of cereal crops, see the article by Goff and Salmeron (2004).

103 The increasingly favourable economics of marker-assisted selection relative to conventional phenotypic selection has been described for wheat breeding by Kuchel *et al*. (2005) who estimate that marker-assisted selection may result in an overall cost saving of 40% in addition to improved genetic gains.

104 Koebner and Summers (2003).

105 Varshney *et al*. (2005).

106 Dreher *et al*. (2003); and Morris *et al*. (2003).

107 Young (1999).

108 Koebner and Summers (2003), p. 63.

109 Some of the advanced journals in this area, all of which have abstracts available online, include *Molecular Breeding, Plant Breeding, Molecular Plant–Microbe Interactions, Transgenic Research, In Vitro Cellular and Developmental Biology: Plant* etc.

110 The technology of TILLING was invented by graduate student Claire McCallum at the Fred Hutchinson Cancer Research Center, Seattle, Washington, USA, following her frustration with attempts to generate mutants by using transgenic knockout methods (McCallum *et al*., 2000a, 2000b). This was very much a basic research project based on an interdisciplinary collaboration between institutes involved in medical and botanical studies including the Howard Hughes Medical Institute and the Botany Department at the University of Washington.

111 Slade *et al*. (2005) describe the application of TILLING to the identification of an allelic series of variants in the granule-bound starch synthase I (GBSSI) gene in hexaploid and tetraploid wheat. GBSSI or Waxy plays a critical role in the synthesis of amylose, which, in addition to amylopectin, comprises the starch fraction of the seed. Reduction or loss of GBSSI function results in starch with a decreased or absent amylose fraction, which is desired for its improved freeze–thaw stability and resistance to staling compared to conventional starch (taken from Moehs, 2005).

112 See Arcadia Biosciences website at: http://www.arcadiabio.com/.

113 Nielsen and Richie (2005).

114 Some of the many potential applications of TILLING are described in the review articles by Slade and Knauf (2005) and Weil (2005).

115 Dedijer (1963).

116 In this analysis, I see no useful purpose in dividing science and scientists into so-called fundamental and applied categories. In this respect I would disagree with colleagues who seek to separate basic science, i.e. the creation of knowledge, from applied science, e.g. the invention of new or improved technologies. For example, Lewis Wolpert argues that: 'technology is not science' (The phrase is the title of Chapter 2 in his book *The Unnatural Nature of Science*, Wolpert, 1993). Wolpert's notion may have once been true, but things are not nearly so simple nowadays. For example, the ancient, empirically based technologies, from fermentation and metallurgy to plant breeding itself, were not informed by any meaningful scientific knowledge until the last few centuries, but after the seventeenth century they became ever more based on reliable scientific knowledge. This remains a gradual process and is still incomplete in the case of plant breeding. Here, as many eminent and successful breeders would readily aver, there are still many intangible factors that inform what they still describe as their 'art'. In the recent words of Donald Duvick, former Senior Vice-President at Pioneer Hi-Bred: 'Breeders universally depend on experience and art more than genetics' (Duvick 1996).

117 States have always commissioned projects that demand the use of new technology and often such a demand can stimulate technological innovation, e.g. the demand for specific types of ships by navies led to greatly improved designs. However, the actual innovators were normally private individuals, possibly acting under contract to the state.

118 State-sponsored crop improvement dates back to the earliest historical records. For example, the Assyrian king Tiglath-pileser I (1114–1076 BCE) founded a botanical garden at Nineveh, in which he planted specimens of various useful plants that he had gathered during his numerous military campaigns. The tradition continued with the imperial botanisers of the Dutch and British maritime empires, and culminated in the extensive series of US government initiatives, most notably the Morill act, during the nineteenth century.

119 'Turnip' Townshend and Thomas Coke were wealthy agricultural improvers in eighteenth century Norfolk.

120 Tusser (1560), chapter LII.

121 To quote from Mark Overton's book on the agricultural revolution in England: ' ... in the late seventeenth century [in England], agricultural production was seen as an activity in which the individual husbandman worked for himself and his family on his own land. By the mid-eighteenth century the husbandman had become the farmer, and, instead of "husbanding" nature, was seen as an entrepreneur, calculating the costs and benefits of alternative courses of action.' Overton (1996).

122 Note that Overton's thesis has been strongly challenged by Robert Allen who asserts that the most rapid phase of agricultural growth occurred much earlier, in the period 1520–1651 (Allen, 1999). Suffice it to say for our purposes that there was a huge increase in crop productivity in England some time after the early sixteenth century and that this was informed by new ideas of entrepreneurship and new science-based knowledge.

123 This sort of appeal for applied scientific research might come from a government department related to agriculture, from a government-funded Academy or institution, or directly from government leaders, as was the case with the appeal from Napoleon III for research into a butter substitute in the 1860s. At this stage, most of the actual research done in response to such appeals would have been carried out by privately funded scientists. This was because scientific research had yet to emerge as a recognised profession beyond the cloistered halls of academia. By the nineteenth century, the offering of cash prizes for technological innovations was a well-established tradition. One of the best known examples in the UK was when, in 1714, Parliament offered the princely sum of £20000 (equivalent to millions of dollars today) to the first person to devise a reliable method to measure longitude. As recounted in the popular book by Dava Sobel, the prize was won by clockmaker, John Harrison, and his device (an accurate portable clock) went on to revolutionise navigation (Sobel, 1995).

124 Murphy (2005).

125 This is also the derivation of the popular name, Margaret.

126 The earliest forms of margarine were derived primarily from animal fat, such as kidney and beef fats. Various vegetable fats were then tried until the process of catalytic hydrogenation enabled almost any oil to be used in the manufacture of margarine (Wilson, 1954). However, the product was not destined to be a great commercial success for several decades. Two technical advances tipped the balance towards the use of plant fats in margarine and allowed it to become an effective competitor with butter. Firstly, improved refining methods allowed the purification of a greater variety of liquid oils and solid vegetable fats that could be blended to make a good spreadable margarine. Secondly, the process of hydrogenation, which was invented in 1901 by English chemist William Normann, allowed the large-scale conversion of relatively cheap plant oils into solid fats. Not only did the hydrogenation process produce a good, inexpensive butter substitute, it also significantly reduced the amount of oxidation-prone polyunsaturates in the solid margarine, which greatly extended its shelf life and therefore its utility to consumers (Murphy, 2005).

127 The value of globally traded vegetable oils is currently about $40–50 billion/year. Almost 83% of world production of traded plant oils is derived from just four major oil crops: soybean, oil palm, oilseed rape and sunflower (Oil World, 2005). About 85% of this 100 million tonnes of annual oil vegetable output is used in the production of margarines, cooking oils and processed foods.

128 Twain (1883).

129 Unlike butter, which is normally distinctly yellow in colour, most animal and vegetable fats are very pale, even white, in their native state. Because consumers are used to yellow fats for spreading, margarine is routinely dyed with added pigments. In the USA shoppers had to buy the dye separately and mix it with the white margarine themselves, which clearly reduced the appeal of margarine compared to butter. For more on the fascinating history of margarine, see the book by van Stuyvenberg (1969).

130 Surprisingly, the US-style protectionist hysteria against margarine is still alive and kicking in another supposed bastion of free enterprise, Canada. As recently as March 2005, the Canadian Supreme Court ruled that the Province of Quebec (a dairy industry stronghold) was justified in continuing its ban, dating from 1886, on yellow margarine (see news story at: http://www.cbc.ca/news/background/food/margarine.html). Interestingly, the major source of the yellow margarine that is so abhorrent to Quebeckers is canola oil from crops grown by their compatriots in the prairie provinces of Western Canada.

131 Klose (1950).

132 For a thorough description of Washington's deep interest in agriculture, see the book by Haworth (1915).

133 Letter from Washington to Arthur Young, dated December 5, 1791, as quoted by Haworth (1915).

134 Despite fighting against the British Crown forces for many years, and eventually winning the independence of the thirteen colonies, George Washington continued to regard himself as, culturally at least, an English gentleman. Like most of the Virginian elite, he had strong ties with the mother country and continued an extensive intercourse with his (often admiring) peers across the Atlantic. Naturally, there was a deep vein of pragmatism for the Americans in maintaining this dialogue. Early nineteenth century Britain was at the vanguard of its industrial revolution and had already, decades before, revolutionised its agriculture. There was much for Washington and his ilk to learn from such a society that could (and would) lead to their own considerable profit.

135 Many of the 'founding fathers' of American independence were keen agricultural improvers. Among the most notable was Thomas Jefferson who was a vigorous proponent of experimentation with new crops and new varieties. Witness the following quotation from a letter from Jefferson to William Drayton in 1786:

> We are probably far from possessing, as yet, all the articles of culture [crops] for which nature has fitted our country. To find out these, will require an abundance of unsuccessful experiments. But if, in a multitude of these, we make one or two useful acquisitions, it repays our trouble.

In another letter to M. Lasteyrie in 1808, Jefferson wrote:

> The introduction of new cultures crops, and especially of objects plants of leading importance to our comfort, is certainly worthy the attention of every government, and nothing short of the actual experiment should discourage an essay of which an hope can be entertained.

(In other words, test the new crop before assuming it has nothing to offer. Note also the plea for government involvement in such endeavours.) A more complete list of the writings of Jefferson is available in Peterson (1984). An extensive list of Jefferson quotations, many of them relating to agriculture, is available from the University of Virginia online at: http://etext.lib.virginia.edu/jefferson/quotations/.

136 The archetypical 'Yankee capitalist' was obviously a New England northerner, but apart from the very small slave owning 'plantocracy' of the Deep South, many southerners and westerners would have shared similar entrepreneurial capitalist values with the 'Yankee' of popular myth. Such was the case from the relatively industrialised regions of northern Virginia to the non-slave owning yeoman class of South Carolina, not to mention the growing numbers of footloose pioneers in the West. For an outline account of the development of capitalism in the antebellum United States, see Chase-Dunn (1980).

137 This was the supreme era of the unregulated entrepreneur in the USA. One of the best examples is the epic tale of swindling and corruption by such colourful characters as Charles Crocker, Leland Stanford and Thomas Durant, who built the first transcontinental railroad. Soon after construction of the railroad was finished at Promontory Summit, Utah on 10 May 1869, it was discovered that these roguish gentlemen had defrauded the public of millions of dollars (Martin, 1873; Lewis, 1938). Nevertheless, they continued to be respectable citizens, with Crocker setting up the major bank that still bears his name, while Stanford established his eponymous university, which is now acknowledged as one of the finest private universities in the world. A good flavour of the shenanigans of these entrepreneurs can be found in the recent popular account of the men who built the railroad by Ambrose (2001).

138 The US government has always played an important role in the conduct of business, from seizing new lands for settlement to providing a complex physical infrastructure for trade and commerce that still gives US farmers a competitive advantage over other countries in the twenty-first century. For example, the Federal government has constructed, at great public expense, a nationwide network of waterways and interstate highways for the efficient movement of goods, such as crops, to their markets. In contrast, the unsubsidised farmers of Brazil, who do not enjoy comparable state largesse, either have to build their own private highways or suffer the additional cost and delay of moving goods along unimproved dirt roads (Diaz, 2004).

139 Letter from Thomas Jefferson to M. de Warville (1786, ME 5:402).

140 Letter from Thomas Jefferson to Benjamin Austin (1816, ME 14:389).

141 A good example of the British use of the food weapon is their refusal to supply flour to the French government in 1789, despite increasingly desperate appeals following a series of crop failures. British Prime Minister William Pitt refused the French request although he had grain stocks in abundance. The famished French populace exploded into rebellion soon afterwards and within a month the Parisian *sans culottes* had stormed the Bastille. There is little doubt that, while it did not directly cause the French Revolution, the withholding of food shipments by the British served to accelerate this process.

142 The US politicians were absolutely correct in their appraisal of the potential power of food as a weapon of realpolitik. Thanks to new crop varieties and the use of new equipment and fertilizers, wheat yields increased seven-fold between 1850 and 1900. Better rail and steamship transportation opened new markets in America's growing cities and across the Atlantic to an industrialising Europe. By the beginning of the twentieth century, America began to come of age as a world agricultural power. Throughout the twentieth century and beyond, the US government continued to use its status as the 'breadbasket of the world' to favour allies (e.g. feeding the British during both World Wars) and to punish foes (e.g. the 1980–1981 grain embargo against the USSR), not to mention earning a very healthy profit from sales of the exported food. In the 1980s, US Secretary of Agriculture, Earl Butz even referred explicitly to what he called the 'food weapon' (quoted by Kloppenburg (1988), p. 49).

143 A good flavour of the essentially amateur nature of the study of plant biology in Europe in the mid nineteenth century can be gleaned from Darwin's book on plant and animal domestication. For example, in Chapter 9, Darwin refers to numerous English enthusiasts with whom he corresponded, such as Buckmann, Hooker and Bentham (Darwin, 1883; Buckmann, 1857). He also refers to many continental authorities, but nowhere is there any sense of an organised, systematic and professional approach to plant research on anything like the scale that was then emerging in the USA.

144 Lest I incur the wrath of the Darwinophiles, amongst whose ranks I would also care to be numbered, I am using dilettante here in its original sense of an amateur enthusiast who takes especial delight (Italian: *dilettare*) in his interests.

145 In the mid nineteenth century, the idea of a university-based research laboratory was only just beginning to emerge, largely thanks to the pioneering work of Justus von Liebig at the University of Giessen, and later Munich, in Germany (Holmes, 1973). Liebig inspired a generation of agricultural chemists, including Evan Pugh and John Bennett Lawes at Rothamsted in England.

146 As happened, for example, when Justus von Liebig and colleagues in Germany developed nitrate fertilisers and eventually invented a process for their chemical synthesis (von Liebig, 1840). This laid the foundations of the immensely successful and world-leading German chemical industry in the early twentieth century. One of the original spurs to Morrill and his colleagues was the publication in 1841 of the first American edition of Liebig's book: *Organic Chemistry and Its Applications in Agriculture and Physiology*. This book helped to stimulate the establishment of several local agricultural colleges in the years immediately preceding passage of the Morrill Act of 1862, e.g. Michigan, 1855, Iowa, 1858, Pennsylvania, 1859, New York, 1860 (Kloppenberg, 1988).

147 The reference to education in military tactics in this passage from the Act should be understood in the context of the hugely destructive civil war that was raging throughout the USA in 1862 and the ever present threat from European powers. Despite this stipulation, the Morrill Act was overwhelmingly agricultural in its focus and there was little, if any, stress on military studies at the Land Grant Universities.

148 Speaking before the Vermont Legislature in 1888, Senator Morrill said:

> Only the interest from the land-grant fund can be expended, and that must be expended, first – without excluding other scientific and classical studies – for teaching such branches of learning as are related to agriculture and the mechanic arts – the latter as absolutely as the former. Obviously not manual, but intellectual instruction was the paramount object. It was not provided that agricultural labor in the field should be practically taught, any more than that the mechanical trade of a carpenter or blacksmith should be taught. Secondly, it was a liberal education that was proposed. Classical studies were not to be excluded, and, therefore, must be included. The Act of 1862 proposed a system of broad education by colleges, not limited to a superficial and dwarfed training, such as might be supplied by a foreman of a workshop or by a foreman of an experimental farm (NASULGC, 1995).

149 In 1862, the USDA had also been established but the Department was not effectively involved in field studies of crops for several more decades.

150 Quoted in Busch and Lacy (1983), p. 9.

151 For a perspective on the future direction of Land Grant Universities, see Campbell (1996).

152 An account of the tradition of US Land Grant Institutions can be found in NASULGC (1995).

153 The US Navy had dispatched earlier plant-collecting expeditions, the first of which was the voyage around the Pacific Ocean of Commander Charles Wilkes from 1838–1842. Also, one of the results of the famous expedition to Japan by Commodore Perry in 1853, where his gunboats forcibly opened Japanese ports to Western trade, was a vast haul of seeds and plant materials that was taken back to the USA (Klose, 1950). Although the foreign

materials were disseminated, these earlier collecting ventures by the Navy were not as effective as the later USDA programme, which was run and staffed by professional botanists of the ilk of Carlton and Meyer.

154 The official USDA plant collection programme began with a single Congressional appropriation of just $20 000 in 1898 (Klose, 1950). The programme continues to this day, and is currently directed under the auspices of the National Germplasm Resources Laboratory of the Agricultural Research Service. Nowadays, such 'bioprospecting' activities are often regarded with suspicion by the host country and USDA therefore attempts to run its collecting expeditions as collaborative joint ventures with botanists from the host country (Kaplan, 1998). Unfortunately, in some countries, such as Brazil, fear of bioprospecting and associated 'biopiracy' have led to the imposition of restrictions that are now hampering genuine scientific work, even by local researchers (Astor, 2005).

155 Blouet and Luebke (1979).

156 Kaplan (1998).

157 For more on Vavilov's plant-collecting expeditions to America, as well as to much of the rest of the globe, see the biography by Popovsky (1984). Many American accessions collected by Vavilov were eventually housed in his purpose-built Research Institute of Plant Industry in Leningrad. During the 900-day siege of the city by the Germans in World War II, the dedicated staff of the institute protected these plants, many of which were edible, from the starving human population of the city, from hordes of voracious rats, and from their own hunger. In her recent novel, *Hunger*, Elise Blackwell explores this turbulent period for Soviet plant sciences that encompassed the persecution of Vavilov, the rise of the charlatan Lysenko, and the climactic siege of Leningrad (Blackwell, 2004). Vavilov also developed the 'centres of origin' concept that lies at the heart of our understanding of crop evolution (Vavilov, 1926, 1935, 1992).

158 Soybeans also used to be the major oil crop in the world but this ceased to be true in 2006, when the emerging new giant of the vegetable oil industry, oil palm, finally outstripped soybean in its annual output. Unlike soy, rape and sunflower, which are the other major oil crops, palm oil comes from its fleshy fruit, or mesocarp, tissue rather than the seed. Hence it is not regarded as an oilseed crop. Just to make things more confusing, however, the seeds (normally called kernels) of oil palm also produce a different type of oil, i.e. palm kernel oil, which is mainly used as a non-edible feedstock in the detergent and oleochemicals industries (Murphy, 2006).

159 In 2004, the US produced about 85 million tonnes of soybeans with a value of more than $16 billion. In comparison, maize production was 269 million tonnes with a value of $28 billion, while wheat production was 60 million tonnes and was valued at $9 billion (USDA website).

160 For a description of the contemporary relationship between the USDA research centres (USDA–ARS) and Land Grant Universities, see the article by Lamkey (2003). Some idea of the extensive network of nationwide ARC research facilities in the USA can be gleaned from the interactive online map at: http://www.ars.usda.gov/pandp/locations.htm.

161 Scott (1962), p. 14.

162 Quoted in Jones (1957), p. 67.

163 For a snapshot of the major trends in US agriculture over the past century, the USDA National Agricultural Statistics Service has an interesting and informative website at: http://www.usda.gov/nass/pubs/trends/introduction.htm.

164 Sir William Bate Hardy was an eminent physiologist and cell biologist who first suggested use of the term 'hormone'.

165 The USA became a pioneer in public sector crop improvement early in the nineteenth century. Indeed, as early as 1799, George Washington had suggested to Congress the

establishment of a National Board of Agriculture. In 1819, New York State legislature set up a State Board of Agriculture, the first organisation of this sort to be found anywhere in the world. But it was not until the latter part of the nineteenth century and early twentieth century that a truly systematic national organisation was put in place.

166 To get a flavour of this complexity, one need do no more than visit the websites of three of the overlapping plant breeding organisations still at Wageningen, namely: the Laboratory of Plant Breeding (formerly the Department of Plant Breeding) of the Agricultural University, Wageningen (http://www.dpw.wau.nl/pv/index.htm); Applied Plant Research – Praktijkonderzoek Plant & Omgeving (http://www.ppo.dlo.nl/ppo/Eng/index_engels.htm); and Plant Research International (http://www.plant.wageningen-ur.nl/default.asp?section=about). As with the seemingly shambolic system in the UK, this apparent lack of a bureaucratically pleasing organisation in the Netherlands has certainly not prevented the country from maintaining its place in the front rank of international plant breeding. For example, the Netherlands still ranks second in the world – after the USA – as an exporter of seed materials.

167 For a more detailed account of the development of plant breeding research in the UK from 1900–1970, see Palladino (1990, 1996, 2002).

168 This was part of a long-established British tradition of encouraging innovation without actually funding it. Hence, King Charles II was pleased to extend his approval to the newly formed Royal Society in 1660, but not to provide it with any financial aid. In contrast, the French equivalent, the Académie des Sciences, was set up as a government institution under the patronage of Louis XIV in 1666. Its greater security and more generous funding, compared to the Royal Society, allowed the Académie des Sciences to build its own laboratories and to provide generous financial support to overseas scientists of the stature of Huygens. The first German Academy was the Prussian Akademie der Wissenschaften, established in Berlin on the French model in 1700. The National Academy of Sciences (NAS) of the United States was a relative latecomer: it was signed into being by President Abraham Lincoln in 1863 and was funded by the US government.

169 Goodman *et al.* (1987).

170 One example of such innovative research was the investigations of the organisation of DNA in wheat and other cereals at the Plant Breeding Institute in Cambridge that started in the 1970s. The research at Cambridge was one of the first steps in using the new science of molecular genetics and DNA manipulation for the benefit of crop improvement.

171 Newman (1852). John Henry Newman was one of the most noted English academics and theologians of the nineteenth century, converting to Catholicism in 1854 and becoming a cardinal in 1879.

172 Mill (1867). John Stuart Mill was a noted philosopher and proponent of utilitarianism. Although one might expect such a person to favour pragmatism and the encouragement of technical innovation, he remained very much a creature of his time and place in his elitist attitude to the role of universities. The quotation is from the inaugural address upon his appointment to the prestigious post of Rector of St Andrews University in Scotland.

173 In today's terms, the bequest to establish the Treforest Mining College would be worth close to $150 million. At that time, early in the twentieth century, the average Welsh miner received a wage of about 20 shillings a week, or £50 a year.

174 A useful text on university–industry relationships over the period from 1850–1970 is Sanderson (1972).

175 Palladino (2002), pp. 86–90.

176 Fedoroff and Brown (2004).

177 As stated on the Rothamsted website (in March 2005) 'Rothamsted Research is a charity and a company limited by guarantee and occupies land and buildings owned by the Lawes

Agricultural Trust. Research at Rothamsted Research is funded from a variety of sources which represented a total income of over £27M in 2002' (http://www.rothamsted.ac.uk).

178 In the 1930s William Lawrence and John Newell, two scientists at the John Innes Horticultural Institute, set out to formulate composts that would give consistently good and reliable results. Their particular motivation was the difficulty they were having in growing Chinese Primrose (*Primula sinensis*) for experimental purposes. Their objective was to obtain much better and more reliable germination and growth among their experimental materials by standardising growing conditions, including the growing medium. Following a series of experiments, they established methods of heat-sterilising the compost to destroy pests and diseases without affecting the development of plants grown in the medium. They also determined the physical and nutritional qualities needed in compost to achieve optimum plant growth. Lawrence and Newell also took into account the need to alter the nutritional status of the compost according to the plant's growth stage. Their research led to the introduction of the two standard composts, one for seed sowing and one for potting, which revolutionised the growing of pot plants (source, John Innes Centre website, http://www.jic.bbsrc.ac.uk).

179 According to the John Innes Centre website, which lists the 2000–2001 corporate accounts, this non-public funding is derived from private charities (14%), industrial companies (8%) and 'others' (7%) (http://www.jic.bbsrc.ac.uk/corporate/About_JIC/corporate_accounts.html).

180 Scotland had to wait until 1921 before the Scottish Plant Breeding Station was founded in Edinburgh. In 1981, the Station was amalgamated with the Scottish Horticultural Research Institute to form the Scottish Crop Research Institute, and the combined centres are now located near Dundee.

181 Palladino (1990).

182 Hides and Humphreys (2000).

183 Palladino (2002), pp. 50–54.

184 During the late-Neolithic period of 4400–3800 BP, there was a large movement of people to upland areas of Britain, which were intensively farmed for over a millennium before their abandonment in about 3000 BP. The causes of this desertion of upland farming across Britain remain unclear, but may be related to climatic changes that rendered it impractical (Simmons and Tooley, 1981). During the late medieval climatic deterioration of the 'little ice age' there was a similar abandonment of arable farming in many areas of Northern Europe. The little-known disaster of the little ice age had a drastic impact on the agrarian economies of the countries of Northern Europe; the combined impact of this climatic shock and the Black Death reduced the population of late fourteenth century Norway by over two thirds. The population of Iceland was more than halved as wheat cultivation became increasingly difficult in the cooler climate, and the Norse settlements on Western Greenland perished altogether (Marcus, 1980).

185 The Station had been funded by various agencies but was increasingly reliant on funds from the UK Agricultural ministry. In 1987, the Station finally became an AFRC Station.

186 The quote is from Palladino (1990). See also Anonymous (1891) and Sanderson (1972).

187 Palladino (2002), pp. 43–46, 73–77.

188 PBI severed its formal links with Cambridge University in 1948, but did not move from the University Farm to its own premises at Trumpington until 1955.

189 Palladino (1996).

190 Many of these issues are explored by Tom Wilkie in his book on postwar British science and politics (Wilkie, 1991).

191 Nehru (1961).

192 Kloppenburg (1988, 2004).

193 For example, in 1905, the Indian Agricultural Research Institute was established at Pusa in Bihar (moving to its present location in New Delhi in 1936 after a devastating earthquake hit Pusa), together with a number of Agricultural Colleges around the country.

194 Kloppenburg (1988), p. 158.

195 In the early sixteenth century, Diego Columbus (the brother of Christopher) noted in his journal that he had walked for almost 30 kilometres through a single and seemingly never-ending field of maize, bean and squash (Visser, 1986).

196 For an account of the causes of the Dust Bowl, see the book by Geoff Cunfer (Cunfer, 2002) or the online article at: http://www.eh.net/encyclopedia/article/Cunfer.DustBowl.

197 Mexico is a particularly fertile and biodiverse country that is a major centre of origin for some of the world's most important crops, including maize, squash, *Cucurbita pepo*, and common beans, *Phaseolus vulgaris*. Mexican agriculture had once sustained the mighty Aztec and Mayan civilisations, and people such as Henry Wallace felt that its fertile soils should be able to produce even greater crop yields than the 'breadbasket' that was the US Midwest.

198 The quotation is from Borlaug's presentation speech given when he received the 1970 Nobel Peace Prize for his work on crop improvement (Borlaug, 1970).

199 The IRRI research headquarters has laboratories and training facilities on a 252-hectare experimental farm on the main campus of the University of the Philippines Los Baños, about 60 kilometres south of the Philippine capital, Manila.

200 More than half of all EMBRAPA staff have Ph.D. qualifications (see website at: http://www21.sede.embrapa.br/).

201 Just two examples of such centres set up during the postwar period are the Palm Oil and Rubber Research Institutes in Malaysia, but there are dozens of others around the world.

202 Quoted from AgBioWorld, see online at: http://www.agbioworld. org/biotech_info/topics/borlaug/mexico.html.

203 Borlaug (1970).

204 Scientific accounts of Borlaug's research on the wheat stem rust disease can be found in Borlaug (1954, 1958).

205 Note, however, in certain countries with particularly intensive input regimes and liberal public subsidies, such as the UK (8 tonnes per hectare) and Ireland (10 tonnes per hectare), even higher average yields are regularly achieved.

206 Evanson and Gollin (2003a).

207 In Borlaug's case, he was able to grow two spring wheat crops a year using different climatic zones in Mexico. We are now able to use shuttle breeding for winter-sown crops as well by growing the crops alternately in the Northern and Southern hemispheres.

208 Ehrlich (1968).

209 The year of 1968 was characterised by an immense intellectual and artistic ferment. This was the year of the Woodstock Festival, the Prague Spring, and the Paris *evenements de Mai*. It was the year of the Youth Revolution and the original 'Summer of Love'. Many of those who were late-teenagers during this period experienced a profound sense of hope and confident optimism for the future. It was therefore shocking and disturbing to read in Ehrlich's decidedly threnodic book that much of humanity apparently faced mass starvation, while we students in the privileged West experimented so earnestly with our long hair and loud music. Ironically, as later decades showed, although most of our over-optimistic aspirations for a better society proved naïvely groundless, on the plus side, so did virtually all the doom-laden prophesies of Ehrlich *et al*.

210 Since Ehrlich's dire predictions in 1968, India's population has more than doubled, but its wheat production has more than tripled, and its gross national product has grown nine-fold.

211 Ehrlich discreetly omitted his predictions about mass famine in India and its lack of ability to feed its people from later editions of *The Population Bomb*. Despite being on the verge of overtaking China with a population well in excess of one billion, food production in India has more than kept pace with its population growth so that the country now enjoys a level of food security that it has not experienced for hundreds of years.

212 As early as August 1968, efforts led by M. S. Swaminathan to breed the new Mexican wheat in conjunction with local varieties looked so promising that the Indian government issued a special commemorative stamp, entitled 'Wheat Revolution' (Food and Agriculture Organisation, 2004a). Evidently, the late 1960s doomsayers such as Paul Ehrlich were either unaware of these developments or did not comprehend their momentous import.

213 India's wheat harvest has continued to improve since the 1960s and, in 2003, production was a record 76 million tonnes.

214 Khush *et al.* (2001).

215 For a recent review of the molecular genetics of crop dwarfing phenotypes, see Hedden (2002).

216 Jain and Kharkwal (2004).

217 The success of the Green Revolution was largely due to genetically improved plant varieties but it was also facilitated by three additional factors: the development of irrigation systems, better use of inorganic fertilisers, and reforms in government regulation and management that facilitated market growth (Khush, 2005).

218 There is no Nobel Prize category for agriculture or plant science, so achievements in these fields are normally placed in such incongruous categories as chemistry, medicine or peace.

219 Borlaug said of his Nobel Prize:

'It was a disaster as far as I'm concerned. You get pushed into so many things. A lot of your energies are cut off from the things you know best. Some of them you have to do. Because you end up being the spokesman for science in general' (Tarrant, 2001).

220 The widely accepted conclusion that the Green Revolution only benefited wealthier farmers has been challenged by several lines of evidence, including studies that imply a scale-neutral diffusion of the technology to farmers (Fischer and Cordova, 1998) and the beneficial effects of cheaper food, especially for the poor (Osmani, 1998). A useful analysis of some of the negative aspects of the Green Revolution, as recognised with the immense benefit of hindsight, can be found in the article by Conway (2003). See also Davies (2003) for an historical perspective on the Green Revolution and its links with today's so-called 'gene revolution'.

221 The single major exception to the beneficial diffusion of Green Revolution improvements in crop yields has been in sub-Saharan Africa, where very little progress was made compared to Asia and Latin America (Lappé, 1998; Pinstrup-Andersen and Schiøler, 2000). Part of the reason for this may have been the over-hasty introduction of high-yield varieties from elsewhere in the world without crossing them with more suitable, locally adapted varieties. Hence the rush for a 'quick fix' ended up delaying the availability of improved varieties for many African farmers (Evanson and Gollin, 2003a, 2003b). Other reasons are discussed by Conway (2003).

222 For example, the immediate cause of the well-known Ethiopian famine of 1984–1985 was drought and crop failure in the north of the country. However, food was available elsewhere in the country and beyond. The government of Ethiopia, which was involved in a civil war in several regions, including the north, deliberately withheld food supplies available in the centre of the country. Meanwhile, the transport of other supplies from relief agencies, that were being shipped in from the coast, was disrupted by warfare in Eritrea. About eight million people suffered food shortages and an estimated one million

died but food was available; it was just not delivered in time to those in need. Nearly all modern famines or instances of severe food shortage are caused by similar human acts of commission and/or omission.

223 The world population has increased by 2.2-fold since 1950, but what is less well appreciated is that food production increased by 3.1-fold over the same period (data from the UN Food and Agriculture Organisation (www.fao.org)). So we now produce 40% more food per capita than in 1950 and most of the world's population is now consuming significantly more food per capita than previously. This does not mean that we are all necessarily eating 40% more calories than in 1950, although given the current global incidence of obesity some people may be doing so. A large proportion of the additional crop calories is being fed to animals as economically better-off populations around the world, e.g. in China and South East Asia, continue to increase their intake of meat.

224 Pingali and Traxler (2002); Raney and Pingali (2004).

225 Evanson and Gollin (2003a).

226 Most of these consultations were held at Bellagio in Italy, and all of them came to be known collectively as the Bellagio Conferences. An idea of the seriousness of this process can be gleaned by looking at some of the international policymakers who contributed to the Bellagio Conferences. They included such senior figures as Addeke Boerma (FAO), Sir John Crawford (Australia), John Hannah (USAID), George Harrar (Rockefeller Foundation), Forrest 'Frosty' Hill (Ford Foundation), Paul Hoffman (UNDP), David Hopper (Canada/IDRC), Robert McNamara (World Bank), Maurice Strong (Canada/CIDA) and Sir Geoffrey Wilson (UK).

227 For a list of the original donors to CGIAR, see the World Bank website at: http://www.worldbank.org/html/cgiar/publications/founding.html.

228 In the words of Forrest Hill, a CGIAR pioneer, the founding objective of the organisation was to 'increase the pile of rice' (http://www.cgiar.org/who/history/).

229 This nationally televised speech entitled 'A time for choosing' was given in support of the presidential candidature of Barry Goldwater. The speech is a powerful critique of public sector involvement in agriculture and launched the political career of Reagan, who became Governor of California just two years later. The entire speech is available online at: http://www.americanrhetoric.com/speeches/reaganatimeforchoosing.htm.

230 These tensions between industry, plant science institutes, and government arose time after time during the twentieth century. Some of the key events in the UK have been chronicled in detail by Palladino (1990, 1996, 2002).

231 The interest of the private sector in these particular crops was largely driven by two factors. First, there had to be a large enough market to enable companies to recapture the considerable investment costs associated with the development and use of new plant breeding technologies such as transgenesis and molecular markers. Second, marketing systems and traits needed to be selected such that farmers would be encouraged, or obliged, to purchase new seed each year as the companies could only derive their income from royalties on the seeds or from trading margins on their sale. Hence, the major seed companies at the time focused on crops that would satisfy these two criteria, as follows. Pioneer focused on maize, and the global distribution of a mainly hybrid crop; Sandoz developed sugar beet, which was mainly triploid and, therefore, infertile. It was also the major sugar crop in Europe and the USA. Limagrain and DeKalb both focused on maize; KWS also developed sugar beet; ICI Seeds developed several crops including maize, sugar beet, sunflower (mainly hybrid) and oilseed rape (as this had the prospect, ultimately realised, of becoming hybrid). Monsanto, on the other hand, was primarily a technology-driven company whose initial aim was to license its technology to seed companies as its major revenue stream. This strategy was unsuccessful and the company

instead began to acquire businesses such as Asgrow (vegetables), PBI (wheat), and Holden's and Dekalb (both maize). (Information from Ian Bartle, personal communication.)

232 Fernandez-Cornejo *et al.*(2004) See especially pp. 41–50 for data on the changing balance public/private R&D for plant breeding in the USA.

233 A particularly useful account of the gradual rise of the private sector in US agribusiness, up to the late 1980s, can be found in Kloppenberg (1988).

234 The famous Adam Smith quote about the 'invisible hand' is as follows:

> ... every individual necessarily labours to render the annual revenue of the society as great as he can. He generally, indeed, neither intends to promote the public interest, nor knows how much he is promoting it. By preferring the support of domestic to that of foreign industry, he intends only his own security; and by directing that industry in such a manner as its produce may be of the greatest value, he intends only his own gain, and he is in this, as in many other cases, led by an *invisible hand* to promote an end which was no part of his intention. Nor is it always the worse for the society that it was no part of it. By pursuing his own interest he frequently promotes that of the society more effectually than when he really intends to promote it. I have never known much good done by those who affected to trade for the public good. Smith (1776).

235 Joseph Alois Schumpeter lived from 1883–1950. The quotation is from *Capitalism, Socialism and Democracy*, published posthumously in 1954 (Schumpeter, 1954).

236 Harlan and Martini (1936).

237 Interestingly, this US Plant Patent Act established a new precedent that went beyond the criteria necessary for the standard utility patents that were granted in other areas of technology. Regular utility patents require that an invention be distinctive, novel, and useful, but the new plant patents were granted simply according to the first two criteria. This meant that a patent could be granted for a variety that was inferior to existing varieties, just as long as it was distinctive and novel.

238 In 1993, Carrol Bolen of Pioneer Hi-Bred identified three factors needed for commercialisation of any crop as follows: 'Pioneer will likely not be interested in any new crop that doesn't have one of the following protection mechanisms: (1) capable of being hybridized; (2) an effective plant patent; (3) vertical integration through contract production' (Bolen, 1993). In my discussion, I have focused on the first two factors as being immediately relevant to the case of transgenic crops, whereas the third factor is a more generic commercial issue.

239 A propagule is part of a plant that can be used to reproduce, or propagate, new copies of the plant. Seeds are the propagules that result from sexual reproduction, but there are other sorts of propagule, such as cuttings, rootstocks and cell cultures, that can be used to effect somatic, or asexual, reproduction of the plant. Whereas a somatic propagule will grow into an identical copy, or clone, of the parent plant, a seed may differ significantly from the parent plant(s), unless it is highly inbred.

240 Sterile seed techniques are part of a wider suite of methods to deter farmer saving that are known as 'genetic use restriction technologies' (GURTs), as described in Various (2004) and Goeschl and Swanson (2003).

241 Louwaars and Minderhoud (2001).

242 For an overview of the process of intellectual property protection of plant breeding technologies, see Sechley and Schroeder (2002).

243 Unlike the earlier Plant Patent Act in the USA, the European UPOV legislation required that, in order to qualify for legal protection, new varieties must be useful, i.e. a demonstrable improvement on existing varieties. For more discussion of European legislation on breeders' rights, see Berlan and Lewontin (1986), Innes (1982, 1984, 1992) and Rangnekar (2000a, 2000c).

244 There is detailed account of the struggle by the US private sector to secure breeders' rights in Chapter 6 of Kloppenburg (1988).

245 Janis and Kesan (2002). This study relates mainly to soybeans and maize but several other studies have reached similar conclusions, i.e. they fail to find support for the hypothesis that PVP has a positive influence on crop yields, in other major crops. Examples include wheat (Alston and Venner, 2002), tobacco (Babcock and Foster, 1991; Perrin *et al.*, 1983), and canola (Carew and Devadoss, 2003).

246 Naseem *et al.* (2005).

247 Pray (1991); Thirtle *et al.* (1998).

248 Srinivasan (2004); Diez (2002).

249 These figures are given as inflation-adjusted 1996 US dollars (Heisey *et al.*, 2001).

250 See Chapter 6 of Kloppenburg (1988).

251 See Kloppenburg (1988), table 6.2, p. 145.

252 Heisey *et al.* (2001).

253 The recent favouring of parts of private sector agribusiness by government incentives, and the potential for overly protective IRP regimes to accelerate industrial concentration and the evolution of quasi-monopolies, has been explored by several agricultural economists, including Lesser (1998) and Phillips (2003a).

254 There are many examples of this type of funding across the industrialised world. One area in which I have been extensively involved is development of non-food crops for uses ranging from biofuels to biodegradable plastics. Government funding for such projects, sometimes paid directly to companies and sometimes to academic-company consortia, have totalled many hundreds of millions of dollars over the past decade.

255 Phillips (2003a).

256 The sorry saga of Calgene and its tomatoes is recounted in the book, *First Fruit*, by ex-Calgene scientist, Belinda Martineau (2001). A shorter, and rather more racy, account can be found in Chapter 10 of Charles (2001).

257 The three agencies are the Food and Drug Administration (FDA), Environmental Protection Agency (EPA), and United States Department of Agriculture (USDA).

258 Fernandez and Smith (2005).

259 For example, Pray *et al.* (2005) describe 'anecdotal evidence that some universities and private sector companies have abandoned patenting, commercialization, or even research programs due to this problem'. See also Pray and Naseem (2005), Heller and Eisenberg (1998), and Shapiro (2001).

260 Note that these reservations about broad claims and over-mighty oligopolies do not invalidate the overall concept of the appropriately regulated patent as a stimulus to technological innovation, which can be efficacious in both the private and public sectors. For example, Pray and Naseem (2005) have discussed the cases of molecular marker and plant transformation technologies and conclude that despite 'some examples of research and GM variety marketing that were slowed down by patents on tools … our preliminary assessment of the evidence suggests that the benefits from patents on tools outweigh the costs.'

261 This conclusion comes from a recent study by William Lesser of Cornell University and the same issue was raised as a key problem at a Crop Science Society of America conference held in 2003 (Lesser, 2005) as well as by earlier authors (e.g. Janis, 2002).

262 For example, by Shano Shimoda, president of BioScience Securities Inc., in the September 1999 issue of *Farm Chemicals* (available online at: http://www.findarticles.com/p/articles/mi_qa3842/is_199909/ai_n8870857).

263 Paradise *et al.* (2005).

264 Jensen and Murray (2005).

265 The quotation is from Jensen and Murray (2005) but refers to an earlier article by Heller and Eisenberg (1998).

266 Nuffield (1999).

267 For a recent review of the patterns of agbiotech patenting in the public and private sectors, see Heisey *et al.* (2005). The broader question of widening access, especially for developing countries, to overprotected agbiotech and pharmaceutical innovations is discussed by Brewster *et al.* (2005).

268 In this judgment, the US Court of Appeals ruled that the Monsanto EST collections did not constitute an invention of demonstrable practical utility (United States Court of Appeals for the Federal Circuit, 2005). The ruling will also affect five further patent applications from Monsanto and as many as 650 000 ESTs in total. According to the written ruling of Chief Judge Michel, the granting of such EST patents without proof of utility 'would discourage research, delay scientific discovery, and thwart progress ... '.

269 Examples of large agrochemical companies that became involved in crop breeding include Monsanto, Dow, DuPont, Zeneca and Novartis. Some independent seed companies that developed interests in advanced crop breeding include Pioneer, Nickerson, Limagrain and Cargill.

270 Some of the better known small agbiotech companies of the 1980s and 1990s include Calgene (USA), Mogen (Netherlands), DNA Plant Technology (USA), Agracetus (USA), Mycogen (USA) and Plant Genetic Systems (Belgium).

271 During the 1990s, another type of GM tomato, with a down-regulated polygalacturonase gene, was developed in the UK by ICI (later to be Zeneca). Whereas the Calgene FlavrSavr® tomatoes were sold as fresh fruit, the ICI/Zeneca tomatoes were only sold in the form of tomato paste in supermarkets such as Safeway and Sainsbury's. The thicker nature of the pulp from the GM tomatoes allowed for cost savings in production and the paste was priced at a significant discount to non-GM alternatives. The reason for this was that there was no need to heat the tomatoes to such an extent as in jam making. In the latter case, heat is required to destroy the pectinases and ensure good 'setting' of the jam. The lower temperatures required in producing paste from GM tomatoes enabled the volatile organic compounds that provide much of the flavour to be retained; the process also used less energy. In contrast to FlavrSavr® tomatoes, the Zeneca GM paste was a modest commercial success in the UK until supermarkets suddenly withdrew the product in 1999 in the wake of the Pusztai affair and in response to a particularly effective campaign by anti-GM pressure groups.

272 According to a 2005 report from the Center of Food Safety (a US organisation that is against GM crops), from 1995–2005, Monsanto had filed 90 lawsuits against American farmers in 25 states that involved 147 farmers and 39 small businesses or farm companies. It is also alleged that Monsanto has set aside an annual budget of $10 million dollars and a staff of 75 devoted solely to investigating and prosecuting farmers. The largest recorded judgment found thus far in favour of Monsanto as a result of a farmer lawsuit is just over $3 million. Total recorded judgments granted to Monsanto for lawsuits amount to $15.3 million. This report is available online at: http://www.centerforfoodsafety.org/Monsantovsusfarmersreport.cfm.

273 Two of the large agbiotech multinationals sold [transgenic seed + herbicide] packages in the first generation of mass-cultivated transgenic crops that was grown from 1996 onwards. Monsanto sold the Roundup Ready® package, which consisted of seeds containing a gene conferring resistance to the herbicide glyphosate (aka. Roundup), plus the herbicide itself. AgrEvo sold the Liberty Link package, which consisted of seeds containing a gene conferring resistance to the herbicide glufosinate (aka. Liberty), plus the herbicide itself.

274 Note that not all transgenic herbicide tolerance traits were linked to herbicides produced by the same company. In the early 1990s, ICI/Zeneca was one of a number of seed companies offering Liberty Link oilseed rape in Canada (incorporating glufosinate resistance from Hoechst) without such ties. As with glyphosate, glufosinate was noted for its efficacy and safety, being broken down on contact with the soil and, therefore, an improvement in environmental terms over such residual herbicides as atrazine that had to be applied prior to crop emergence. Hence, there was an environmental benefit from the use of these herbicides, in addition to the commercial benefit to both the farmer and the seed/agrochemical companies.

275 Phillips (2003a).

276 In 1998–1999, there was a dramatic food scare in the UK, following allegations on a television programme that rats fed on transgenic potatoes suffered developmental abnormalities that were due to the presence of the transgenes themselves. Within a few days, all UK supermarkets withdrew any GM food products and soon afterwards the European Union imposed what was effectively a moratorium on the use of transgenic crop varieties. Although the original study on rats is now widely regarded as flawed, European consumers and supermarkets have still not recovered their confidence in foods from transgenic organisms, although ironically they have universally embraced pharmaceutical products from transgenic bacteria and yeast. I will return to this topic in a forthcoming book, now in preparation, on the impact of agbiotech controversies on scientists and the scientific establishment. The importance of identifying and listening to your consumers is summarised as follows in the words of business guru, Bo Bennett 'Not being in tune with your customers is like living in an alternate reality; the way you think your customers feel about your product is not always the same as what your customers really think about your product.'

277 The impact of patenting on the US seed industry in the years immediately prior to the development of commercial agbiotech have been examined by Butler and Marion (1985).

278 The stifling effect of over-broad biotech patents has been discussed in the broader European and North American contexts by Anonymous (1995), Stone (1995), van Wijk (1995) and Packer and Webster (1996), and in a specifically Canadian context by Phillips (1999) and Carew (2000).

279 Webster and Packer (1996).

280 Values are inflation-adjusted 1996 US dollars (Heisey *et al.*, 2001).

281 Sometimes these findings were publicised simply as press releases announcing a new finding, although the more sophisticated companies and their research collaborators also used conference papers and publication in reputable journals as more credible ways of both validating and publicising their work. The latter practice has led to some devaluation of scientific language as even rather mundane findings become labelled as 'breakthroughs' in order to hype up both the significance of the work and the achievement of the scientist.

282 Phillips (1999).

283 In his book *The Rape of Canola*, Brewster Kneen presents a robust case for a public-good ethos in canola breeding that has now largely disappeared (Kneen, 1992).

284 Lindner (2004a, 2004b).

285 The evidence for a policy of planned obsolescence in UK wheat breeding has been presented by Rangnekar (2000b), although most private sector breeders would dispute this analysis. Note also that, while such a short-termist strategy may seem to be against the public good, there is also a counter-argument that planned obsolescence can act as a driver for technological progress, as witnessed for example by the computer industry (Fishman *et al.*, 1993).

286 Srinivasan *et al.* (2003).
287 Rangnekar (2000b).
288 It is difficult to substantiate claims of deliberate obsolescence, especially given that the approval of any new variety is normally conditional on criteria that should ensure both novelty and utility. The main criteria are DUS (distinctness, uniformity and stability), i.e. the variety must be genuinely new, and VCU (value for cultivation and use). The latter is usually based on a yield assessment but milling and malting varieties could pass VCU provided the combination of yield and quality was seen as an overall improvement over existing varieties. In England and Wales new varieties were then entered into the NIAB Recommended List trials (now carried out by the HGCA). This screening process further reduced the number of varieties likely to be commercialised as farmers normally base their purchasing decisions on this list of approved varieties.
289 Phillips (1999).
290 Alston and Venner (1998).
291 Rangnekar (2000a, 2000b, 2000c).
292 This is not to imply that seed companies necessarily conspired with agrochemical companies to promote disease susceptible varieties, but the fact remains that private sector breeders tended to de-emphasise broad resistance traits in favour of more immediately profitable yield-related traits that included enhanced responsiveness to agrochemical inputs.
293 See Bingham *et al.* (1991) and Bingham and Payne (1993).
294 Bayles (1991).
295 Commission on Intellectual Property Rights (2002).
296 Doyle (1985).
297 Leonard (1987).
298 NRC (1972).
299 Data are from the World Resources Institute (http://www.wri. org/) and can also be found in the online version of the *AAAS Atlas of Population and Environment* (Harrison and Pearce, 2001).
300 According to Kent Whealy of the US Seed Savers Exchange '*Of the 230 (seed) companies that we inventoried in 1984, 54 had gone out of business or been taken over by 1987. From 1981 to 1994, we lost 84 percent of all the non-hybrid vegetables that are available in this country*' (http://www.alternet.org/envirohealth/11417/).
301 In addition to the formal public seed banks, there is an increasing number of private citizen initiatives, including a range of public-good, seed savers networks in countries around the world. Some of these initiatives are aimed at conserving plant diversity in general, while others are more focused on agriculturally related species.
302 There are also many minor local varieties of banana and plantain grown in tropical countries but these tend to have much smaller fruits with relatively short shelf lives and are therefore not suitable for large-scale cultivation for export to the major markets in industrial countries. Commercial export bananas all belong to Cavendish and Gros Michel subgroups of triploid clones and have a monospecific *Musa acuminata* origin. Triploids were formed when normal haploid gametes fused with aberrant non-reduced diploid gametes (Raboin *et al*, 2005).
303 The controversial possibility that transgenic methods might be used to ameliorate the genetic deficiencies of commercial bananas has been discussed by John Innes Centre crop scientist, James Brown (Brown, 2000).
304 Reif *et al.* (2005)
305 Public sector breeding centres have not been entirely immune to this tendency to concentrate on a narrow set of parental genetic material. For example, in 1993, a report

on a twenty-year study of 27 rice-breeding programmes across Asia identified strong similarities in genealogies across the released varieties (NRC, 1993). In another study, it was found that, in some CGIAR rice breeding programmes, a set of semi-dwarf females was so extensively used as to raise concern regarding genetic similarity across all of the released varieties (Busch *et al.*, 1995) See also Flower and Mooney (1990).

306 Jain (1991).
307 Johnson (1986).
308 This issue is explored in detail by Raney and Pingali (2004).
309 Johnson (1986).
310 The quotation is from Webb and Bruce (1968), as cited in Kloppenburg (1988).
311 Wallace and Brown (1956) p. (111). See also the earlier book on maize and its cultivation by Wallace and Bressman (1949).
312 Rasmussen (1982); Goodman *et al.* (1987).
313 Thereby hangs a considerable tale, as eloquently recounted in Jim Hightower's memorable book, *Hard Tomatoes, Hard Times* (Hightower, 1973). One of the most controversial episodes of tomato 'improvement' was the displacement of a large force of Mexican fruit pickers. As late as 1977–1980, when I was working as a plant researcher at UC Davis, the role of the university in these events was still a lively topic of debate and recrimination, both on campus and further afield.
314 The FlavrSavr® tomato was developed as a variant of the conventional thick-walled varieties in which the cell-wall pectin was broken down less rapidly and, therefore, the ripening and subsequent rotting processes were slowed down giving the tomatoes a longer shelf life. However, Calgene failed on all counts. Their new tomatoes were not significantly tastier than conventional Californian varieties and neither were they much longer lasting. My favourite anecdote about the whole Calgene saga comes from 1992 when I was attending a scientific conference in Paris with some of their researchers. As we sampled our lunchtime salad in a restaurant just off the Boulevard Saint-Germain, our American colleagues expressed astonishment at the exquisite taste of the fresh plum tomatoes that had been served. One of them went on to remark: 'I guess there won't be much of a market for our FlavrSavr® tomatoes here in Europe.' How true!
315 A detailed historical account of the development of hybrid maize in the USA can be found in Chapter 5 of Kloppenburg (1988).
316 The quotation is from the first edition of Simmonds (1979), but note that this text has subsequently been revised and updated as Simmonds and Smartt (1995).
317 Lewontin (1982).
318 Berlan and Lewontin (1983).
319 Simmonds (1979), p. 159.
320 The private sector is also a major force in the breeding of wheat and alfalfa, although there is still a significant public effort as well for these two crops (Duvick, 2004).
321 In 2003, for the first time in half a century, a lack of funds caused CIMMYT breeders in Mexico to abandon their wheat breeding trials for that year. It was reported that more than half of the fields at the Ciudad Obregón field station were lying fallow and a new cohort of trainee plant breeders could not be taught (Knight, 2003).
322 This quotation from Sallust has regularly been used to counsel against a lack of balance between wealth and resources, not only between public and private sectors, but also between different groups within society.
323 Kenney *et al.* (1983).
324 Sullivan (2004).
325 Pirie (2002).
326 For an account of the theory of privatisation, see Pirie (1988).

327 Pray (1996).

328 NSDO was the marketing arm for public sector-bred crop varieties in the UK. Hence, it marketed seeds of grass and forage crops from Aberystwyth, potatoes from SCRI in Dundee, Scotland, and cereals etc. from PBI Cambridge. In the case of cereals, NSDO produced basic improved seed varieties in relatively small quantities and it was the private sector that took the additional risk in multiplying them to meet anticipated demand.

329 Boden *et al.* (2004).

330 Sharp (2005).

331 The term 'near-market' research was first used in the 1988 Barnes Review, as discussed by Webster (1989). The term was then used throughout the 1990s by research policymakers in the UK public sector but is less frequently encountered nowadays. It has been defined as 'innovative work aimed at generating or partially generating a specific product, artefact or idea for the commercial market'. However, as cautioned in the 1993 UK government White Paper *Realising Our Potential* 'the "near-market" guideline cannot be applied unthinkingly or too rigidly. An important test here will be the extent to which all the benefits of support can be captured by an individual firm or venture, which might therefore be expected to pay for the work itself, or by a wide group of firms or society generally. The approach must take careful account of the circumstances of each individual case.' (UK Government, 1993).

332 For a wider perspective on the effects of privatisation on science in general during the 1980s, see Wilkie (1989).

333 UK Parliament (1998).

334 Vickers and Yarrow (1988).

335 There was a delicious irony to this privatisation because, in the end, the UK government was unable to recover directly any of the proceeds of the sale of PBI. This was because PBI was legally a charitable trust and was not therefore actually owned by the government. This fact had been overlooked when PBI had been auctioned off by the government in 1987. In 1988, the Charities Commission ruled that the sale was illegal and the proceeds were duly transferred to that part of PBI that had remained in the public domain (Webster, 1989; Pray, 1996). This was the so-called 'Cambridge Laboratory', which I joined as a department head as it relocated to Norwich early in 1990. The government responded by withholding our salaries and running costs. This obliged us to use the proceeds from the PBI sale, which had accrued to our charitable trust, in order to meet all of these costs. We went on like this for several years, until all the privatisation funds had been exhausted.

336 Webster (1989).

337 As in the case of several other privatisations, PBI was artificially held back from some opportunities to develop by the government in the years before its sale. Hence, during the early 1980s, it was refused permission to establish any experimental stations in Europe. In contrast, soon after Unilever purchased PBI in 1987, it was able to set up stations in France and Germany without let or hindrance from the UK government (Pray, 1996).

338 The evidence suggests that the UK government was significantly under-investing in PBI research in 1986. Even with the inclusion of the costs associated with less applied but scientifically noteworthy molecular biology research, PBI was still generating about $3 million more in revenues than it cost to run. This meant that when Unilever bought PBI in 1987, they would have access to a notional extra $3 million per annum that they could use for expanded research operations, or simply pocket as additional profit (financial data are from Lazard Brothers & Co., as quoted in Pray, 1996).

339 The gene in question encoded a 5S ribosomal RNA (Gerlach and Dyer, 1980).

340 The two most important UK crops are wheat and barley, which respectively make up 35% and 27% of total arable cultivation. The two next most important crops are oilseed

rape at 9% and peas at 6% total arable acreage (data from DEFRA wesite at: http://www.defra.gov.uk/farm/crops/index.htm#arable).

341 Teich (1986).

342 PBI was described as a 'jewel in the AFRC's crown' by Roger Gilmore, Chief Executive of Agricultural Genetics Company in 1987 (Webster, 1989).

343 The non-privatised part of the PBI was transferred to the Institute of Plant Science Research, which included the John Innes Institute, Norwich and the Nitrogen Fixation Laboratory, Sussex.

344 The move to Norwich was completed in April/May 1990. The three departments and their Heads were respectively: Brassica and Oilseeds – Denis Murphy; Cereals – Mike Gale; and Molecular Genetics – Mike Bevan. Following the relocation to Norwich, the ex-PBI researchers officially maintained a separate identity as the 'Cambridge Laboratory' under Head of Laboratory, Colin Law (who had previously been Director of PBI in Cambridge), for several more years.

345 See the *Preferred Options Report – Cambridge Southern Fringe Area Action Plan* from the South Cambridgeshire District Council, which is available online at: http://egov.scambs.gov.uk/ldf/representation.php?docid=45&chapter=3#1879.

346 Quoted in Knight (2003).

347 See CSREES website at: http://www.csrees.usda.gov/.

348 Farmers pay for ADAS crop trials via a levy on their harvest, which funds bodies such as the Home Grown Cereals Authority (which deals with oilseeds as well as cereals). It is the breeders and HGCA that directly fund ADAS crop trials.

349 These figures are taken from a US-based economic study of the consequences of the PBI and ADAS privatisations (Pray, 1996).

350 Chapman and Tripp (2003). However, one might comment that, once he had been broken, not even 'all the king's horses and all the king's men' were able to put Humpty Dumpty together again. Similarly with ADAS perhaps?

351 Pray (1996).

352 Pray (1996).

353 Webster (1989).

354 These companies included Zeneca, Pioneer and Nickersons (Pray, 1996).

355 Thirtle *et al.* (1991).

356 Some of the latest cuts in institute research funding occurred as this book was being finalised in early 2006, with a £2 million withdrawal in DEFRA funding of IGER in Wales, with several dozen associated redundancies to plant scientists. At the same time, the Natural Environment Research Council announced the closure of four research centres with the loss of 200 science posts.

357 By mid-2006, all of the major multinational companies, such as Monsanto, Unilever, and Syngenta had closed down or withdrawn their plant breeding and agbiotech research divisions from the UK. One of the final casualties was French-owned Biogemma, based in Cambridge, which closed in June 2006 with the loss of 25 scientific posts.

358 Quoted in McGuire (1997).

359 McGuire (1997).

360 I am indebted to David Crystal who brought this passage to my attention in his fine book on the English language (Crystal, 2005). The quotation is sampled from a message to Bishop Waeferth dating from about 890 CE, and found in the Preface to a translation of Gregory's *Curia Pastoralis*. The text of the original Anglo Saxon, and its modern translation (which is more preoccupied with matters godly than botanical), can be found online at: http://www.towson.edu/~tinkler/prose/1oe.html.

361 This saying originated from one of the maxims of the early Roman author, Publius Syrus: 'Solamen miseris socios habuisse doloris – It is a consolation to the wretched to have companions in misery' (Syrus, 42 BCE).

362 The privatisation of the Canadian canola breeding programme has been analysed by Phillips (1999).

363 See Figures 2 and 3 in Heisey *et al.* (2001). For a useful overview of recent changes in US public sector plant breeding, see Coors (2002).

364 Frey (1996).

365 Tracy (2003).

366 Day Rubenstein and Heisey (2005).

367 New Zealand Crown Research Institutes related to plant breeding include AgResearch Ltd (http://www.agresearch.co.nz/default.asp) and the New Zealand Institute for Crop & Food Research Ltd (http://www.ccmau.govt.nz/crop-and-food.html).

368 The GRDC is the main funding body for crop breeding in Australia. It disburses an annual $5 million, which is raised from a combination of industry levies and government funds. According to its website: 'The GRDC's mission is to invest in research and development for the greatest benefit to its stakeholders – grain growers and the Commonwealth. The Corporation links innovative research with industry needs. The GRDC's vision is for a profitable, internationally competitive and ecologically sustainable grains industry' (http://www.grdc.com.au).

369 See newspaper article: http://www.abc.net.au/news/newsitems/ 200511/s1495178.htm.

370 See 'GM Canola – the Canadian facts' online at: http://www.grdc. com.au/growers/cd/ west/western_region05011.htm.

371 Data from FAO: http://www.fao.org.

372 In global terms, sunflower is second only to oilseed rape as a commercial oilseed crop. Although soybean and oil palm produce more oil than these two oilseed crops, they are not themselves classified as oilseeds. Soybean is a protein-rich legume that produces a modest amount of oil as a by-product, while oil palm is a perennial tree crop that produces most of its oil in its fleshy fruits, rather than its seeds.

373 While Ukraine was the centre of Soviet grain production, Georgia was assigned the role of producing wine, citrus fruit, tea and vegetable crops, Plant breeder Taiul Berishvili (director of the Department of Plant Breeding of the Botanical Institute in Tbilisi) related efforts to maintain the precious but sadly neglected collection of Georgian cereal germplasm, which at one stage he found being eaten by mice (see: http://www.slowfood. com/img_sito/PREMIO/vincitori2003/pagine_en/Georgia_03.html).

374 A research centre carries out primary research and development into agricultural improvement while a service centre is a more direct link with farmers. Service centres may be involved in multiplying, storing and distributing seed stocks, giving advice and training, on-farm testing of new varieties in different regions, and monitoring of individual farmer compliance with national standards.

375 Turner *et al.* (1999). This study from the ICARDA seed unit, which focuses on the Near East and North Africa, makes the point that privatisation is far from being a panacea for the structural problems of many publicly funded seed centres. While a greater private sector role might improve the dynamics and economic performance of seed provision, there is also a risk that the private sector will only focus on the most profitable crops and leave an impoverished and demoralised public sector to deal with a mass of less glamorous 'orphan' crops. Part of a possible solution is to have a mixed economy with vigorous and accountable public and private partners that work together both for mutual reward and for the good of the country. See also Ramirez and Quarry (2004).

376 Price (1999); Tripp and Byerlee (2000); Pingali and Traxler (2002); Graff *et al.*, (2003); Jones (2003); Gerpacio (2003); Tracy (2003); Lindner (2004a, 2004b).

377 The delights of the academic ivory tower were summarised as far back as the early seventeenth century by Joseph Hall (1574–1656) in his poem: *Discontent of Men with Their Condition*

> 'Mongst all these stirs of discontented strife,
> O, let me lead an academic life;
> To know much, and to think for nothing, know
> Nothing to have, yet think we have enow.

378 The contentious term 'wealth creation' was still in use in scientific policy documents some 20 years later, but its meaning seems to have changed somewhat – at least in the following example from a BBSRC document dated December 2005: '*We do believe that "wealth creation" should be taken in its broadest sense and include non-financial values. We accept that this term can be confusing to the public and indeed have found this to be the case ourselves during a recent policy dialogue activity*' (BBSRC, 2005). During the 1980s and 1990s, it was not only the general public who were confused about 'wealth creation', but in the UK research community, we were clear that it was meant in the narrower, more literal context of monetary value.

379 Naturally, there are many reasons, other than the merely utilitarian, to pursue research into plant science. However, since the vast majority of such research is funded from the public purse, it is legitimate to expect that some serious consideration be given to its possible application. This is especially true for relatively expensive research areas like plant development and genetics where possible applications for crop improvement are frequently trumpeted, but much less often followed through, by publicly funded scientists. This deficiency was recognised in a report published in mid-2004 on BBSRC funded plant science research in the UK (BBSRC, 2004).

380 Check (2004); Murphy (2005), pp. 18–21.

381 'Freely available' here means available without restriction to any academic scientist. There are sometimes charges made for the supply of such resources, e.g. stocks of mutant plants or specific DNA sequences, but these are normally to cover handling costs.

382 The extent to which other countries have taken up the UK model of research assessment is discussed in detail in appendix G of the report by Roberts (2003).

383 The Research Selectivity Exercise, as it was originally known, was first held in 1986 and was repeated in 1989. The process was renamed the Research Assessment Exercise for the third exercise in 1992. The fourth and fifth exercises took place in 1996 and 2001 respectively. The next exercise, known as RAE 2008, will be held in 2007–2008.

384 Following the last RAE in 2001–2002, 95% of funds distributed by the Higher Education Funding Councils (HEFCs, successors to the University Grants Committee) were allocated on the basis of the RAE grade of each university department or research unit. For a list of the RAE scores across the UK, see the RAE website at: http://www.hero.ac. uk/rae/.

385 I have personally participated, both as an assessor and as an assessee, in several research assessment exercises, for both institutes and universities. The overwhelming focus of these exercises has always been on research quality as measured by peer-reviewed published output. This has led to criticisms that such a focus may be too narrow and that it fails to take into account, or give enough merit to, more applied research outputs, such as an improved plant variety (Roberts, 2003; UK Parliament, 2004b). For this reason, the RAE for 2008 may take a wider view of research quality and may not be so obsessed with notoriously unreliable indicators such as the 'impact factors' of research journals (see the

document: *Guidance on submissions, June 2005*, available online at: http://www.rae.ac.uk/pubs/2005/03/).

386 This relentless pressure to publish more or less anything brings to mind Lord Byron's wonderfully sardonic couplet: 'Tis pleasant, sure to see ones's name in print; A book's a book, although there's nothing in't.' This is taken from English bards and Scotch Reviewers, published by Byron in 1809.

387 The possible adverse consequences of the RAE for applied subjects such as engineering and medicine have been highlighted in the academic press (Banatvala *et al.*, 2005; Fazackerley, 2004 & 2005).

388 However, see note 385 about the changes planned for future research assessments, including RAE 2008, which may address some of these criticisms.

389 Universities UK originated in the nineteenth century as a body that represented the interests of universities, e.g. to government. Formerly known as Committee of Vice-Chancellors and Principals of the Universities of the United Kingdom (CVCP), Universities UK adopted its present name in 2000. The Universities UK report on the RAE and research concentration can be found as Adams and Smith (2003).

390 Adams and Smith (2003, p. 7)

391 An example of concerns expressed about over-concentration of research in a few large centres is the article by Richard Bateman (Keeper of Botany at the Natural History Museum, London) in the *Times Higher Education Supplement* entitled 'Creativity crushed by super-fuelled juggernaut' (Bateman, 2005).

392 As of mid-2006, there are indications that the UK government may abandon the RAE after the 2008 exercise, although the proposed metrics-based system to replace RAE may end up even worse for many less fashionable areas of research (MacLeod, 2006).

393 The better performance of US researchers cannot be simply attributed to higher funding levels, more high-quality facilities, or more critical mass. For example, in plant science one of the most successful initiatives is the harnessing of researchers across the country to work on interrelated aspects of *Arabidopsis* biology. In addition to the usual high-profile centres of expertise, such as Stanford and Harvard, this initiative includes a host of researchers, many of them in relatively small laboratories in lesser known institutions such as Oklahoma State University, Stillwater OK, University of Pennsylvania, Philadelphia, PA, Ohio State University, Columbus, OH, Rice University, Houston, TX, Miami University, Oxford, OH, Kansas State University, Manhattan, KS, and Florida Gulf Coast University, Fort Meyers, FL.

394 Buhler *et al.* (2002).

395 The dangers of relying on short-term funding were graphically illustrated in April 2005 when the John Innes Centre admitted to a shortfall of over $5.5 million that would necessitate the loss of 10% of its scientific staff. Such pressures have also led to yet another proposed merger between two UK research institutes, i.e. the John Innes Centre and the Institute for Food Research, mainly on the grounds of cost saving rather than for any compelling scientific reasons (Adam, 2005).

396 Whereas research assessments are always carried out ad hominem, the actual funding that proceeds from such an assessment is invariably directed to the entire institute or to a university department rather than to individual scientists.

397 This kind of research funding is therefore programme focused where the overall topic area is determined by policymakers rather than the researchers themselves. In contrast, much of the basic science funding from the UK Research Councils is allocated to the best projects submitted by researchers with no preconceived topic areas – this is known as 'response mode' funding. One of the clear tendencies over the past 20 years, especially in

plant science, has been the reduction in the proportion of 'response mode' funding and the increase in more targeted programme-focused research.

398 The dubious economic case for biodiesel is discussed by Murphy (1999a), while the equally dubious case for the supposed energy efficiency of biofuels in general is critically examined by Pimentel and Patzek (2005). Moreover, as many advocates of renewable energy acknowledge, the mass conversion of land in developing countries to biodiesel crops is fraught with risks for future food production. In the usual hyperbolic language of the genre, UK environmental writer George Monbiot has commented: 'if biofuels take off, they will cause a global humanitarian disaster' (Monbiot, 2004).

399 This topic has been examined recently in an article by Carl Pray where he enumerates the many strategies, and potential pitfalls, that face research managers, scientists and policymakers who wish to obtain funding from the private sector (Pray, 1998).

400 The firm conclusion from experience over the past two decades is that public sector institutions cannot come anywhere near to achieving financial self-sufficiency by marketing their expertise, e.g. by a vigorous use of IPRs for fund raising, or otherwise relying on market mechanisms for such funding (Lesser, 2005).

401 In addition to the examples from the UK and USA mentioned in the text, there have been dozens of similar strategic partnerships set up between commercial companies and public sector research institutions around the world. For example, alliances in plant genomics have been formed between the Max Planck Institut für Züchtungsforschung in Germany and a consortium of companies including AgrEvo (now part of Bayer), and also between the Institute of Molecular Agrobiology (Singapore) and Rhône-Poulenc (now also part of Bayer).

402 For a fairly recent review of public–private collaborations in agricultural research, see the book edited by Fuglie and Schimmelpfennig (2000).

403 Busch *et al.* (2004).

404 Lawler (2003).

405 The US examples are quoted from Pray (1996).

406 See press release: John Innes Centre (2002).

407 See press release 'Zeneca invests £50M in wheat laboratory and research collaboration', available online at: http://www.syngenta. com/en/media/article.aspx?pr=091698&Lang=en.

408 John Innes Centre (2002).

409 Soon after its withdrawal from the UC Berkeley and John Innes Centre alliances, Syngenta also underwent a major reduction in its own in-house research operations. This involved closure of major crop biotech facilities at the Torrey Mesa Research Institute in California and at Jealotts Hill in the UK.

410 Adam (2005); John Innes Centre (2005).

411 As often happens in such cases, the technology fees either ended up elsewhere in the institute, or were siphoned off by the relevant government department. Either way, the Chinese researchers who initially developed the assets that were licensed or sold off gained no benefit from the arrangement, but suffered all the adverse consequences of the resulting international isolation. This is neither an isolated occurrence, nor is it limited to developing countries like China.

412 Note that this example of restrictions on germplasm sharing is by no means limited to China. As we will see in Chapter 16, one of the major culprits is the US government, which still restricts the sharing of germplasm with crop breeders from selected 'unfriendly' countries.

413 Unlike the previous US experience with Chinese hybrid rice, the new joint venture has been successful. In December 1999, the US partner, RiceTec Inc. of Alvin, Texas, began selling the first hybrid rice seed in the US. Although the hybrid seed is three times more

expensive than conventional varieties, this additional marginal cost is dwarfed by the 30% yield improvement of the hybrid rice (Andrews, 2001).

414 In order to cope with such eventualities, public research institutions, and some universities, have been obliged to include so-called 'supervening circumstances' clauses in staff contracts, e.g. for postdoctoral researchers, that allow for abrupt termination of the contract in the event of a default or withdrawal by a sponsoring company.

415 This figure is from the BBSRC budget for 2005, which is available from the website (http://www.bbsrc.ac.uk/). According to the same source, out of its total 2005 budget of $550 million, BBSRC spent $340 million on university research, which means that about one third of its funds went to the research institutes while two thirds went to university research projects.

416 Data from the National Audit Office, Session 2002–2003, Reaping the Rewards of Agricultural Research, HC 300, p. 21. See also Tenth Report of the Environment, Food and Rural Affairs Select Committee, Session 2002–03, Horticulture Research International, HC 873.

417 See parliamentary announcement online at: www.parliament.uk/ documents/upload/ Horticulture%20Research%20International. pdf.

418 See BBC news story: 'Research centre to make job cuts', available online at: http://news. bbc.co.uk/1/hi/wales/mid/4655250.stm.

419 Foot and mouth disease (known as hoof and mouth disease in the USA) is a virulent, but not fatal, viral disease that spread across the UK in 2001. Following scientific advice, which has since been questioned, the government implemented a drastic cull that resulted in the slaughter of six million animals at an estimated cost to the economy of $15 billion, plus the closure of much of rural Britain to all visitors for almost a year (see the DEFRA FMD site online at: http://footandmouth.csl.gov.uk/).

420 UK Parliament (2004a).

421 UK Parliament (2006).

422 Heisey *et al.* (2001).

423 This reminds one of the unfortunate person, who waits interminably at a bus stop, until suddenly three buses arrive together. We had waited for 15 years while the predicament of UK plant breeding research steadily worsened and then suddenly, within a few months in 2004, there arrived all manner of reviews and reports from Parliament, the BBSRC, the AEBC, and the Institute of Biology.

424 BBSRC (2004).

425 BBSRC (2004), p. 16.

426 Dale (2004).

427 AEBC (2005a). The remit of the AEBC was to provide the UK government with independent, strategic advice on developments in biotechnology and their implications for agriculture and the environment. In 2004, AEBC established a subgroup to examine how research agendas in agricultural biotechnology are determined, and whether they address policy needs and public aspirations (for more details, plus the latest AEBC consultative document, see online at: http://www.aebc.gov.uk/aebc/subgroups/re-search_agendas.shtml). For reasons that are not entirely transparent, AEBC was wound up in 2005 and it is uncertain whether it will be replaced by another advisory mechanism for agbiotech.

428 The immediate response of BBSRC to the AEBC report was published in December 2005 (BBSRC, 2005), but so far only short-term palliative measures have been announced in the form of research funding initiatives in some crop-related areas. While these measures are most welcome, they do not address the more fundamental and long-term structural problems that confront applied plant science in the UK.

429 This initiative includes a collaboration between three of its research institutes (IGER, Rothamsted and John Innes), known as MONOGRAM. The BBSRC policy document is available online at: www.bbsrc.ac.uk/about/pub/policy/bbsrc_delivery_plan.pdf.

430 AEBC (2005b).

431 For example, in its 2005 report the AEBC made the following statement: 'Due to Government policy decisions to pull out of research in this area and the sale of PBI, several of those we spoke to expressed a concern that there are now very few places to train plant breeders in the public sector in the UK.' This information in the AEBC report was obtained from John Macleod (British Beet Research Organisation) and Chris Leaver (Oxford University Plant Sciences Department), in February 2005 (AEBC, 2005b, p. 11).

432 We could extend this metaphor and say that much of the expensively produced water from the UK sprinkler system is now being diverted to nourish other gardens (e.g. via the export of expertise and scientists to the USA) instead. And rather than replace our hosepipe so that we can use our own water in the UK, we are now buying water (because we no longer generate many of our own varieties, we must import seed from abroad) from our neighbours to save the garden from expiring altogether. And all this trouble is caused for want of a mere hosepipe.

433 Stiglitz (1993).

434 National Academy of Sciences (2000).

435 See also the useful policy document (available online) by Zephaniah Dhlamini (formerly of the International Atomic Energy Agency) on the use of non-GM biotechnologies in developing countries (Dhlamini, 2006).

436 In 2003, the US, Canadian and Australian governments made a formal complaint to the World Trade Organization against the European Union for refusing to allow the sale of genetically modified (GM) food or crops. In early 2006, the WTO panel delivered a preliminary ruling in favour of the plaintiff countries but the litigation is still ongoing. A leaked version of the draft WTO ruling was published by Friends of the Earth online at: http://www.foeeurope.org/biteback/WTO_decision.htm.

437 Analogous observations have been made about the supposed biotech revolution in biomedicine, a good example of which is the essay by Michael Fitzpatrick, entitled: 'Body politics. Francis Fukuyama and Gregory Stock take different sides in the biotechnology revolution. What revolution?' Available online at: http://www.spiked-online.com/Printable/00000006D919.htm.

438 The demise of Calgene as an independent company was largely due to its failure to realise the key importance of cultivar selection for its FlavrSavr® tomato, in addition to serious management problems (Martineau, 2001; Charles, 2001).

439 See press release online at: http://www.cimmyt.org/english/wps/ news/irriAlliance.htm, and a commentary by Normile (2005).

440 Knight (2003).

441 This quote was from a notorious commentary on the AgBioView discussion site that equated anti-GM activists like Mae-Wan Ho and Vandana Shiva with the perpetrators of the 9–11 suicide attacks in the USA. It demonstrates that shallow and ill-founded ad hominem invective is by no means the sole preserve of the anti-GM lobby (http://www.gmwatch.org/archive2.asp?arcid=5019). Such personal attacks against anti-GM activists are neither warranted nor effective. It is far better in the long run to rely on the tools of evidence-based reason and logical argument to advance the cause of science.

442 For example, since 1989, I have published numerous articles in books and journals that have highlighted the vast potential of transgenic oil crops to produce novel oils for food and industrial uses (Hills and Murphy, 1990; Murphy, 1991, 1994, 1998, 1999b, 1999c).

However, although several GM oilseed varieties with novel oil profiles have been produced, none of them have been commercially successful (Murphy, 2002, 2003a, 2006).

443 According to CIA estimates, the combined gross domestic product of the 25 countries of the EU in 2006 was about $13.74 trillion, while that of the USA was $13.22 trillion (http://www.cia.gov/cia/publications/factbook/geos/ee.html).

444 Quoted from Busch (2004), p. 27. Busch was asked by UC Berkeley to head a group to report on the highly controversial partnership between that university and the agbiotech company Novartis (later Syngenta). The subsequent report by Busch *et al.* (2004) was critical of the overbroad scope of the partnership and its lack of significant achievements.

445 For an up-to-date assessment of the economic impact of GM crops in the Americas (where 94% were grown in 2005), see Taxler (2006).

446 There are many books about GM foods and agbiotech, some of which are so distorted and error ridden as to be of little value. An example is the book by Jeremy Rifkin that excoriates modern biotechnology, and especially GM crops in a particularly polemical and acerbic manner, such as: 'Pestilence, famine, and the spread of new kinds of diseases throughout the world might yet turn out to be the final act in the script being prepared for the Biotech Century.' (Rifkin, 1999, pp. 90–91). Some of the better written, more informative, and more balanced popular books on GM crops include the following: McHughen (2000), Pinstrup-Anderson and Schiøler (2000), Charles (2001), Priest (2001), Lurquin (2001 & 2002), Ruse and Castle (2002), Winston (2002), Pringle (2003), Fedoroff and Brown (2004). A few of the above books veer towards a selectively pro-agbiotech perspective in places, but are in general factually accurate. A useful presentation of an unabashedly pro-agbiotech case is Miller and Conko (2004). Some of the more interesting anti-agbiotech books are Lambrecht (2001), Bailey and Lappé (2002), and Rowell (2003). Many books in this latter genre tend to be somewhat lacking in objectivity and rigour, but most are worth reading, albeit with a generous measure of scepticism.

447 Many agronomic traits are regulated by a combination of genetic, epigenetic and environmental effects, but even the latter are often mediated via inherited factors, as discussed in Chapter 1.

448 Lev-Yadun *et al.* (2002).

449 Murphy (1991); Murphy *et al.* (1994).

450 I will be exploring the effects of agbiotech, and its associated controversies, on individual scientists, and on the overall conduct of the scientific process, in a separate book, now in preparation.

451 Gaskell and Bauer (2001); Bauer and Gaskell (2002).

452 There are many non-transgenic varieties of herbicide tolerant crops, including maize, wheat and oilseed rape. These are resistant to widely used herbicides such as triazine and imidazolinone.

453 This topic is discussed in detail in my forthcoming book *People, Plants, and Genes* (Murphy, 2007).

454 Genes are also being exchanged between bacteria, such as *E. coli* and *Synechocystis*, at the relatively rapid rate of 16 kb per million years (Martin, 1999). It is also estimated that no less than 18% of the entire *E. coli* genome may be of relatively recent foreign origin.

455 Horizontal gene transfer is reviewed in the book edited by Syvanen and Kado (2002).

456 Davis *et al.* (2004, 2005) have described the transfer of genes to an endophytic parasite in the Rafflesiaceae from its obligate hosts in the genus *Tetrastigma*; as well as gene transfer in the opposite direction, from members of the parasitic angiosperm order, the Santalales (which includes the sandalwoods and mistletoes) to one of their non-angiosperm host plants, the rattlesnake fern (*Botrychium virginianum*). Meanwhile, Mower *et al.* (2004) described evidence for gene transfer from parasitic members of the *Bartsia* and *Cuscuta*

genera to their host species in the unrelated genus, *Plantago*. Hence, gene transfer can occur either from host to parasite, or vice versa, among plants and can cross the boundaries of genus, order, family, and even kingdom.

457 For more on virus-mediated gene transfer, and other mechanisms, see Bergthorsson *et al.* (2003, 2004) and Martin (2005). In one case, the genome of the non-parasitic, tropical shrub *Amborella trichopoda* was found to contain no less than 26 foreign genes (Bergthorsson *et al.*, 2004).

458 Fungal–plant gene transfer in the context of mycorrhizal associations (which affect as much as 90% of soil-growing plants) is potentially a very powerful and uniquely pervasive mechanism for horizontal gene flow between almost all plant species (Davis *et al.*, 2004).

459 As discussed in Syvanen and Kado (2002).

460 Ghatnekar *et al.* (2006).

461 The gene transferred from *Poa* to sheep's fescue encoded the enzyme phosphoglucose isomerase; in this case an entire functional gene was transferred between these divergent plant lineages (Ghatnekar *et al.*, 2005).

462 The gene transferred from rice to *Setaria* spp. encoded a *Mu*-like transposable element, or MULE (Diao *et al.*, 2006).

463 Won and Renner (2003).

464 This report was from Habetha and Bosch (2005), who discovered that the nuclear genome of *Hydra viridis* contained a plant gene called ascorbate peroxidase. This hydra is a member of the phylum Cnidaria, which includes the hydras, jellies (or jellyfish), sea anemones and corals. The gene probably originated from the alga, *Chlorella vulgaris*, which can associate symbiotically with animals such as hydra. The plant ascorbate peroxidase is fully functional in the modern hydra and is expressed specifically during oogenesis.

465 This phenomenon of DNA removal has been well studied in some insects, where the rate of DNA removal seems to be correlated with the size of the genome. Hence, the fruit fly, *Drosophila melanogaster*, has been able to lose exogenous DNA at forty times the rate of the Hawaiian cricket, *Laupala cerasina*, with the result that the cricket now has an eleven-fold larger genome than the fruit fly (Petrov *et al.*, 2000). So far we do not know whether a rapid rate of DNA loss is more adaptive than a slower loss, but the existence of such variations in the rate in both plants and animals suggests that rapid loss may not always be advantageous.

466 San Miguel *et al.* (1998).

467 Note the deliberate use of emotive adjectives in this section. Examples are: toxin, unnatural, abnormal, DNA-disrupting, artificial, and chemical. Such terms are frequently applied to GM technology, but can be used with equal validity for so-called conventional breeding. For other examples of the use of emotive language in the GM debate, see Chapters 3 and 12 and Cook (2004).

468 The notion of a 'species' is itself an artificial concept that has been adopted by scientists largely for convenience of identification, but is full of exceptions and may be of only limited value in many aspects of plant biology. Hence a recent article in *Nature* began as follows: 'Many botanists doubt the existence of plant species, viewing them as arbitrary constructs of the human mind, as opposed to discrete objective entities that represent reproductively independent lineages or "units of evolution"' (Rieseberg *et al.*, 2006). It may also be salutary to reflect that our most definingly 'human' biological character, namely our DNA, is well over 90% non-human (and probably most viral) in origin. In this context, concepts like 'species integrity' would appear to have little value, other than some sort of religio-mystical significance, for those that believe in them.

469 Genetic engineering is frequently criticised because small stretches of DNA are inserted randomly into a plant genome, possibly disrupting existing genes in the process. This is quite true, but transgenic plants that have serious disruptions to important genes will exhibit abnormal phenotypes and can be eliminated by selection during subsequent rounds of breeding and backcrossing. The likelihood of serious genetic abnormalities is far higher in the progeny of wide crosses, where thousands of genes, rather than one or two, are effectively transferred to the crop genome. An additional risk comes from the large amount of non-coding DNA, which can be transferred to a crop genome during a wide cross. Such DNA might include retroviral-like elements that might end up proliferating out of control in the crop genome during later generations. Curiously, such risks have never been raised in relation to non-transgenic modes of plant breeding.

470 The 'interested public' in this context means individuals and groups who may not work directly in plant and crop research but who still have an interest or a stake in such activities. It may include policymakers, politicians, environmentalists, consumer groups, journalists, academics and individual citizens.

471 Dubcovsky (2004).

472 Goodman (2004).

473 This quotation from Major M. Goodman and Martin L. Carson, of North Carolina State University, is cited in Cox (2002).

474 For example, when people around the world were offered just one of these two stark choices, and no others, by George W Bush Junior after the 9–11 disaster: 'nations are either with us or against us!' (Bush has repeatedly made speeches to this effect, e.g. in the White House Rose Garden on 24 June 2002, see online commentary at: http://www. nixoncenter.org/publications/articles/Kemp/073002nextstpes.htm).

475 The allusion is, of course, to the well-known Panglossian statement in Voltaire's, *Candide*: 'In this best of all possible worlds, My Lord the Baron's castle was the finest of castles, and My Lady the best of all possible Baronesses. "It is demonstrated" he said, "that things cannot be otherwise, for, everything being made for an end, everything is necessarily for the best end."' (Voltaire, 1759).

476 Pollan (1998).

477 This quote is from pp. 158–159 of Charles (2001).

478 Quoted in Flint (1998).

479 The description of Bob Shapiro's tenure as Monsanto CEO, the book *Lords of the Harvest* by science reporter Daniel Charles, makes an entertaining read (Charles, 2001).

480 The speech by Hendrik Verfaillie is available from the Monsanto website at: http://www. monsanto.com/monsanto/layout/media/speeches/11-27-00.asp.

481 As recently as September 2005, a Monsanto researcher published a generally well-written article on the GM debate in Europe, which quite rightly draws attention to the lamentable state of affairs there, but the writer seems to have no conception of the pivotal role played by her own company in bringing this situation into being in the late 1990s (Tencalla, 2005).

482 Mira (2006).

483 The contrasting PR skills of Monsanto and Greenpeace can be judged by two anecdotes recently related to me by a senior colleague at a Swiss university. The first example came when Greenpeace planned a demonstration against GM crops at his institute. A group of well-drilled professional protestors were bussed in from outside the region with maximum publicity to attract a police response. The protestors exclusively filmed (preventing others from doing so) their own actions, making their film available to media outlets just in time to gain peak evening TV viewing audiences and newspaper deadlines. The result was a highly publicised, colourful, and very effective account of what appeared to be a

spontaneous popular action against faceless corporate enterprises. In actual fact, the transgenic plants that were the target of the protest were part of an academic research project by Ph.D. students. The second example came in 2005 when Swiss supermarket giant, Migros, invited Monsanto representatives to discuss ways in which they might resolve the GM foods impasse that had seen Migros banning such foods since the 1990s. Migros is the largest supermarket chain in Switzerland with a 2005 turnover of almost $16 billion. The Monsanto response was to the effect that Switzerland was too small a unit for them to bother having a dialogue with, and the US company declined the offer of talks. The impact of the ensuing publicity both in Switzerland itself, and across Europe in general, can well be imagined.

484 The figure of $500 million comes from a 2003 estimate by Neil Harl, a professor of economics at Iowa State University, and represents the amount that the company has paid out to farmers, food processors and grain handlers since the problem was first identified in 2000 (Jacobs, 2003).

485 Knight (2003).

486 Thatcher (2004).

487 The company did not specify where these R&D and other cutbacks would fall (http:// www2.dupont.com/Media_Center/en_US/news_releases/2005/article20051107.html).

488 King and Schimmelpfennig (2005).

489 This figure comes from the former CEO of Monsanto, Richard Mahoney, who is now at Washington University in St Louis, Missouri (Mahoney, 2005). The dominance of Monsanto had reportedly grown to almost 98% of the global acreage of transgenic crops by late 2006 (Davoudi, 2006).

490 In 2001, Monsanto CEO Hugh Grant reported a negative cash flow totalling $8.3 billion from 1996–2000 (Phillips, 2003b).

491 Phillips (2003a).

492 Pollack (2006).

493 Duvick (2004).

494 For more details of this article, see USDA website at: http:// www.ers.usda.gov/ AmberWaves/November03/DataFeature/.

495 For example, see Binenbaum *et al.* (2000), Binenbaum (2004), Binenbaum and Pardey (2004), and Graff and Zilberman (2004).

496 It was precisely in order to dispel this poor public image that the big biotech companies eventually agreed, amid great publicity, to waive their rights to exploit the transgenic 'golden rice' in developing countries. However, the major owner of some of the technology, Syngenta, has always maintained that it will retain the rights to profit from this rice in the rest of the world.

497 Phillips and Dierker (2001).

498 Dierker and Phillips (2003).

499 Brennan *et al.* (2005).

500 For example, an economic analysis of the effect of monopoly conditions on the pricing and welfare effects of agbiotech innovations gives a mixed picture where in some cases even developing countries may not necessarily be adversely affected by monopoly or near-monopoly conditions (Acquaye and Traxler, 2005).

501 As judged by their share of total agbiotech industry utility patents, Dow and BASF are approximately half as active as the 'big four' (King and Schimmelpfennig, 2005).

502 The increased BASF commitment to agbiotech was announced on 11 April 2006, at the BIO 2006 conference in Chicago. Entitled 'A Whole New Field of Dreams: The BASF Vision For Green Biotechnology', the full text of the press release is available online at: http://www.corporate.basf.com/en/produkte/biotech/?getasset=file1&name=

BPS_BIO_Release_April_06.pdf&MTITEL=BASF+Plant+Science+Debuts+at+BIO+2006&suffix=.pdf& id=kVEva8TXfbcp0m6.

503 The challenges for improved innovation and efficiency in the agbiotech industry have been reviewed by Pray *et al.* (2005).

504 In this study, Ashikari *et al.* (2005) found that a combination of a known semi-dwarfing gene, and a newly discovered gene, *Gn1a*, that regulated the growth hormone cytokinin, resulted in shorter rice plants that carried almost 50% more grains on each branch. The higher yielding rice has a defective version of the *Gn1a* gene, which results in the formation of more grains. Because all cereals have somewhat similar genomes (Devos and Gale; 2000; Delseney, 2004), it is now possible to disrupt the *Gn1a* gene in a wide range of cereal crops and thereby obtain higher yielding varieties. This could be done either by a direct transgenic approach or by the tried and tested mutagenesis/selection method.

505 This could end up as a 'deal breaker' for the present-day agbiotech sector if companies are unable to provide useful customer-orientated traits at an appropriate cost to growers together with some certainty about their consumer acceptability. As concluded in a recent review by Jefferson-Moore and Traxler (2005):

> At least three major challenges face the industry as it tries to replicate the rapid market growth that has characterized first generation GMOs. First, genetically modified VECs (value-enhanced crops) must offer large per-acre profits compared to alternative crops. The VEC product will have to generate significant price premiums for farmers due to yield drag. Second, the adoption of VEC GMOs may also be slower than for input trait GMOs, because they introduce uncertainty into the production process, as it will take time for farmers to form yield expectations, thus increasing subjective production risk. Finally, it will take time, coordination, effort, and investment for the appropriate marketing arrangements to evolve. These transaction costs will require that some of the surplus created by VEC innovations be shared downstream in the marketing chain, making it more difficult to pay farmers a significant adoption premium.

506 I was one of many such scientists who expressed concerns to scientific colleagues at Monsanto. Interestingly, the impression we got from these researchers was that most of them, especially those who regularly travelled to Europe, shared our concerns. Nevertheless, the policy to force GM crops into Europe in the mid-1990s had already been decided at the top levels in the company and the researchers had no influence over such decisions. Colleagues in most of the other agbiotech companies were generally dismayed by the Monsanto decision and correctly predicted that if things went wrong, it would probably set the entire industry back by at least a decade. It must be stressed that these and similar market signals from well-informed sources were conveyed to Monsanto well in advance of its decision to enter the hypersensitised European market, so nobody should be able to claim that the subsequent reaction in Europe took them by surprise. I should add here that many researchers have a very high regard for our scientific colleagues in Monsanto and other agbiotech companies. In a lot of cases, they have shared our frustrations with the actions of their company hierarchy, in the same way that public sector researchers may on occasion be frustrated at the actions of our own national policymakers, governments, and local senior management.

507 The organic movement gets its name from its assertion that 'organic' (i.e. biologically derived) materials, such as manure, are preferable to non-organic fertilisers, such as nitrates. This ignores the fact that plants assimilate nitrogen in the form of non-organic nitrates and nitrites; even legumes fix atmospheric nitrogen into nitrates. The nitrogen in manure is useless to crops until it has broken down to inorganic nitrogen. Meanwhile, the use of manure might reduce risks of fertiliser runoff into watercourses, but it increases risks of manure-borne pathogens, such as *E. coli* O157:H7 and *Cryptosporidium parvum*, entering the human food chain (Comis, 1999).

508 Organic practices often have a more favourable environmental impact than conventional farming, albeit at the price of overall yield, but there is very little rigorous evidence that the organic system per se (rather than just good husbandry) results in better quality crops.

509 In the words of a recent review of the use of antibiotic resistance markers in transgenic food crops 'We conclude that, although fragments of DNA large enough to contain an antibiotic-resistance gene may survive in the environment, the barriers to transfer, incorporation, and transmission are so substantial that any contribution to antibiotic resistance made by GM plants must be overwhelmed by the contribution made by antibiotic prescription in clinical practice' (Gay and Gillespie, 2005).

510 The quoted title is from Bryant and Leather (1992); in the same year a joint paper from public and private sector scientists (including Robert Fraley from Monsanto) raised similar issues, as noted in Flavell *et al.* (1992).

511 Calgene Inc. (1990).

512 For example, Dale and Ow (1991) described how the *Cre/lox* recombination system could be used to remove unwanted DNA, such as antibiotic resistance genes. Meanwhile, by the early 1990s, several alternative methods both either positive and negative selection were available that avoided the need for using antibiotic resistance genes altogether (for example, Perl *et al.* 1993; Béclin *et al.* 1993).

513 Badosa *et al.* (2004).

514 Some of the more considered comments on the Bt10 affair are from *Nature Biotechnology*, 'Syngenta's gaff embarrasses industry and White House Accidental release of Bt-10 in US farms does not bode well for the agbiotech industry' (Herrera, 2005) and *Nature* 'Stray seeds had antibiotic-resistance genes. Accidental release of genetically modified crops sparks new worries' (Macilwain, 2005).

515 There are now dozens of often ingenious systems, such as for removing selectable markers and other unwanted DNA from transgenic plants, or use of alternative marker systems, that do not give rise to any of the concerns that surround the use of antibiotic resistance or herbicide resistance markers. A few examples include Hare and Chua (2002), Puchta (2003a, 2003b), Sun and Zuo (2003), de Vetten *et al.* (2003), Erikson *et al.* (2004), Rommens (2004), Mentewab and Stewart (2005), and Rea (2005).

516 Meganucleases are DNA-cleaving enzymes. Cellectis has developed proprietary technology to use such enzymes to remove DNA from genomes (Epinat *et al.*, 2003; also, see website at: www.cellectis.com).

517 Recent examples of reality-disconnect in the areas of food and health are legion, but some cases from the UK include the MMR vaccine case (Fitzpatrick, 2004), BSE and beef safety (Phillips *et al.*, 2000), homeopathic medicines, the Pusztai GM potatoes affair, and the alleged nutritional benefits of organic food.

518 Gressel (1992); Crawley *et al.* (1993); Raybould and Gray (1993).

519 Huppatz *et al.* (1995); Fail and Schmid (2002).

520 See BRACT website and reference to clean gene technology at: http://www.bract.org/cleangenetechnology/cleangenetechnology.html.

521 Rea (2005).

522 The important role of the use of specific types of language in the public discourse about transgenic crops is examined in the book *Genetically Modified Language* (Cook, 2004).

523 Examples of irregular and unpredicted patterns of transgene fragmentation and insertion into plant genomes can be found in Porsch *et al.* (1998) and Afolabi *et al.* (2004). In the latter paper, the authors report the presence of fragmented transgene inserts in 70–78% of 62 independently transformed plant lines (numbering over 4000 plants) that they analysed. See Kohli *et al.* (2003) and for a review of transgene integration into higher plant genomes. The problems of the transfer of 'superfluous' DNA into plant genomes

during transformation, and the subsequent potentially adverse evolutionary consequences, plus methods of avoiding such problems are discussed in the reviews by Smith *et al.* (2001) and Latham *et al.* (2006).

524 A polycistronic transgene contains several tandemly linked genes under the control of a single promoter; for more details, see the comprehensive review by Halpin (2005).

525 Shapira *et al.* (1983); Folger *et al.* (1982).

526 For example, transgene insertion frequencies via homologous recombination of only 0.1% (Kempin *et al.*, 1997), <0.1% (Halfter *et al.*, 1992), and 0.08% (Miao and Lam, 1995) have been reported in various plant systems. It has also been found that extrachromosomal DNA integrates almost exclusively at random, non-homologous sites, although the presence of protein complexes, such as Rad50/Mre11/Xrs2, can facilitate targeted transgene insertion (Mengiste *et al.*, 1999; Gherbi *et al.*, 2001).

527 The Japanese group was from the well-reputed National Institute for Basic Biology in Okazaki (Terada *et al.*, 2002). For recent reviews on gene targeting research, see Britt and May (2003), Hanin and Paszkowski (2003), Maliga (2004), and Cotsafis and Guiderdoni (2005).

528 Cotsafis and Guiderdoni (2005).

529 Out of the 60000 computer-predicted genes in the rice genome, roughly half of them have been assigned an as-yet uncertain role on the basis of their DNA sequences, while only about one hundred genes have a known and verified function (Cyranoski, 2003).

530 The potential for gene transfer from crops to wild relatives has been examined in the book *Dangerous Liaisons?* (Ellstrand, 2003a).

531 The unauthorised transfer of transgenic traits into local crop varieties is a problem in several countries, most notably in India. Here, transgenic insect resistant cotton, containing one of the Bt genes produced by Monsanto, was crossed into local cotton varieties and the pirated seed widely sold on the black market. Farmers favour Bt cotton because they save on pesticide costs and often get higher yields, but are unwilling or unable to pay the price demanded by Monsanto for the seeds, especially as this might include a contract that locks them into a longer term obligation with the company. If the Monsanto GM varieties produced sterile or otherwise non-propagable seeds, this piracy problem could be avoided.

532 Various (2004).

533 For a review of gene containment methods for crops, see Daniell (2002).

534 Maliga (2002); Daniell *et al.* (2005).

535 Apomixis is the plant equivalent of parthenogenesis, or virgin birth: seeds arise from unfertilised female tissues and are therefore genetically identical to the parent plant. Apomixis occurs spontaneously in some plants but has not yet been engineered as a trait in transgenic crops (Spillane *et al.*, 2004). Male sterility normally means infertile or absent pollen. It is already used in several crops, including transgenic oilseed rape. Sterile seed traits are common in many crops, including cereal crops, and are now being developed as gene containment systems in some transgenic crops. As with inbred-hybrids and male sterile systems, seed sterility also prevents seed saving for repropagation by farmers. Cleistogamy means pollination only within a closed flower, which prevents pollination by other plants. There are recent encouraging indications that cleistogamy in some of the major crops may be controlled by a very small number of genes, hence making manipulation of this potentially valuable trait much easier (Takahashi *et al.*, 2001; Turuspekov *et al.*, 2004).

536 Sangamo BioSciences was founded in 1995, and went public in 2000, see website at: http://www.sangamo.com.

537 Halpin (2005).

538 In areas like genomics, a degree of such outsourcing has already been commonplace for the past few years, with large agriculture-based firms like Cargill, Dow and Monsanto signing agreements with (among many others) such smaller technology-rich, gene-discovery companies as Mendel Biotechnology Inc. (http://www.mendelbio.com/) and Ceres Inc. (http://www.ceres-inc.com/) in the USA.

539 These data are from the study by Vain (2005) which analyses published research output in plant transgenic science. The authors found no significant increase in papers relating to technology development from 1995–2003, and predict that this lack of growth is likely to hamper the further evolution of basic and applied transgenic research and the development of transgenic crops. However, the study also notes an increase in papers on the application of GM technology and in areas such as GM crop performance and risk assessment.

540 Use of insect-control sprays containing a pro-toxin-producing *Bacillus thuringensis* suspension has been common for over 30 years in organic farming, but use of Bt toxins in transgenic crops is much more recent. The Bt toxins are a family of so-called crystalline (cry) proteins that are converted into their active form during digestion in the gut of a range of insect larvae, resulting in a disruption of potassium ion transport that rapidly becomes lethal (USDA Server, www.nal.usda.gov/bic/BTTOX/bttoxin.htm). Mammals do not convert the pro-toxins into their active forms and are therefore unaffected by them. Early indications (Briggs and Koziel, 1998) suggest that transgenic Bt crops are effective in controlling insects and improving yields (by 7%), while also reducing the need for spraying with more toxic and less desirable pesticides that often affect beneficial organisms such as insectivorous birds (saving growers some $40/hectare).

541 Tabashnik *et al.* (2003); Goldberger *et al.* (2005); Kranthi (2005); Romeis *et al.* (2006).

542 See Sundstrom *et al.* (2002) for a review on identity preservation issues in agbiotech.

543 Macilwain (2005).

544 Herrera (2005); Macilwain (2005).

545 In my own research area of oilseed crops, I have repeatedly pointed out the serious management challenges involved in the design and enforcement of stringent systems of transgenic crop segregation, followed by the identity preservation of all the products that are derived from such crops (Murphy, 1991, 1994, 1995, 1996, 1999b, 2003a, 2004c; see also comment online at: http://www.gmsciencedebate.org.uk/topics/forum/0002.htm). Recent publications on transgenic crop segregation and identity preservation in other crops include Wilson and Dahl (2002), Huffman (2004), and Lapan and Moschini (2004).

546 Edible oilseed rape varieties make up about 90% of the UK crop. The oil is rich in the monounsaturated fatty acid, oleic acid, and is used for a host of food applications as well as being sold directly as a salad oil or in the form of margarine. About 10% of the UK oilseed rape area is used to grow another variety that looks identical to the eye but has a totally different seed oil. In this case, the oil is rich in erucic acid, which is used to make polymers and non-stick surface coatings. Despite the fact that bees can carry oilseed rape pollen for several kilometres, UK farmers have grown the two crops on a large scale and have successfully kept the crops, seeds and oil completely segregated from each other every season for over three decades, with no problems of cross contamination. As we have seen in the case of the StarLink and Bt10 scandals, parts of the agbiotech industry do not have a similar record of good management.

547 These transgenic virus resistant crops express complete or partial proteins from a particular virus (typically part of the viral coat protein complex), which causes the plants to become sensitized to subsequent infections with the same virus. When such plants are attacked by the viral pathogen, they mount a successful defence response in a manner that is analogous to immunisation in animals that have been injected with an attenuated virus, although the exact mechanism of viral immunity in plants remains to be explained.

A recent example of a transgenic virus resistant crop is a papaya variety developed in Hawaii and in Australia. The papaya ringspot virus is a major threat to the cultivation of papayas in tropical countries like Hawaii. Transgenic papayas that express the ringspot virus coat protein, which on its own is harmless to the plants, are considerably more resistant to infection with the active virus than are non-transgenic papayas. (International Service for the Acquisition of Agri-biotech Applications Server http://www.isaaa.org.)

548 Nematodes are the most destructive animal parasites of crops, causing over $100 billion annual losses to world agriculture. Biological resistance to nematodes is relatively restricted in the major crops, so chemical control methods are used instead. These agents, such as carbamates, include some of the most toxic pesticides in widespread use; they are both costly to the farmer and environmentally damaging. The development of transgenic crops with nematode resistance could therefore have appreciable economic and environmental benefits. Some of the research approaches include the induction of so-called 'suicide genes' in plant cells that become infected with a nematode, or the expression in plants of protease inhibitors that inhibit nematode growth (Atkinson *et al.*, 1995). To date these studies are still some way from commercial application, but they remain a promising option for the future.

549 The broader question of engineering plants with increased disease resistance, both regarding what genes to use and how to ensure that they are expressed in the right place and at the right time, is examined in two back-to-back reviews by Gurr and Rushton (2005a, 2005b).

550 Zhai *et al.* (2000).

551 Shah *et al.* (1995).

552 Gao *et al.* (2000).

553 *Phytophthora infestans* is the pathogen that caused the devastating Irish famine in the 1840s.

554 Altman (1999). The potential impact of current changes in global climates were also highlighted at the CGIAR annual meetingin December 2006 (CGIAR, 2006).

555 For example, see the seminal review by Boyer (1982).

556 A useful overview of abiotic stress tolerance and its possible amelioration via genetic engineering is the review by Wang *et al.* (2003).

557 Vinocur and Altman (2005).

558 Mittler (2006)

559 Yamaguchi-Shinozaki and Shinozaki (2001).

560 Flowers (2004).

561 Tarczynski *et al.* (1992).

562 Nuccio *et al.* (1999).

563 Zhang *et al.* (2001).

564 Yamaguchi and Blumwald (2005).

565 As stated in Flowers (2004):

> It is surprising that, in spite of the complexity of salt tolerance, there are commonly claims in the literature that the transfer of a single or a few genes can increase the tolerance of plants to saline conditions. Evaluation of such claims reveals that, of the 68 papers produced between 1993 and early 2003, only 19 report quantitative estimates of plant growth. Of these, four papers contain quantitative data on the response of transformants and wild-type of six species without and with salinity applied in an appropriate manner. About half of all the papers report data on experiments conducted under conditions where there is little or no transpiration: such experiments may provide insights into components of tolerance, but are not grounds for claims of enhanced tolerance at the whole plant level. ... After ten years of research using transgenic plants to alter salt tolerance, the value of this approach has yet to be established in the field.

566 Masle *et al.* (2005).

567 Moore *et al.* (1995)

568 Anderson *et al.* (1998).

569 For reviews on the use of plants in 'bioremediation', see Prasad and de Oliveira Freitas (1999) and Gressel and Al-Ahmad (2005).

570 Prasad and de Oliveira Freitas (1999). There have also been efforts to overexpress another class of glutathione-derived metallothioneins, termed phytochelatins, by transferring a bacterial γ-glutamylcysteine synthase gene into poplar trees (Arisi *et al.*, 1997). Preliminary data indicate that the transgenic trees contain higher levels of glutathiones, but it is not yet clear whether this leads to increased phytochelatin accumulation or tolerance to heavy metals.

571 Recent studies have shown that metallothioneins are highly expressed during fruit ripening (Moriguchi *et al.*, 1998; Abdullah *et al.*, 2002), leaf senescence (Buchanan-Wollaston and Ainsworth, 1997), wounding and viral infection (Choi *et al.*, 1996) and fungal and bacterial infection (Butt *et al.*, 1998).

572 Abdullah *et al.* (2002).

573 Ye *et al.* (2000).

574 For an example of the arguments against 'golden rice', see the Greenpeace Server online at: http://www.greenpeace.org/~geneng/. Interestingly, commercial rights for 'golden rice' in the USA and Europe have now been acquired by Syngenta. It is possible that this could lead to the marketing of 'vitamin-enhanced' food products derived from golden rice, e.g. vitamin A enhanced breakfast cereals, which might be more acceptable to the public than the current generation of food from input trait modified GM crops.

575 Early versions of golden rice contained only 1.6 micrograms of β-carotene per gram of rice (Ye *et al.*, 2000).

576 Paine *et al.* (2005).

577 Other partners in the HarvestPlus programme include CIMMYT, Mexico; IITA, Nigeria; the University of Illinois, Urbana-Champaign; and Wageningen University, Netherlands. Monsanto Company is also a partner.

578 For more information on VITAA and the new high vitamin A sweet potatoes, see the website at: http://www.cipotato.org/vitaa/.

579 Low *et al.* (2001).

580 Cahoon *et al.* (2003).

581 In one study, tocols were found in unrefined, or 'virgin', palm oil at levels in excess of 1500 ppm, which is comparable to some commercial vitamin supplements (Han *et al.*, 2004). Red palm oil is also an excellent source of provitamin A.

582 Carotenoids are the pigments that have been engineered into transgenic 'golden rice', but they also occur in many coloured vegetables, such as the eponymous carrots, and in most unrefined plant oils. Lycopenes are the red pigments that are so prominent in tomatoes, cayenne and bell peppers, red grapefruit and a few vegetable oils, most especially from the oil palm. Although both carotenoids and lycopenes are available as vitamin supplements, by far the most reliable way to ensure their efficient uptake in the body is by consuming them as part of an original food product, such as a fruit or an unprocessed plant oil, rather than in isolated capsule form.

583 The iron-rich transgenic rice was produced by inserting three groups of transgenes; a ferritin gene from the bean, *Phaseolus vulgaris*, doubled the overall iron content of the grain, a phytase gene from the fungus, *Aspergillus fumigatus*, decreased phytate levels, and an endogenous cysteine-rich metallothionein was added to stimulate dietary uptake of iron (Lucca *et al.*, 2002).

584 Zimmermann *et al.* (2002).

585 White and Broadley (2005).
586 Haas *et al.*, (2005); Graham (2003).
587 Murphy (1994, 1999b, 2003a).
588 The market for oils that contain reduced or zero levels of *trans*-fatty acids is currently driven by health concerns that have led to the imposition of labelling requirements revealing whether a product contains over a given threshold of these fatty acids. Such labelling requirements were introduced in the USA in 2006 and are also likely to be required in the European Union in the near future (Murphy, 2006).
589 One of the problems encountered in trying to produce novel oils in transgenic plants is that, although the plant cells can usually make the new fatty acids, the latter are sometimes then broken down for as-yet unknown reasons, as we discovered when attempting to engineer transgenic oilseed rape to accumulate an industrially useful isomer of oleic acid, called petroselenic acid (Murphy *et al.*, 1999).
590 High-oleic soybean varieties with as much as 83% oleate and less than 3% α-linolenate have been developed (Rahman *et al.*, 2001) and are now being marketed by major seed companies, including Monsanto, which launched its VISTIVE™ soybeans in 2004. Despite almost two decades of effort to produce high-oleic soybeans via transgenic methods, the VISTIVE™ varieties were obtained by conventional, non-transgenic approaches. Breeders have also developed other lines of soybeans with high levels of stearic acid (Rahman *et al.*, 2003) and other nutritionally relevant fatty acids. Several high-oleic canola/rape varieties have been developed with about 70–80% oleate, 15% linoleate, and only 3% α-linolenate. Major agbiotech companies such as Cargill, Dow Agrosciences and Bayer are now developing such varieties for various end-use markets, both edible and non-edible. By 2004, high-oleic rape/canola was already being planted in Canada on about 250,000 ha, or 5% of the total area of total canola cultivation (AgCanada, 2004). For further discussion of oil crop modification, see Murphy (2006).
591 Kleingartner (2002).
592 Monsanto used non-transgenic breeding to produce a high-oleic variety of soybean, called VISTIVE™ (see note above). The PR dilemma facing the company can be seen by the way they promote VISTIVE™ beans as '*produced through conventional breeding*', although the commercial varieties also carry the Roundup transgene and so are technically classed as 'GM'. Hence, low-*trans* fatty acid oils derived from VISTIVE™ beans are unlikely to be used by European retailers, notwithstanding any potential nutritional benefits.
593 Oils rich in these omega-3 fatty acids include the so-called 'fish oils' (or more correctly 'marine oils'), which are characterised by relatively high levels of very long chain polyunsaturated fatty acids (VLCPUFAs) such as eicosapentaenoic acid (20:5v-3, EPA) and docosahexaenoic acid (22:6v-3, DHA). These compounds are part of the group of omega-3 fatty acids that are essential components of mammalian cell membranes, as well as being precursors of the biologically active eicosanoids and docosanoids (Funk, 2001; Hong *et al.*, 2003).
594 Crawford *et al.* (1997), Benatti *et al.* (2004), Spector (1999).
595 An example of another fatty acid supplement is gamma linolenic acid, or GLA, which is usually obtained from evening primrose, *Oenothera biennis*, or borage, *Borago officinalis*, oils. GLA supplements are taken for almost every ailment under the sun, but a healthy person with a balanced diet should be able to synthesise enough GLA for all normal metabolic functions without recourse to such nutritional supplements.
596 The most serious challenges to the engineering of VLCPUFA production in plants is the number of enzymes that are needed for the conversion of a typical plant C18 PUFA, such as linoleate or linolenate, to the C20 and C22 VLCPUFAs with up to six double bonds that are characteristic of fish oils. Other key challenges are similar to those that have

confronted previous attempts to engineer transgenic oilseeds, namely to ensure seed-specific expression of the transgenes and to channel the novel fatty acids towards oil accumulation and away from membrane lipids (Murphy, 2006). During 2004 and 2005, there were several reports that encourage the view that the economic production of VLCPUFAs in transgenic plants might be possible (Abbadi *et al.*, 2004; Qi *et al.*, 2004; Wu *et al.*, 2005).

597 Wu *et al.* (2005)

598 Murphy (2002, 2003a).

599 The economic argument against low- and intermediate-value products from transgenic crops is made in Murphy (2002). The major problem with relatively high-volume intermediate-value products, such as bioplastics, is their high land-use requirement, which could begin to compete significantly with food crops, plus their low profitability compared with existing conventional sources of such products.

600 See reviews by Freese (2002), Rodgers (2003) and Stoger *et al.* (2005)

601 Mahoney *et al.* (2005).

602 The capital cost of conventional fermentation-based bioreactor production of therapeutics is estimated to be about ten times greater than that of plant-based systems (Pew Initiative, 2004).

603 Many therapeutic proteins, such as vaccines, enzymes and antibodies, are post-translationally modified, e.g. by glycosylation or acylation, when expressed in their normal host species. However, the bacterial expression systems that are currently used to produce most recombinant proteins are often unable to incorporate such modifications, and the resulting proteins may not have the required activity. However, because they are eukaryotes just like animals, plants are often able to make appropriate post-translational modifications and thereby produce active versions of therapeutic proteins (Joshi *et al.*, 2004). In several cases, plants produced proteins, including follicle-stimulating hormone (Dirnberger *et al.*, 2001), measles virus (Huang *et al.*, 2001), and acetylcholine esterase (Mor *et al.*, 2001), which were therapeutically active, but not necessarily glycosylated in the pattern that is normal in humans. In other cases, e.g. an immunoglobulin produced in transgenic alfalfa (Bardor *et al.*, 2003), the glycosylation pattern was, encouragingly, exactly the same as in humans.

604 For a detailed risk analysis of vaccine production in plants, see Kirk *et al.* (2005). Similar principles apply to other classes of biopharmed crops.

605 Vandekerckhove *et al.* (1989) reported the expression of enkephalins in transgenic plants. Enkephalins are pentapeptides that act as ligands for opiate receptors in the human brain. These small peptides normally bind endorphins but can also act as extremely potent painkillers. In the same year, Hiatt *et al.* (1989) reported the production of mammalian immunoglobulins and assembly of functional antibodies in transgenic tobacco plants.

606 The public health implications of biopharming have been examined at length in a review by Law Professor, Rebecca Bratspies, of the City University of New York. In the article, she is especially critical of the '*laissez-faire*' attitude of US regulatory agencies, such as USDA and FDA, towards the containment of biopharmed crops (Bratspies, 2004). Similar criticisms were voiced in a report by the USDA's own auditor in late 2005 (USDA, 2005).

607 Fox (2003).

608 Hileman (2003).

609 Jones and Brooks (1950).

610 Harl *et al.* (2003).

611 One of the most seriously affected food companies was the fast food franchise, Taco Bell, which was reimbursed $60 million for sales lost after the recall of Taco Bell taco shells that contained StarLink maize. Taco Bell franchises maintained that their sales were hurt

by the recall, even though the affected taco shells were supplied only to supermarkets (Taylor and Tick, 2002).

612 Aventis CropScience was a division of the much larger multinational pharmaceutical firm Aventis (now called Sanofi-Aventis), which is one of the largest of the 'big-pharma' companies with global sales of $27 billion (Economist, 2005e). By 2001, Aventis was planning to sell off its underperforming agbiotech division anyway, in order to focus on its more profitable pharmaceutical activities, but this decision was greatly reinforced and accelerated by the StarLink disaster.

613 Fox (2003).

614 Ellstrand (2003b).

615 For the biopharming industry perspective on risk management, see Biotechnology Industry Organization (2002). A more impartial perspective can be found in the paper from the Pew Initiative (2002), while a more anti-agbiotech view from the Union of Concerned Scientists in the USA is available from Andow (2004).

616 In 2003, Monsanto abandoned its biopharming R&D programme after deciding that the potential for making profits from the venture was too far into the future.

617 It is not only small companies like ProdiGene that are using mainstream crops like maize for biopharming; Dow Chemical Company is also using maize for the expression of drugs such as appetite suppressants.

618 The ProdiGene case for the use of maize as a production system for human and animal vaccines can be found in the paper by Streatfield *et al.* (2003).

619 Lee and Lau (2004).

620 Anheuser-Busch is one of the biggest brewers in the world and makes Budweiser, the most popular brand of beer in the USA. At first, Anheuser-Busch was strongly opposed to the Ventria trials but in April 2005 the two companies came to an agreement that the transgenic rice would be grown at least 120 miles away from the prime rice growing areas of Southeastern Missouri. This agreement came too late for the 2005 growing season, but the company now plans to go ahead with field trials in 2006 (http://www.checkbiotech.org/root/index.cfm?fuseaction=search&search=ventria&doc_id = 10217&start = 1 &fullsearch = 0).

621 In 2006, Ventria explored sites in North Carolina and Kansas (Downing, 2006). Eventually, in 2007, the USDA gave preliminary approval for cultivation of 1215 hectares of the transgenic rice in Kansas.

622 Einsiedel and Medlock (2005).

623 Brasher (2006).

624 Bailey (2005) & Hoffert (2005).

625 The relative efficacies of different types of plant production, ranging from tobacco leaves to potato tubers, are compared in Fischer *et al.* (2004).

626 Another possibility, which is being investigated, is to link the transgene to a promoter that is only activated once the leaves are harvested from the parent plant. Senescence-induced or wound-induced promoters may fit the bill here (Cramer *et al.*, 1999), as long as the crop is not excessively damaged while still in the field, e.g. by wind or herbivores, or is not left out until it begins to senesce anyway.

627 Padidam (2003); Hunzicker *et al.* (2004).

628 Examples of recent reviews of safety in biopharming include: Commandeur *et al.* (2003), Andow (2004), Mascia & Flavell (2004), Peterson and Arntzen (2004) and Elbehri (2005).

629 Details of Icon Genetics are at: http://www.icongenetics.com/ html/home.htm and Large Scale Biology Corporation at: http://www.lsbc.com/. The use of inducible gene promoters in transgenic tobacco has been pioneered by Carole Cramer and colleagues (at Virginia Tech and Arkansas State Universities) as reported in Cramer *et al.* (1999) and

Medina-Bolivar *et al.* (2003). After a shock announcement in December 2005, Large Scale Biology Corporation was shut down, making its staff redundant. This demonstrates the high risk and volatility of agbiotech R&D, especially for small companies (Agres, 2006).

630 Gleba *et al.* (2005); Marillonnet *et al.* (2005).

631 The virus on its own is not very efficient at infecting plant cells, so the Icon team has used the soil-borne plant pathogen, *Agrobacterium tumefaciens*, to deliver viral replicons into the plant cells (Marillonnet *et al.*, 2005).

632 Something similar happens to each of us when we get a cold or any other form of viral infection. During such an infection, virus particles enter some of our cells, which are then transiently recruited to make large quantities of new viral proteins so that the virus can replicate and spread. For a few days these cells become genetically engineered biopharming factories making millions of viral proteins and nucleic acids. Luckily, we normally recover and our partial genetic transformation by the virus only lasts for a few days.

633 The full-scale production phase of the anti-caries antibody was announced in July 2005 and is a joint venture with Plant Biotechnology, based in California, as reported online at: http://www.in-pharmatechnologist.com/news/printNewsBis.asp?id = 61277.

634 Yamamoto *et al.* (2001).

635 Cell cultures of the yew *Taxus brevifolia* have been reportedly scaled up to 70 000 litres for commercial-scale taxol production (Venkat, 2001). Several companies are using this and other *Taxus* species for cell culture production of taxol, including *Taxus cuspidata*, and *Taxus canadensis* (Yukimune *et al.*, 1996; Ketchum *et al.*, 1999).

636 For example, transgenic root cultures of the plant *Hyoscyamus muticus* expressing the *h6h* gene accumulated 100-fold more scopolamine than non-transformed controls (Jouhikainen *et al.*, 1999; Oksman-Caldentey (2000).

637 See Mahoney *et al.* (2005).

638 Turner (1993) Morphine is obtained from the opium poppy, *Papaver somniferum*, while artemisinin comes from the herb, *Artemisia annua*.

639 Canter *et al.* (2005).

640 The lax regulatory system in the USA was the subject of an editorial entitled: 'Drugs in crops – the unpalatable truth', in the journal *Nature Biotechnology* (Anonymous, 2004). Weaknesses in USDA regulation of pharmaceutical crops were also highlighted in a critical report by the Office of the Inspector General in December 2005 (USDA, 2005).

641 Kirk *et al.* (2005).

642 Ma *et al.* (2005).

643 In this quotation, Swift argues that the chronic hunger that beset the poorer classes in eighteenth century Ireland was principally due to the policies carried out by the English over a prolonged period, rather than being the immediate consequences of a good or a bad harvest (Swift, 1725). Likewise, the persistent malnutrition and hunger that we see today in some parts of the world owe their origins to more complex economic and social causes, instead of being simply due to a localised failure in agricultural production.

644 The case for agbiotech as a panacea for world hunger has been presented by most of the relevant companies. For example, the Council for Biotechnology Information (a trade organisation) has stated the following regarding hunger: 'few advances offer a more feasible way to help put an end to this suffering than biotechnology. With the help of biotechnology, more food – food even richer in necessary vitamins – can be produced, and crops in underdeveloped countries can be protected from the ravages of insects, diseases and even inclement weather.' (quoted on p. 142 of van Wijk, 2002). Similar perspectives, on 'feeding the world', in this case from Monsanto scientists, can be found in the articles by Cockburn (2004) and Mackey and Montgomery (2004).

645 For example, see 'Malthus foiled again and again' (Trewavas, 2002). It should be noted that, although Malthus initially believed that populations would reach a limit once the available fertile land was used, he went on to apply a more sophisticated view to his analysis of the causes of famines. He realised that famine and food availability can co-exist. In some of his later work, Malthus correctly observed that famine was due as much to the distribution of income as to the availability of food. In other words, he made the case for the contribution of poverty to famine as being at least as important as any lack of food production (Wrigley, 2004).

646 In addition to Paul Ehrlich's book, as previously discussed in Chapter 6 (Ehrlich, 1968), this notion was most powerfully enunciated in the 1972 report from the Club of Rome, entitled *The Limits to Growth*, which outlined a series of scenarios, most of which led to global catastrophe within a few decades due to 'fundamental limits to growth in global population, agriculture, resource-use, industry and pollution' (Meadows *et al.*, 1972). Although this book has subsequently been updated on two occasions (in 1992 and 2004), it was the original 1972 edition that had the greatest effect on policymakers and public alike in helping to set out the received wisdom of seemingly inevitable ecological disaster.

647 For example, in May and June 2003, US President George W. Bush accused European governments of impeding efforts to fight 'world famine' and 'global hunger' because of their trade policy of blocking new transgenic crops (Driscoll, 2003). In April 2006, Bill Clinton told the Biotechnology 2006 conference in Chicago that the solution to feeding the world's poor lay in GM crops (Daly, 2006). During the 1990s, Bill Clinton had also supported the link between transgenic crops and the battle against hunger, albeit in more moderate language. Meanwhile, although they accept the flawed evidence of the Malthusian doomsayers, many US politicians do not accept some of the better-founded evidence about climate change. Ironically, in this respect, their inconsistent and self-serving actions resemble those of many environmental protest groups.

648 For example, Cassman *et al.* (2003) have raised doubts about whether the yields of the major cereals can continue to increase via conventional crop breeding.

649 Duvick (2004).

650 Borlaug and Dowswell (2001).

651 Examples of such organisations include the FAO and the Pew Initiative, while respected and well-informed individuals who have made similar statements include dozens of plant breeders of the calibre of Norman Borlaug and Donald Duvick.

652 Food and Agriculture Organisation (2004b).

653 A useful outline of some (but not all) of the key issues in the relationship between agbiotech and world hunger can be found in the discussion paper released by the Pew Initiative (2004).

654 According to the UN Population Division, the global population was predicted to grow from 6.5 billion in early 2005 to 9.1 billion in 2050, which is a 40% increase (2004 Revision, http://esa.un.org/unpp/). However, it is also estimated that there is an 85% probability that the world population will stabilise, and may even begin to fall, by the end of the twenty-first century (Lutz *et al.*, 2001).

655 For example, the ratio of births per woman is 5.0 in Pakistan but only about 1.5 in Europe (www.iiasa.ac.at/Research/LUC/Papers/glch1/chap1.htm). Relative economic security and wellbeing are the major drivers of reduced family size, although other factors like the availability of contraception also play a part. Hence, it is largely the rapid growth of the Chinese economy since the 1980s, rather than the single-child policy, that will ensure that the population of China will soon be outstripped by that of India. Meanwhile, the region of the greatest relative population growth will be sub-Saharan Africa. Much of this region has experienced negative growth over the last twenty years so,

despite its gruesome AIDS pandemic, it is likely that its population will not stabilise until near the end of the 21st century (Lutz *et al.*, 2001).

656 Fry (1998).

657 Not all of our increased food crop production is used directly as human foodstuffs. In the case of crops such as maize, the majority of US and European production of this edible crop is used for animal feed, as is a large proportion of the soybean harvest.

658 The Wartime diet that many of our parents and grandparents had to endure during the dark days of World War II may seem unappealing to modern tastes, but its low fat, high fibre and sensibly sized portions meant that the British population as a whole enjoyed a level of health and fitness unsurpassed since 1945. This is despite the fact that it was very starch and energy rich, it had a higher salt content, and most of the fat was highly saturated. The Wartime diet was also low in fruit, especially in the winter, but did not contain the hundreds of added chemicals found in the pre-prepared convenience foods of today. The UK Wartime diet has now been popularised as a model for today's overweight Britons, as exemplified by publication of *The Ration Book Diet* (Brown *et al.*, 2004).

659 *World Agriculture: Towards 2015/2030* is available from the FAO website (Food and Agriculture Organisation, 2004b).

660 Food and Agriculture Organisation (2003, 2004a, 2004b).

661 Numbers of hungry people in the world, excluding sub-Saharan Africa, are predicted to fall by over a half, from 576 to 257 million, by 2030 (Food and Agriculture Organisation, 2004a, 2004b).

662 This point was recognised in the 2006 report of the World Trade Organization (WTO, 2006). The wider issue of the need for thriving and stable markets as a prerequisite for agricultural improvement in Africa is examined in detail by Djurfeldt *et al.* (2005). See also the article on the delayed Green Revolution in Africa in the Economist (2006).

663 Some recent examples of this are Western campaigns that have been launched against Malaysian and Indonesian oil palm and Brazilian soybeans because of fears about the environmental impact of such crops in biodiverse tropical regions. While some of the concerns about unnecessary habitat destruction are valid, these only relate to a small fraction of the areas of both crops, which provide hundreds of thousands of jobs and are key export earners for the producer countries.

664 Panayotou (1993).

665 For more detailed discussion of modern interpretations of the ECK hypothesis, see Arrow *et al.* (1995), Chevé (2000) and Bimonte (2001).

666 This topic is discussed in more detail in the article entitled: 'Refining plantation technologies for sustainable production: the path to eco-economy' (Chan *et al.*, 2003). See also Basri *et al.* (2004) for recent development in sustainable oil palm research.

667 Some plantations, such as those managed by EPA Management Sdn. Berhad, have been using a zero-burn technique for oil palm since as long ago as 1984 (Leng, 2000).

668 Even in a prosperous country like the UK, it was only in 1993 that the government made it illegal to burn crop residues. Prior to this, late summer in many rural areas had been marked by huge clouds of choking smoke that were not only environmentally unsound, but also the cause of numerous traffic accidents. To this day, the open burning of agricultural residues is permitted in many parts of the USA. This is especially marked in California, where I directly experienced the unfortunate side effects of the ensuing smoke clouds at several abortive summertime barbecues, while working at UC Davis in the heart of the agricultural Central Valley.

669 See discussions in Panayotou (1997) and Aznar-Marquez and Ruiz-Tamari (2005).

670 Many of these sustainability issues are discussed by Tilman *et al.* (2002).

671 Conway (1997).

672 Harrison and Pearce (2001).
673 Data from UN FAO (http://www.fao.org).
674 Data from UN FAO (http://www.fao.org).
675 It used to be thought that large-scale cereal cultivation in North Africa was abandoned for mainly climatic reasons but this view has been challenged more recently by a more nuanced perspective involving the interaction of social and climatic factors. Cereal cultivation in the region was highly organised and largely dependent on elaborate irrigation systems and terraces, the remains of which are still evident throughout the area. Once this infrastructure was destroyed by invaders such as the Vandals, it would have been very difficult to replace. At the same time, the increasing difficulty in shipping the grain to the main markets in Italy, coupled with a decline in demand as Rome itself became depopulated, removed any incentive to rebuild the cereal infrastructure. This may have been exacerbated by drier climatic conditions but the latter was not the main cause of the disaster. Once the infrastructure had been abandoned, irreversible changes such as soil erosion and salinisation occurred, which largely precluded any future attempts to resume large-scale cereal cultivation.
676 www.europa.eu.int/comm/agriculture/publi/caprep/prospec ts2004a/grains.pdf.
677 A useful description of Amerindian settlements in the Hudson Valley and Western Connecticut can be found in Russell Shorto's immensely readable account of the rise and fall of the Dutch colony of New Amsterdam, *The Island at the Centre of the World* (Shorto, 2005).
678 A useful summary on the future of South American agriculture can be found in the paper by Featherstone and Conforte (2002).
679 The estimate of 100 million hectares of new arable land is taken from Schnepf *et al.* (2001). In 2005, Silvio Crestana, director of EMBRAPA, estimated that at least 30 million hectares of new land could be developed immediately without impacting on environmentally sensitive areas (Economist, 2005g). Note that the Brazilian *cerrado* is a savannah and not a rainforest ecosystem. Although a small part of the recent expansion of soybean cultivation in parts of Brazil has been at the expense of primary rainforest, the majority has involved the colonisation and development of non-rainforest habitats, such as the *cerrado*, which are much easier to clear and are far more productive in the longer term than forest.
680 Harrison and Pearce (2001).
681 During early 2005, soybean cultivation in some parts of Brazil was badly hit by a drought, which reduced national grain production. However, the drought mainly affected the existing major growing area of Rio Grande do Sul in the south of the country. In contrast, the Matto Grosso and Parana areas, where much of the new *cerrado* cultivation is planned, had good rainfall and would have made up for shortfalls in other parts of the country (see USDA report at: http://www.fas.usda.gov/pecad/highlights/2005/03/BrazilSoy/).
682 Warnken (1999); Schnepf *et al.* (2001).
683 Diaz (2004).
684 Schnepf *et al.* (2001).
685 Economist (2005b).
686 Brazil is also belatedly improving its railroad system. For example, in 2005 work began on a new extension of the *Ferronorte* line from Matto Grosso to the port of Santos, cutting the cost of shipping soybeans for export by 20–25% (Economist, 2005g).
687 Thailand is now an agricultural superpower, ranked eighth in the world in terms of its $11 billion annual export volume (data from Food and Agricultural Organisation, www.fao.org).

688 Denmark is the most intensive hog producer, with 25 million pigs (5 pigs per person), followed by the Netherlands and Canada with 15 million pigs each. The largest producer is the USA, with over 110 million pigs per year, almost all reared on intensive lots.

689 Tilman *et al.* (2002).

690 Economist (2005f) Note that the farmland areas quoted in this article are mistakenly converted from acres to hectares. The true values are 2.6 million hectares in 1993 and 4.2 million hectares in 2005.

691 Wahid *et al.* (2004) and MPOB website at: www.mpob.gov.my.

692 See Fearnside (2000) for a perspective on the possible environmental threat posed by soybean farming in Brazil.

693 For example, in the region of the UK where I live (South Wales), two centuries of relentless mineral exploitation and industrial production have totally removed virtually all vestiges of any native flora and fauna. But the resulting wealth sustained the growth of Britain as a world power, and was instrumental in creation of the high living standards in the UK today. One wonders how people in nineteenth century South Wales would have felt if they were approached by groups of rich Indian and Chinese environmentalists and told that they should not seek to emulate the wealth of the Eastern nations lest they disturb the habitat of oak forests, red squirrels, badgers and other priceless wildlife.

694 Brock (2003), Diaz (2004), Morais (2005) , McVey *et al.* (2000).

695 Quoted from p. 98 of Economist (2005g).

696 Oil World (2005).

697 Murphy (2006).

698 Tinker (2000), Murphy (2004c).

699 Chan *et al.* (2003).

700 These exceptional oil palm trees are located on the Jendarata Estate plantation of United Plantations Berhad in Malaysia (website at: http://www.unitedplantations.com/).

701 Olives also produce most of their oil in their mesocarp, i.e. the fleshy part of the fruit, but the overall oil yield of an olive tree is much inferior to that of an oil palm tree. Another one of the very few crops with oily fruits is the avocado, but avocados are normally consumed as whole fruits, rather than serving as a source of extracted vegetable oil.

702 Murphy (2004c).

703 Murphy (2006).

704 Ho *et al.* (1996).

705 Ho *et al.* (1996).

706 Some of the complex arguments relating to the economics of demand for food and the possible role of agbiotech in ameliorating food supplies are explored by Krattiger (2000).

707 Wambugu (1999).

708 Data are from the USDA Production Estimates and Crop Assessment Division, Foreign Agricultural Service (see website at: http://www.fas.usda.gov/pecad/highlights/2004/12/Ukraine%20Ag%20Overview/).

709 For example, a US government delegation to the Ukraine in 2001 reported that the region had less than one third of the combining equipment needed to ensure a timely harvest. As a result, over 15 million tonnes of grain are avoidably lost each year, which is more than the entire annual wheat production of the UK. The harvesting equipment used to be manufactured locally but the supply system broke down post-1990. Hence this is more of a management issue than one of affordability (report from USDA Foreign Agricultural Service, see online at: http://www.fas.usda.gov/remote/soviet/2001/fsu_trip_report/fsu_report.htm).

710 Food and Agriculture Organisation (2003) pp. 19 and 22–23.

711 Between 1993 and 2004, it is estimated that the proportion of Vietnamese people living in poverty fell from 58% to 20%. What makes this decline in poverty even more impressive

is that it occurred despite a population rise of 15% (from 72 to 83 million) over the same period (Economist, 2005h).

712 From 1993 to 2005, the Vietnamese economy has sustained an average annual GDP growth rate of 7.2%, and the once iconic vision of the bicycle-clogged streets of Ho Chi Minh City has given way to streets filled with motorcycles and, more recently, by cars (Economist, 2005h).

713 ISAAA newsletter, 17 June 2005, 'Vietnam collaborates with U. S. and India on Biotech', available online at: http://www.isaaa.org/kc/.

714 Rosegrant and Cline (2003).

715 Note that, although agricultural over-production is closely linked to government subsidies, the two phenomena are not necessarily dependent on each other. For example, Wise (2004) has argued that, in the case of maize, the reduction of subsidies to US growers is unlikely to help Mexican maize farmers because large US companies will continue to dump below-cost maize into Mexican markets.

716 See USDA report online at: www.fas.usda.gov/wap/circular/ 2003/03–08/wap%2008–03.doc.

717 One of the most frequent justifications for subsidies in Europe and Japan is the desire to maintain a rural way of life that is seen by some to be emblematic of the country at large, e.g. the French family wine producer, or the Japanese small rice farmer. Quite why this sector of the rural economy should be privileged in such a way, while the rest of the economy has to 'sink or swim' is less easy to understand.

718 For more information on the movement against agricultural subsidies, see the Global Policy Forum website at: http://www.globalpolicy.org/socecon/trade/subsidies/ and the article by Vidal (2002).

719 Evans (2005).

720 The unnecessary and over-subsidised production of sugar beet in the European Union is an even greater scandal than US-subsidised cotton crops. The production cost of sugar from EU beet crops is double that of sugar cane from the tropics. The EU currently spends almost $2 billion each year to subsidise EU sugar beet farmers, many of whom run relatively wealthy agribusiness enterprises. In 2005, this practice was roundly condemned by the Appellate Body (the highest court) of the WTO, following a case brought by Brazil, Thailand and Australia (WTO Arbitration Ruling of 25 October 2005 is at: http://www.wto.org/english/tratop_e/dispu_e/265_266_283_arb_e.pdf). The reform of this gross misuse of public funds would benefit not only tropical sugar cane farmers, but also European consumers who could see reductions of as much as 40% in sugar prices.

721 Similar considerations regarding unfair subsidies of US cotton apply to other Sahelian countries like Benin, Mali and Chad.

722 The WTO 1986–1994 Uruguay Round negotiations culminated in a signed agreement at the Marrakech ministerial meeting in April 1994. The full text is available online at: http://www.wto.org/english/docs_e/legal_e/legal_e.htm.

723 See the text of the WTO agreement on cotton at: http://www.wto. org/english/news_e/news04_e/sub_committee_19nov04_e.htm.

724 This widely reported speech is described on the World Bank website under the title 'Wolfowitz urges Burkina Faso to persevere' at: http://web.worldbank.org/WBSITE/EXTERNAL/COUNTRIES/AFRICAEXT/0,,contentMDK:20545378~ menuPK:258649 ~pagePK:146736~piPK:146830~ theSitePK:258644,00.html.

725 WTO (2006).

726 See Bhagwati (2005) and the article in the Economist (2005c).

727 It has been pointed out that, while increased food prices will benefit the rural sector of developing countries, such price rises would adversely affect the urban poor (Economist, 2005c). But with greater economic growth from manufacturing and services, as well as

from agriculture, the income of urban poor in most developing countries is likely to increase sufficiently to offset such costs.

728 According to Tilman *et al.* (2002): 'There is general consensus that agriculture has the capability to meet the food needs of 8–10 billion people … but there is little consensus on how this can be achieved by sustainable means.'

729 See Heady (1982) and comments on p. 13 of Goodman *et al.* (1987).

730 Conway and Toenniesson (1999).

731 Duvick (2004) has made this point for yields of maize and wheat. I have made a similar point regarding selection for genetically improved yields in oil palm earlier in this section and also in Murphy (2006). In the case of rice, Tran (2001) has reviewed the sometimes conflicting views of various experts and concludes that in certain regions and with certain types of rice there is still significant scope for increasing crop production.

732 This point was made by Huang *et al.* (2002) in a commissioned review article in *Nature*. However, they also stated in the same article that: 'interviews with 22 leading scientists and observers of international research indicate that there is still the potential to increase productivity by conventional breeding'.

733 The tropical plantains should not be confused with the quite unrelated common plantains, *Plantago lanceolata* and *Plantago major*, which are broadleaved species native to Europe that are commonly found as garden weeds.

734 Contrast this paltry performance with the yield increases of several hundred-fold for the Green Revolution crops, as described in Chapter 6.

735 Varro's *Rerum Rusticarum* is an especially illuminating insight into the complexities of agricultural management in Italy, over two millennia ago. But its sophisticated focus on practical aspects of agronomy and commerce, rather than matters botanical, reflects the lack of knowledge about crop breeding at this time (Hooper, 1999).

736 According to the World Bank Millennium Development Goals, malaria kills 1.1 million and affects 300–500 million, 90% of whom are in sub-Saharan Africa; tuberculosis kills 2 million and infects a further 8 million each year and is most prevalent in sub-Saharan Africa; worst of all, 38 million people live with AIDS, two thirds of them in sub-Saharan Africa, but only 7% of them have access to AIDS medication (see Millennium Development Goals website at: http://www.developmentgoals.org/Hiv_Aids.htm).

737 The shortcomings of the current pharmaceutical industry paradigm for addressing global health issues have been recognised by the decision of one of the richest couples on earth (Bill and Melinda Gates) to donate hundreds of millions of dollars of their private fortune into research aimed at finding cures for orphan diseases such as malaria. To some extent this gesture echoes the actions of earlier philanthropists, such as the Ford and Rockefeller Foundations, in laying the groundwork for the agricultural Green Revolution. One can only hope that the Gates Foundation has similar success in the sphere of global health.

738 For a useful summary of 'Why should plant breeding be supported by taxes', see the review by Tracy (2003).

739 Heisey (2001).

740 Dreher *et al.* (2003); Dubcovsky (2004); Heisey (2001); Moreau *et al.* (2000).

741 Jain (2000).

742 Maredia and Byerlee (2000).

743 Normile (2004).

744 For example, during the 1990s, while the public and private sectors in industrial countries spent almost $10 billion on agbiotech, the CGIAR network invested just $25 million (CGIAR, 2005b).

745 Amounts are expressed in inflation-adjusted 1993 US dollars (Pardey and Beintema, 2005).

746 Eicher (1999).

747 Heisey *et al.* (2002).
748 Fowler (2003).
749 Lantican *et al.* (2005).
750 The range from $1–6 billion is due to different stringencies used in the economic assessment models employed during the study (Lantican *et al.*, 2005, p. 3).
751 CIMMYT (2005).
752 CIMMYT (2005), p. 52.
753 Conway (2003). For more on breeding for Africa, see de Vries and Toenniessen (2001).
754 An early account of the potential for a more participatory role by farmers can be found in the publication edited by Lori Ann Thrupp for the World Resources Institute (Thrupp, 1996). For more recent accounts of research into participatory barley breeding, see Ceccarelli *et al.* (1997, 2000), Toomey (1999), Almekinders and Elings (2001), Vernooy (2003), and Morris and Bellon (2004).
755 Ashby and Lilja (2004).
756 Apart from the most southerly regions, sub-Saharan Africa largely missed out on the Green Revolution gains for a complex series of reasons. These include a whole host of infrastructural deficits, from roads to credit facilities, as well as a focus on crops that were ill adapted to the drier soils cultivated by most of the poorer African farmers (Conway, 2003). Nevertheless, even when these challenges are taken into account, there was still a failure in the uptake of crop varieties specifically designed for African farmers for the reasons outlined in this section.
757 The increasing separation between farmers and scientific breeders is explored in Simmonds (1990) and by Cleveland and Soleri (2000).
758 Dalton (2004); Dalton and Guei (2003); Lilja and Dalton (1997).
759 Steele *et al.* (2002 and 2004); Virk *et al.* (2003).
760 Witcombe *et al.* (2003).
761 Qualset and Shands (2005).
762 The largest CGIAR seed repositories are in India (ICRISAT, ca. 113 000 accessions), Syria (ICARDA, ca. 122 000), Mexico (CIMMYT, ca. 179 000), and Philippines (IRRI, ca. 99 000) (Koo *et al.*, 2004).
763 The UN Population Division estimates that, from 2005–2050, the population of the industrialised countries will remain virtually unchanged at 1.2 billion but that the population of developing countries will increase by 2.5 billion to 7.8 billion (2004 Revision, http://esa.un.org/unpp/).
764 Data from Qualset and Shands (2005).
765 The Norwegian seed bank, also known as the Svalbard Global Seed Vault, was announced in 2006 and will store frozen seed. While this is a laudable venture, many crops, especially from the tropics, cannot be stored in this way so the scope of the seed bank is limited. The new Arctic seedbank will be part of the larger Nordic Genebank (for website, see: http://www.nordgen.org/ngb/).
766 A detailed analysis of the long-term costs of seed conservation within the CGIAR network has recently been published by Koo *et al.* (2004).
767 The Iraqi threats to bomb Tehran, where the Iranian seed bank was located, came after Iranian forces had bombarded the city of Baghdad with rockets, causing considerable numbers of civilian casualties.
768 Qualset and Shands (2005); Whitney (2005).
769 At this time, the US and Iran were at daggers drawn after the fall of the Shah and the US embassy kidnappings of the late 1970s. Throughout the 1980s, the Reagan administration supplied weapons and intelligence support to the Iraqis. Hence the actions of the US

breeders in assisting their Iranian colleagues were in a sense contrary to the foreign policy of their own government.

770 For a discussion of the importance of wild relatives of crops, see Waines (1998).

771 Abu Ghraib is also the site of a notorious prison that housed many opponents of Saddam Hussein until it was taken over by US-led forces in 2003. The prison then achieved even greater notoriety for the abuse of Iraqi prisoners by their US captors.

772 In 2005, over 20 tonnes of cereal and legume seeds were shipped from ICARDA in Syria to farmers in Iraq. Unlike the 'generic' seed sometimes provided in assistance programmes, this seed, some of which had been rescued from the Iraqi gene bank before its destruction after the 2003 invasion, was carefully selected for its genetic suitability to the growing conditions in Iraq.

773 Anonymous (2005).

774 For more details of the 2004 World Food Prizes, see the Press Release at: http://www.worldfoodprize.org/Laureates/04laureates/jones.htm. For more details on the uptake of nerica by African farmers, see Hersch (2004).

775 See the news release of 3 December 2004, 'WARDA Board of Trustees Responds to the Ivorian Crisis', available online at: http://www.warda.org/warda1/main/newsrelease/newsrel-botresponds-Dec04.htm.

776 Shanahan and Cockburn (2005).

777 Waage (2002), p. 12.

778 The July 2005 report entitled 'Tsunami-surviving rice may have salt-tolerance genes' is available online at: http://www.scidev.net/content/news/eng/tsunami-surviving-rice-may-have-salt-tolerance-genes.cfm.

779 CGIAR (2005a).

780 Buerkert *et al.* (2006).

781 Agrobiodiversity is defined as that component of biodiversity that contributes to food and agriculture production. The term agrobiodiversity encompasses intraspecies, species and ecosystem diversity. In simpler terms, it is a measure of the number of different varieties, races, subspecies and species of useful plants and animals. For a brief discussion of the concept, see Brookfield and Stocking (1999), while the book edited by Brush (2000) provides a more detailed, technical survey. See also Kothari (1994).

782 For example, none of the food crops grown in the UK is a native species and most of them came from either the Near East or Central/South America. Equally, hardly any US crops are indigenous. One of the few exceptions is sunflower, but even here the commercial high-oil varieties were developed by breeders in Ukraine and then later imported by the USA, mainly for cultivation in the Great Plains region.

783 Wilkes (1983). The issue is also discussed in detail by Kloppenburg (1988), pp. 161–166.

784 The article by Engels and Visser (2003) is a useful review of the concept of a common heritage of global plant germplasm resources.

785 Letter from US Secretary of State Warren Christopher to George J. Mitchell, US Senate, dated 16 August 1994, urging rapid ratification of the Convention on Biological Diversity. The Convention was never ratified by the USA despite arguments by many people of vision that it was in the nation's national interest (for example, see following posting on the discussion site of the Center for International Development at Harvard University: http://www.cid.harvard.edu/cidbiotech/comments/comments199.htm).

786 Shands (1986); Shands and Stoner (1997).

787 Shands and Stoner (1997).

788 It is important to note that, although access to US-held plant germplasm and related resources is restricted in some instances according to high-level policy considerations, I have personally found individual researchers at such germplasm centres to be only too

willing to share scientific materials. And this is, of course, reciprocated as I have sent seed-derived materials to colleagues at Fort Collins. Many US researchers disagree with aspects of the restrictive policy on germplasm sharing, which is a political, not a scientific policy decision.

789 This extract of the USDA/ARC letter is quoted on p171 of Kloppenburg (2004).
790 Mooney (1983), p. 29.
791 This quote is from an unattributed editorial in *Nature* (Anonymous, 2005).
792 Abu Hureyra in the middle Euphrates region of modern Syria is the site of some of the earliest finds of domesticated cereal grains, dating from 12 000 years ago (Cauvin, 2000; Moore *et al.*, 2000; Hillman *et al.*, 2001).
793 These concerns are well summarised in a review by UC Davis breeding researcher Paul Gepts (Gepts, 2004).
794 Waage (2002). This report from Imperial College at Wye, UK, was one of the factors that eventually led in 2004 to the establishment of The Global Crop Diversity Trust. This body is currently worth more that $50 million, with an additional $60 million of raised funds in negotiation. The objective is to provide funding for national and international crop collections around the world, but most especially in the developing countries where germplasm resources are some of the richest but funding is extremely limited. For example, several important seed collections in Africa have already been lost because their governments could not pay the costs of refrigeration.
795 Waage (2002), p. 10.
796 The Gatsby Charitable Foundation is a UK-based grant- awarding trust (see website at: http://www.gatsby.org.uk).
797 Anonymous (2005). The cost of a permanent endowment to maintain the seed banks alone is estimated at a 'mere' $149 million (Koo *et al.*, 2004), which is a very modest one-off sum for such a worthy objective. Bill Gates take note!
798 This point is well made in the review article by Hammer (2003).
799 There are elements of exclusive modishness and self-absorption in industrial country obsessions as apparently unconnected as organic farming and genetic engineering. These very different phenomena have attracted similar elements of almost messianic proselytising by their respective proponents. Both seem to involve an element of intolerance of other 'faiths'. Both started small but are now underpinned by major commercial interests. And they each seek to lobby and influence public policy to advance their respective causes. Both are private sector enterprises and yet they have both profited greatly from the public sector and seek to manipulate it in their favour. There are, of course, big differences as well; genetic engineering tends to be presented as a more evidence-based technical approach to conventional crop breeding, while organic farming is more concerned with farming systems as a whole, but often appears to be based at least as much upon faith as on physical reality. It is useful to bear such factors in mind when weighing evidence presented by proponents of these two essentially limited (in their different ways) approaches to agriculture.
800 As we have seen in Chapters 9 and 10, the academic focus of the public sector was itself deliberately created by policymakers who withheld funding for 'near-market' research after the late 1980s.
801 I am alluding here to the so-called 'knowledge hierarchy' that was first introduced (although not explicitly as such) in the poem entitled 'The Rock' by T. S. Eliot in 1934:

Where is the Life we have lost in living?
Where is the wisdom we have lost in knowledge?
Where is the knowledge we have lost in information?

802 Zeleny (1987); Ackoff (1989).

803 Peter Alekseyevich Kropotkin was born a Russian prince, although he later renounced the title. He became a noted libertarian and anarchist thinker and wrote prolifically on economics and social justice. Some of his ideas have an eerie resonance today, well over a century after he enunciated them. For example, in his *Appeal to the Young* he says: 'More than a century has passed since science laid down sound propositions as to the origins of the universe, but how many have mastered them or possess the really scientific spirit of criticism? A few thousands at the outside, who are lost in the midst of hundreds of millions still steeped in prejudices and superstitions worthy of savages, who are consequently ever ready to serve as puppets for religious impostors' Kropotkin (1880). We have only to think to the rise of religious fundamentalism and associated intolerance across the globe, the new attacks on the teaching of evolution in our schools and universities, and the campaigns against research into stem cells to feel the contemporary resonance of Kropotkin's words.

804 Kropotkin (1880).

805 Some American observers have gone even further in their claims for the effect of plant breeding on world events. Hence, in 1946, L. E. Stadtler, testified to the Senate Committee on Military Affairs that it was the increased production due to hybrid maize that had largely paid for the Manhattan Project and the development of nuclear weapons in the USA (Shull, 1946, p. 550). Even more dramatically, Paul Manelsdorf credited hybrid maize with containing the spread of communism in the immediate aftermath of World War II by ensuring an adequate food supply to a devastated and hungry Western Europe (Glass, 1955, p. 3; Manelsdorf, 1951). These assertions may seem somewhat hyperbolic today, but they give an idea of the importance that was ascribed to plant breeding as a vital and potent instrument of national policy in the USA in the 1940s and 1950s.

806 This Commission was established by the Association of American Agricultural Colleges and Experimental Stations. It favoured a less superficial and more sophisticated approach to agricultural research, but one that was resolutely applied to the needs of farming in general (Kloppenberg, 1988, p. 76).

807 Hambridge and Bressman (1936).

808 BBSRC (2004).

809 AEBC (2005b).

810 As a keen practitioner of curiosity-driven research, I am of course in favour of its continued generous funding. But we must remember that these funds come from the public purse and there must be a better balance between blue-skies research and more applied activities. Moreover, the virtual abandonment of the latter activity in UK plant science means that there is no way of connecting basic discoveries with a mechanism to exploit them in this country. Unless this link is restored, we face the danger that the basic research itself will then be cut back as it is not perceived to have a socially useful function.

811 Having been something of a sceptic about the value of research assessment exercises (Murphy, 2003c, 2004d), I was rather surprised to be invited to sit on the panel that will be assessing biological sciences for the 2008 exercise in the UK. However, it was immensely reassuring to discover that others, including many eminent colleagues, also shared my concerns (e.g. Adams and Smith, 2003). As a result, in the 2008 assessment we are enjoined to ignore the impact factor of papers submitted by researchers and instead to focus on their intrinsic worth within a particular discipline. This will empower those scientists who work in less fashionable areas, like plant breeding, where papers in *Nature* or *Cell* are hard to come by. In addition RAE 2008 will seek to assess all the different forms of scientific research including interdisciplinary research, applied research, basic/strategic research, practice-based research – again this is good news for those working in less academically fashionable areas of research (RAE 2008 Guidance to panels, see: http://www.rae.ac.uk/news/2006/criteria.htm). The result should be a much fairer process, albeit not a perfect

one. Given the importance of the RAE for the future funding of UK research, I take the view that it is better to be on the inside trying to constructively optimise the process rather than complaining about it from the sidelines.

812 For reviews of public attitudes to biotechnology in Europe, see Durant *et al.* (1998), Gaskell and Bauer (2001) and Bauer and Gaskell (2002).

813 As of 2005, the Tesco, Wal-Mart, Morrison and Sainsbury supermarket chains controlled 80% of the UK grocery retailing market. The largest of these chains, Tesco, held 30% of the entire national market with an annual turnover of more than $60 billion.

814 The first soybean shipments containing transgenic varieties were exported to Europe in 1996. These cargoes only contained about 2–5% of the transgenic soybeans with the remaining 95–97% of the beans being non-transgenic, but all this seed or flour was mixed together as a single batch.

815 Given the positive UK consumer reception of labelled GM tomato paste (which was cheaper and tastier than other brands), GM soybeans might still have been accepted if they had been labelled. There were two problems with this. First, the Monsanto business model was always based on non-segregation of the GM beans; segregation would have been too expensive to guarantee an adequate return on the investment in GM technology. Second, unlike GM tomato paste, GM soybeans had no direct consumer benefits, giving little motivation for shoppers to choose to purchase them instead of the more familiar conventional soybeans.

816 In November 1996, the Greenpeace barge 'Beluga' attempted to stop the bulk tanker 'Ideal Progress' with 44 600 tonnes of new crop soybeans from docking in Hamburg. This action was unsuccessful so the protestors then attempted to blockade the crushing mill operated by Oelmuehle Hamburg (Oelag), a leading European processor. Similar action was taken in Belgium at Antwerp port against the Polish-registered freighter, 'Ziemia Zamojska' and at Gent against the 'MV Sirius'. At the same time, Greenpeace activists also attempted to blockade several grain silos and Mississippi port facilities in Louisiana. A few weeks later, similar action was taken in the UK, when Greenpeace activists blockaded the Polish bulk carrier 'Orleta Lwowskie' with 44 600 tonnes of soybeans as it attempted to dock at the port of Liverpool.

817 There is little doubt that there was a highly organised and well-funded international campaign against GM crops by professional bodies such as Greenpeace in the late 1990s. However, the fact that the campaign largely succeeded in Europe was more to do with the consumer-related deficiencies of the GM business model than to the efforts of the anti-GM lobby. Indeed, it was partially the weakness of the GM paradigm that attracted the attentions of the activists in the first place. Unlike the family automobile or imported consumer electronics, GM crops had few friends among consumers, retailers, or politicians, and would therefore be an easier target than oil companies (cf. Brent Spar) etc.

818 The Pusztai affair began with a TV interview on 10 August 1998 when he disclosed data from rat feeding studies showing adverse effects of an experimental variety of GM potatoes. Some of the data eventually appeared in the *Lancet* in 1999 (Ewen and Pusztai, 1999), amid considerable controversy that such incomplete and potentially misleading work could be accepted for publication in a 'scientific' journal (Gatehouse, 1999; Kuiper *et al.*, 1999).

819 I will examine the reaction to the Pusztai study by the UK scientific establishment and the consequences for researchers in agbiotech and related subjects, in a forthcoming book on scientific controversies.

820 There is a good case for the view that it was the precipitate action of UK (and subsequently other European) supermarkets in banning GM foods in early-mid 1999 that did more than anything else to colour the public perception against GM foods. One reason why this ban was so hasty and retailer led (rather than consumer or producer driven) was the lack of benefit for such retailers from GM food to offset risks to their

brand integrity and quality reputation. A key problem for these input trait GM foods was the failure of the agbiotech industry to recruit their downstream customers in the food industry as additional stakeholders in the GM enterprise. The supermarket bans on GM food are discussed further by Kalaitzandonakes and Bijman (2003).

821 These three weeds are the most serious plant pests in UK winter wheat fields, together causing crop losses of almost 5%. Other serious weeds in wheat cultivation are charlock, thistle, bindweeds, chickweed and knotgrass (data from Rothamsted and the UK Home Grown Cereals Authority).

822 Note that, although I am rather critical about the wider context of the farm-scale evaluation process itself, and in particular its essentially non-scientific and very political genesis and consequences, much of the associated science was of the high quality that one would expect from the talented and exceedingly well-resourced team that carried out the field studies. A good flavour of the perspective of the participating scientists can be found in the personal evaluation by Joe Berry from Rothamsted as follows: http://www. rothamsted.bbsrc.ac.uk/pie/sadie/joe_general_work_GM_FSE_page_3_1.htm#I_Before %20the%20FSE:%20pre-Spring%201999%20label. My concern in criticising the study is not that a great deal of data were collected and analysed in a rigorous manner, it is whether this has actually shed any useful light on the environmental consequences of transgenic crop cultivation, and whether it was even remotely worth the diversion of more than $9 million of funds from the UK agricultural science budget.

823 Which was nice

824 As I commented publicly at the time of publication of the FSE results: 'These papers clearly demonstrate that different transgenic crops can have very different effects on plants and animals in farming ecosystems. It's hardly surprising that using more effective weed control measures may often result in reduced weed numbers, and associated populations of the invertebrates that feed on them.' (See: http://www.sciencemediacentre. org/press_releases/10-16-03_FSEreaction.htm.)

825 For a recent study of the impacts of agricultural intensification on birds, see the review by Wilson *et al.* (2005).

826 According to the UK government strategy on sustainable development, as adopted in 1999, fifteen 'Headline Indicators of Sustainable Development for the UK' were established. Number thirteen on this list is 'wildlife', which is defined as populations of wild birds, including farmland species (see: http://www.sustainable-development.gov.uk/indicators/headline/in-dex.htm). This has fed back into policy as a heightened awareness of farmland birds because of their impact on the overall UK quality of life and sustainability scores. As stated on the DEFRA website: 'Following the Government's Spending Review in 2000, Defra adopted as one of its Public Service Agreement (PSA) targets a commitment to reverse the long-term decline in the number of farmland birds by 2020, as measured annually against underlying trends. The PSA target will have been achieved when the long-term trend in the index (i.e. the smoothed indicator) and the associated lower confidence limit (using a 95% confidence interval) are both positive' (see http://www.defra.gov.uk/news/2005/050404f.htm). This shows how any study that related to farmland birds and their associated food chain (i.e. invertebrates and weeds) would have a special resonance with the public and politicians alike. This may partially explain the seemingly extraordinary decision to devote over $9 million of public funds to the farm-scale evaluations of 1999–2004.

827 The area of the New Forest National Park is about 57 000 hectares, Dartmoor is 95 400 hectares, and Exmoor is 69 000 hectares. However, much of this land is relatively unproductive for arable farming, so it would probably be necessary to toss in another Park, such as the Yorkshire Dale (176 200 hectares) to make up for the losses due to weeds. Hopefully, the absurdity of the pro-weed argument is now apparent.

828 Calgene is an excellent example of such a company. During its heyday in the early 1990s, Calgene scientists acquired a richly deserved reputation as some of the most innovative plant biochemists and molecular geneticists, especially in my own main research area of plant lipids. In particular, their atypical openness in discussing their findings, and their willingness to publish new results, earned them the respect of the research community. In a recent book on *Plant Lipids*, I have noted that: 'In retrospect, the most notable achievement of the Calgene group (the company was acquired by Monsanto in 1998 and no longer has a separate identity) has been their contributions to basic lipid research rather than to the commercialisation of transgenic crops' (Murphy, 2005, p.16).

829 For a recent overview of the status of innovation in the agbiotech industry, see the special issue of *AgBioForum* as edited by Oehmke *et al.* (2005).

830 For example, Ceres, DuPont and Monsanto has made part of their maize genomic database available to researchers (http://www.maizeseq.org/PDF/csila.pdf) while Monsanto has done the same for part of its rice database (http://www.rice-research.org). The catch is that researchers must agree in advance that any commercially useful knowledge derived from such access will be owned by the company in question.

831 Such industry funding can be direct or indirect. Examples of indirect funding include check-off funds from commodity organisations or from royalties (Duvick, 2004).

832 These three firms are currently the major global players in development of commercial GM crops.

833 Of course, an individual wealthy philanthropist like Bill Gates should be able to use his personal profits from his companies in any way he chooses. But it must be stressed that such activities should not be part of the mission of a commercial shareholder-owned enterprise like Microsoft.

834 Thro (2004).

835 Examples of company endowments, gifts etc. to public sector plant breeders in the USA are legion. Hence, Pioneer endowed the Raymond F. Baker Center for Plant Breeding at Iowa State University (http://www.plantbreeding.iastate.edu/default.htm), and Monsanto contributed significantly to the $150 million endowment to set up the Donald Danforth Plant Science Center at Washington University in St Louis (www.philanthropy.iupui.edu/Million$report-by-year.pdf). Private companies also help to support major communal resources such as the National Plant Germplasm System, which holds hundreds of thousands of accessions from around the world in several dozen regional repositories across the USA (http://www.ars-grin.gov/npgs/).

836 Frey (1998).

837 According to the Syngenta Press release of 11 April 2006, the fund will focus on US companies and will be based in Boston: see 'Syngenta to invest in new venture fund focused on plant science', available online at: http://www.syngenta.com/en/media/article.aspx?pr=041106&Lang=en.

838 The Seven Sisters (from a phrase first popularised by Italian oil tycoon Enrico Mattei) refers to the seven major oil corporations, Exxon, Mobil, Chevron, Texaco, Gulf, British Petroleum and Royal Dutch/Shell. These companies acted as an effective oligopoly that dominated global oil production after World War II (Sampson, 1975).

839 Average incomes in Venezuela have fallen significantly in the past thirty years, despite the hundreds of billions of dollars that their government has earned from its nationalised oil industry. The Nigerian government has also received billions of dollars in oil revenues, virtually none of which has percolated down to its citizenry.

840 Oil-service companies include Schlumberger, Halliburton and Diamond Offshore; and examples of countries that have used such services to avail themselves of the latest oil-related technologies include China and Mexico.

841 Economist (2005d).
842 Malaysian public sector researchers have outsourced several key aspects of R&D on oil palm, which is the major cash crop and an extremely important export earner for the country. In some cases, large hardware- and expertise-intensive projects have been contracted out to private sector service companies in the USA. In other cases, joint projects involving technology transfer have been established with university groups in Europe and the USA. For a broader perspective on private–public parthships in developing countries, see Horsch and Montgomeny (2004).
843 This case has been argued eloquently by Donald Duvick, who worked for 40 years as a plant breeder, eventually rising to be Senior Vice-President for Research, at Pioneer Hi-Bred. Following his retirement from Pioneer, Duvick has worked as an affiliate professor at Iowa State University where he has contributed extensively to the debate on plant germplasm resources and their utilisation in plant breeding (Duvick, 1996, 2003).
844 Brewster *et al.* (2005).
845 This topic is also discussed in Chapter 13.
846 Finkel (1999). The implications of commercial IPR restrictions for public sector scientists are explored in detail by Dunwell (2005), who states 'Although such issues are often considered of little interest to the academic scientist working in the public sector, they are of great importance in any discussion of the role of "public-good breeding" and of the relationship between the public and private sectors.'
847 The researchers were aware that some of the key steps in their production of transgenic 'golden rice' were already patented, but the full extent of this patent coverage still took them completely by surprise.
848 Kryder *et al.* (2000). Note that the figure of 70 individual patents and 31 companies is misleadingly high since many of these patents were simply upgrades of existing patents and many of the companies subsequently merged with or were bought out by each other.
849 See USDA publication online at: www.fas.usda.gov/icd/stconf/pubs/scitech2003/part3.pdf).
850 See press release at: http://www.uspto.gov/web/offices/com/speeches/01-01.htm.
851 United States Court of Appeals for the Federal Circuit (2005).
852 Broothaerts *et al.* (2005).
853 The use of non-*Agrobacterium* vectors for gene transfer into plants has been briefly reviewed by Chung *et al.* (2005).
854 The acronym CAMBIA originally stood for 'Center for the Application of Molecular Biology to International Agriculture'. However, the public-good mandate of the organisation has since expanded to include all forms of biological innovation for the betterment of agriculture. In particular, CAMBIA seeks to develop new genetically based technologies and to make these available freely to the participating commons. According to its website, CAMBIA states that: 'Our institutional ethos is built around an awareness of the need and opportunity for local commitment to achieving lasting solutions to food security, agricultural and environmental problems. We envision a situation in which the broadest community of researchers and farmers are empowered with dramatic new technologies to become innovators in developing their own solutions to the challenges they face – solutions for which they feel ownership' (http://www.cambia.org/mission.html).
855 Broothaerts *et al.*, (2005), p. 632.
856 Economist (2005a); Salz (2005).
857 For more information on BIOS, see the website at: http://www. bios.net/daisy/bios/15.
858 See 'Open Source Biotechnology alliance for international agriculture', online at: http://www.biologynews.net/archives/2005/12/07/open_source_biotechnology_alliance_for_international_agriculture.html.
859 The BIOS prospectus is available online at: http://www.bios.net/ daisy/bios/3.

860 The MASwheat website is at: http://maswheat.ucdavis.edu/.

861 For example, in the section on Retained Rights in its exclusive license template, Stanford University asserts that: 'Stanford retains the right, on behalf of itself and all other non-profit academic research institutions, to practice the Licensed Patent and use Technology for any purpose, including sponsored research and collaborations. Licensee agrees that, notwithstanding any other provision of this Agreement, it has no right to enforce the Licensed Patent against any such institution. Stanford and any such other institution has the Technology or a Licensed Patent.' The full text of this document is available online at: http://otl.stanford.edu/industry/resources/exclusive.pdf.

862 Brewster *et al.* (2005).

863 See Egelyng (2005) for an analysis of the improved engagement of international agricultural research centres with IPR issues.

864 Some options for public–private partnerships for agricultural research are presented in the consultation paper by Hartwich *et al.* (2003).

865 In 2004, the global area of transgenic herbicide tolerant crops was estimated at 65.4 million hectares, or 81% of the total transgenic crop area (James, 2006). While some herbicide tolerant crops were from non-Monsanto varieties, such as LibertyLink® from Bayer, well over 75% were the Monsanto Roundup Ready® varieties.

866 http://pewagbiotech.org/newsroom/summaries/display.php3?Ne wsID = 622.

867 This action was taken because Monsanto was unable to recoup its technology fees for the transgenic herbicide tolerant varieties. The decision caused a storm in Argentina because the alternative older soybean varieties were much lower yielding which meant a potential loss to farmers of as much as $78 million (Burke, 2004). In December 2004, it was reported that 'the government of Argentina and biotechnology firms agreed on the key elements of a "technology compensation fund" to compensate companies such as Monsanto for the illegal use of their GM seeds' (http://www.ictsd.org/biores/04-12-20/story3.htm). But this agreement did not last long and by early 2006, Monsanto was reportedly using legal injunctions to blockade the unloading of three Argentinian soybean shipments, valued at several million dollars, in the European ports of Bilbao, Santander, and Liverpool (Mira, 2006).

868 The figure of $21 million in royalties from transgenic soybean varieties does not include the technology fees top-sliced under an agreement with Monsanto, whereby the company gets a proportion of fees from sales of EMBRAPA varieties that contain its proprietary transgenic traits, most notable Roundup Ready®.

869 Thro (2004).

870 James (2004); Cohen (2005).

871 For a recent perspective on the uptake of GM technologies by developing countries, see Fukuda-Parr (2006).

872 The Younger Dryas was so-named by the Scandinavian palaeobotanist, Jansen, who noticed that there were unusual accumulations of the arctic-alpine herb, *Dryas octopetala*, at two strata in organic sediments. These accumulations suggested that the otherwise mild climate had undergone a return to relatively frigid conditions on two occasions that Jansen termed the Oldest Dryas at pre-14 700 BP and the Younger Dryas from 12 800 to 11 600 BP, as originally reported in Jansen (1938). For many years, the Younger Dryas was regarded as a solely European phenomenon but improved analytical tools and dating techniques now suggest that it was a global event as discussed by Various (2002). Some of the main hypotheses that seek to explain the causes of the Younger Dryas are reviewed by Broecker (2003).

873 For an idea of the vast range of potential new crops, and some of the many products that can be obtained from them, see the study by Frey (1997) and the website of the Center for

New Crops and Plant Products at Purdue University at: http://www.hort.purdue.edu/newcrop/default.html.

874 A similar logic applies to the domestication of animals. Witness, for example the groundbreaking Russian researchers who, in just a few decades, were able to domesticate wild foxes to the point that they had many of the more appealing behavioural characteristics of dogs (Trut, 1999; Hare *et al.*, 2005).

875 Some of this work has already shown how modern genetic techniques can complement basic breeding methods to develop reasonably domesticated varieties of some hitherto wild crops. One of the best examples is the work of Steve Knapp and colleagues at Oregon State University (and more recently Georgia University) who developed non-pod-shattering, high-oil yielding varieties of the oilseed species, meadowfoam (*Limnanthes alba*) and *Cuphea* spp. Such crops can act as renewable producers of specialty fatty acids and hydrocarbons, and will be our only sources of oils when non-renewable fossil hydrocarbons eventually run out (Knapp, 1990, 1998; Knapp and Crane, 1999).

876 The National Institute of Standards and Technology website is at: http://www.nist.gov/. and details of the Advanced Technology Program are at: http://www.atp.nist.gov/.

877 In 2002, Sangamo also received a research grant for over $300 000 from the National Institutes of Health.

878 The Small Business Innovation Research Program website is at: http://www.sba.gov/sbir/indexsbir-sttr.html.

879 Celiac disease, or gluten intolerance, is an inherited genetic disorder that affects 4% of Europeans and 1% of Americans (about 20 million people). Removal of the disease-causing gluten proteins from wheat would greatly improve the dietary range and overall health of such people.

880 Soybean isoflavones, also known as phytoestrogens, are phytochemicals that have been shown to reduce menopausal symptoms in women, as well as decreasing the risk of heart disease, osteoporosis, breast cancer and prostate cancer.

881 To quote from the Arcadia press release: 'Diets of troops in some areas are often constrained by the ability to deliver fresh food on a regular basis. Development of longer-lasting produce can help expand the dietary options and improve the morale of soldiers by providing them with good-tasting, nutritious fresh food, said Patrick Dunne, senior science advisor to the US Army Natick Soldier Center DoD Combat Feeding Program.' (14 September 2005: http://www. arcadiabio.com/media/press_DODNRFINAL.pdf).

882 See press release at: http://www.lsbc.com/pdfs/louisville.pdf.

883 Large Scale Biology Corporation filed for bankruptcy in December 2005, following a failure to raise enough funds to continue operations, but shareholders and others are hoping to continue some of the research described in the text, possibly by selling company assets, including IPR, to other biotech firms.

884 Ventria is also funded by the NIH Small Business Innovation Research programme.

885 In December 2005 Ventria withdrew from the agreement with Northwest due to a loss of construction funds from the state of Missouri for the new building.

886 In October 1998, major US tobacco companies settled a pending lawsuit with the States of Minnesota, Florida, Mississippi and Texas, after public disclosure that these companies intentionally targeted youth in marketing and knowingly withheld information on the addictive nature of nicotine. These four States won $40 billion over 25 years. The original 1998 tobacco settlement involved four major tobacco companies R. J. Reynolds Tobacco Co., Brown & Williamson Tobacco Corp., Lorillard, and Philip Morris USA, whose parent company recently changed its name to Altria Group. More than 40 smaller tobacco firms now participate in this settlement. The settlement led 46 states to drop a massive lawsuit against cigarette companies in exchange for sharing $206 billion over 25 years. Much of

this huge sum has been ring-fenced by States for use in health-related research, including biopharming and other aspects of the medical uses of plants.

887 Shepard (2004).

888 EU research consortia must be international and there is always pressure to include participants from less research experienced regions, such as Southern and Eastern Europe. While this may be admirable in terms of overall social policy and the long-term stimulation of nascent research capacity in such regions, it is hardly calculated to give rise to the most highly productive and competitive research. This already bad situation is often exacerbated by stipulations that each participant can only have one postdoc funded by a grant; so an excellent cutting edge lab ends up with the same funding as the weakest member of the consortium. EU grants are also notoriously bureaucratic and, with their low overheads, can even result in a net financial loss to universities in more developed regions. To make an already bad situation worse, more recent EU funding rounds have de-priorititised the already paltry component allocated to plant and genomic research (Hughes, 2006). Small wonder, then, that little sustained innovatory research (especially compared with the USA) comes from this direction.

889 Hughes (2006).

890 Hughes (2006).

891 Murphy (1999b, 2006).

892 Crop yields have actually increased much more than seven-fold because many of our current crops are grown, not for human consumption, but to feed those animals that increasingly affluent societies across the world tend to prefer over plant-based foodstuffs.

893 The article was written to mark Delmer's elevation to become a member of the US National Academy of Sciences (Delmer, 2005).

894 Jana *et al.* (2006)

895 The US–EC Task Force on Biotechnology Research report is available online at: http://europa.eu.int/comm/research/biotechnology/ec-us/docs/ec-us_tfws_2005_june_arlington_ abstracts.pdf.

896 CIMMYT (2005), p.52.

897 Busch (2004).

898 Busch (2004).

899 US–EC Task Force on Biotechnology Research (2005).

900 Wright and Pardey (2006).

901 For example, see the review by Anderson and Jackson (2006) discussing EU policies on GM crops, and their effects on developing countries.

902 Such a low-cost, publicly subsidised agribusiness paradigm has existed for many decades in California, where production costs are kept artificially low by a combination of extremely cheap water for irrigation and the availability of a largely immigrant (often illegal) labour force. That is why semi-arid California produces rice and maize crops and has itinerant labourers living in primitive trailers, all less than ten miles from the powerhouse of high-tech agricultural research that is UC Davis.

References

Abbadi A., Domergue F., Bauer J., Napier J. A., Welti R., Zahringer U., Cirpus P. and Heinz E. (2004) Biosynthesis of very-long-chain polyunsaturated fatty acids in transgenic oilseeds: constraints on their accumulation, *Plant Cell* **16**, 2734–2748.

Abdullah S. N. A., Cheah S. C. and Murphy D. J. (2002) Isolation and characterisation of two divergent type 3 metallothioneins from oil palm, *Elaeis guineensis, Plant Physiology and Biochemistry* **40**, 255–263.

Ackoff R. L. (1989) From data to wisdom, *Journal of Applied Systems Analysis* **16**, 3–9.

Acquaye A. K. A. and Traxler G. (2005) Monopoly power, price discrimination, and access to biotechnology innovations, *AgBioForum* **8**, 127–133, available online at: http://www.agbioforum.org/v8n23/v8n23a09-acquaye.htm.

Adam D. (2005) £3m shortfall at John Innes centre, *The Guardian*, 12 April 2005, available online at: http://www.guardian.co.uk/uk_news/story/0,,1457252,00. html.

Adams J. and Smith D. (2003) Funding research diversity, *Technical Report, Universities UK*, London, available online at: www.UniversitiesUK.ac.uk.

AEBC (2005a) *Research Agendas Workstream Information Paper, from the Agriculture and Environment Biotechnology Commission*, available online at: http://www. aebc.gov.uk/aebc/subgroups/research_agendas.shtml.

AEBC (2005b) What shapes the research agenda in agricultural biotechnology? *Report by the Agriculture and Environment Biotechnology Commission, April 2005*, available online at: http://www.aebc.gov.uk/aebc/reports/reports.shtml.

Afolabi A. S., Worland B., Snape J. W. and Vain P. (2004) A large-scale study of rice plants transformed with different T-DNAs provides new insights into locus composition and T-DNA linkage configurations, *Theoretical and Applied Genetics* **109**, 815–826.

AgCanada (2004) The United States canola industry: situation and outlook, *Agriculture and Agri-Food Canada Bi-weekly Bulletin*, 27 February 2004, Volume 17, Number 4, available online at: http://www.agr.gc.ca/mad-dam/e/bulletine/ v17e/v17n04_e.htm.

Agres T. (2006) LSBC closing signals pharming trouble Fall of pioneer in using transgenic plants for pharmaceuticals has analysts worried field may dry up, *The*

Scientist, 17 January 2006, available online at: http://www.the-scientist.com/news/display/22969/.

Ahloowalia B. S., Maluszynski M. and Nichterlein K. (2004) Global impact of mutation-derived varieties, *Euphytica* **135**, 187–204.

Allen R. C. (1999) Tracking the agricultural revolution in England, *Economic History Review* **52**, 209–235.

Almekinders C. J. M. and Elings. A (2001) Collaboration of farmers and breeders: participatory crop improvement in perspective, *Euphytica* **122**, 425–438.

Alston J. M. and Venner R. J. (1998) The effects of U.S. Plant Variety Protection Act on wheat germplasm improvement. In: *CIMMYT Symposium on IPR and Agricultural Research Impact, 5–7, March 1998, El Batán, Mexico*.

Alston J. M. and Venner R. J. (2002) The effects of the US plant variety protection act on wheat genetic improvement, *Research Policy* **31**, 527–542.

Altman A. (1999) Plant biotechnology in the 21st. century: the challenges ahead, *Electronic Journal of Biotechnology* **2**, (2), available online at: http://www.ejbiotechnology.info/content/vol2/issue2/full/1/index.html.

Ambrose S. E. (2001) *Nothing Like it in the World. The Men who Built the Transcontinental Railroad, 1863–1869*, New York: Touchstone.

Ambrosoli M. (1997) *The Wild and the Sown: Botany and Agriculture in Western Europe 1350–1850*, Cambridge: Cambridge University Press.

Anderson C. W. N., Brooks R. R., Stewart R. B. and Simcock R. (1998) Harvesting a crop of gold, *Nature* **395**, 553–554.

Anderson K. and Jackson L. A. (2006) Transgenic crops, EU precaution, and developing countries, *International Journal of Science, Technology, and Globalization* **2**, 65–80, available online at: http://bcsia.ksg.harvard.edu/research.cfm?program = STPP&project = STG&pb_id = 537.

Andow D., editor (2004) *A Growing Concern: Protecting the Food Supply in an Era of Pharmaceutical and Industrial Crops*, Union of Concerned Scientists, available online at: www.ucsusa.org.

Andrews R. D. (2001) The commercialization and performance of hybrid rice in the United States. In: Peng S. and Hardy B., editors, *Rice Research for Food Security and Poverty Alleviation, Proceedings of the International Rice Research Conference, 31 March–3 April 2000*, Los Baños, Philippines: *International Rice Research Institute*, available online at: http://www.irri.cn/textonly/science/abstracts/rice%20research%20for%20food%20security.htm.

Anonymous (1891) *Cambridge University Reporter*, 12 November 1891, pp. 193–196.

Anonymous (1995) Sweeping patents put biotech companies on the warpath, *Science* **268**, 656–657.

Anonymous (2004) Drugs in crops – the unpalatable truth, *Nature Biotechnology* **22**, 133.

Anonymous (2005) Seeds in threatened soil. US hostility towards Syria is undermining the stability of an important seed bank for dry areas, *Nature* **435**, 537–538.

Arcioni S. and Pupilli F. (2004) Somatic hybridization. In: Thomas B., Murphy D. J. and Murray B., editors, *Encyclopedia of Applied Plant Sciences*, pp. 1423–1431, Oxford: Elsevier Academic Press.

Arisi A. C., Noctor G., Foyer C. H. and Jouanin L. (1997) Modification of thiol contents in poplars (*Populus tremula* × *P. alba*) overexpressing enzymes involved in glutathione synthesis, *Planta* **203**, 362–372.

Arrow K., Bolin B., Costanza R., Dasguputa P., Folke C., Holling C. S., Jansson B. O., Levin S., Maler K. G., Perrings C. and Pimentel D. (1995) Economic growth, carrying capacity, and the environment, *Science* **268**, 520–521 [reprinted in: *Ecological Economics* **15**, 9195].

Ashby J. A. and Lilja N. (2004) Participatory research: does it work? evidence from participatory plant breeding, *4th International Crop Science Congress, September 2004, Brisbane, Australia*, available online at: http://www.regional.org.au/au/cs/2004/symposia/4/1/1589_ashbyj.htm.

Ashikari M., Sakakibara H., Lin S., Yamamoto T., Takashi T., Nishimura A., Angeles E. R., Qian Qian, Kitano H. and Matsuoka M. (2005) Cytokinin oxidase regulates rice grain production, *Science* **309**, 741–745.

Astor M. (2005) Fears of biopiracy hampering research in Brazilian Amazon, *Associated Press*, 20 October 2005, available online at: www.enn.com/today.html?id=9065.

Atkinson H. J., Urwin P. E., Hansen E. and McPherson M. J. (1995) Designs for engineered resistance to root-parasitic nematodes, *Trends in Biotechnology* **13**, 369–374.

Aznar-Marquez J. and Ruiz-Tamari J. R. (2005) Demographic transition environmental concern and the Kuznets curve, *Université Catholique de Louvain, Département des Sciences Economiques Working Paper*, 200500, available online at: http://www.ires.ucl.ac.be/DP/IRES_DP/2005-1.pdf.

Babcock B. A. and Foster W. E. (1991) Measuring the potential contribution of plant breeding to crop yields: flue-cured tobacco, 1954–1987, *American Journal of Agricultural Economics* **73**, 850–859.

Badosa E., Moreno C. and Montesinos E. (2004) Lack of detection of transfer of ampicillin resistance genes from Bt176 transgenic corn to culturable bacteria under field conditions, *FEMS Microbiology and Ecology* **48**, 169–178.

Bailey A. (2005) AmeriFlax reiterates opposition to pharmaceuticals from flax, *Grand Forks Herald*, 13 June 2005, available online at: http://www.grandforks.com/mld/agweek/11880857.htm.

Bailey B. and Lappé M., editors (2002) *Engineering the Farm*, Washington, DC: Island Press.

Banatvala B., Bell P. and Symonds M. (2005) The Research Assessment Exercise is bad for UK medicine, *Lancet* **365**, 458.

Barclay A. (2004) Feral play, *Rice Today*, January 2004, pp. 15–19, available online at: http://www.irri.org/publications/today/pdfs/ 3-1/feral.pdf.

Bardor M., Loutelier-Bourhis C., Paccalet T., Cosette P., Fitchette A. C., Vézina L. P., Trépanier S., Dargis M., Lemieux R., Lange C., Faye L. and Lerouge P. (2003) Monoclonal C5–1 antibody produced in transgenic alfalfa plants exhibits a N-glycosylation that is homogenous and suitable for glyco-engineering into human-compatible structures, *Plant Biotechnology Journal* **1**, 451–462.

Barton K. A., Binns A. N., Matzke A. J. M. and Chilton M. D. (1983) Regeneration of intact tobacco plants containing full length copies of genetically engineered T-DNA, and transmission to R1 progeny, *Cell* **32**, 1033–1043.

Basri M. W., Abdullah S. N. A. and Henson I. E. (2004) Oil palm – achievements and potential. In: *New Directions for a Diverse Planet, Proceedings of the 4th International Crop Science Congress, 26 September–1 October 2004, Brisbane*, available online at: www.cropscience.org.au.

Bateman R. (2005) Creativity crushed by super-fuelled juggernaut, *Times Higher Education Supplement*, 11 February 2005, p. 18.

Bauer M. W. and Gaskell G., editors (2002) *Biotechnology – The Making of a Global Controversy*, Cambridge: Cambridge University Press.

Bayles R. A. (1991) Research note: varietal resistance as a factor contributing to the increased importance of *Septoria tritici* Rob. and Desm. in the UK wheat crop, *Plant Varieties and Seeds* **4**, 177–183.

BBSRC (2004) *Review of BBSRC-Funded Research Relevant to Crop Science. A Report for BBSRC Council, April 2004*, Swindon: BBSRC, available online at: http://www.bbsrc.ac.uk/news/reports/crop_sci_review12_05_04.pdf.

BBSRC (2005) *BBSRC Response to the Agriculture and Environment Biotechnology Commission 'What Shapes the Research Agenda', December 2005*, Swindon: BBSRC, available online at: http://www.bbsrc.ac.uk/about/pub/reports/aebc_2005_dec.htlm.

Béclin C., Charlot F., Botton E., Jouanin L. and Doré C. (1993) Potential use of the *aux2* gene from *Agrobacterium rhizogenes* as a conditional negative marker in transgenic cabbage, *Transgenic Research* **2**, 48–55.

Benatti P., Peluso G., Nicolai R. and Calvani M. (2004) Polyunsaturated fatty acids: biochemical, nutritional and epigenetic properties, *Journal of the American College of Nutrition* **23**, 281–302.

Bender P. (2002) Plant epigenetics, *Current Biology* **12**, R412–R414.

Bergthorsson U., Adams K. L., Thomason B. and Palmer J. D. (2003) Widespread horizontal transfer of mitochondrial genes in flowering plants, *Nature* **424**, 197–201.

Bergthorsson U., Richardson A. O., Young G. J., Goertzen L. R. and Palmer J. D. (2004) Massive horizontal transfer of mitochondrial genes from diverse land plant donors to the basal angiosperm Amborella, *Proceedings of the National Academy of Sciences USA* **101**, 747–752.

Berlan J. P. and Lewontin R. (1983) *Hybrid Corn Revisited*, unpublished ms., cited in Kloppenburg (2004), p. 382, Berlan J. P. and Lewontin R. (1986) Breeders' rights and patenting life forms, *Nature* **322**, 785–788.

Bhagwati J. (2005) Reshaping the WTO, *Far Eastern Economic Review*, January/February 2005, available online at: http://www.columbia.edu/~jb38/index_paper01.html.

Bimonte S. (2001) Model of growth and environmental quality. New evidence of the environmental Kuznets curve, *Department of Political Economics, University of Siena, Working Paper 321*, available online at: www.econ-pol.unisi.it/quaderni/321.pdf.

Binding H. and Nehls R. (1978) Somatic cell hybridization of *Vicia faba + Petunia hybrida*. *Molecular and General Genetics* **164**, 137–143.

Binenbaum E. (2004) The intellectual property strategy of international agricultural research centers, *Australasian Agribusiness Review* **12**, paper 3, available online at: http://www.agrifood.info/review/2004/Binenbaum.html.

Binenbaum E. and Pardey P. G. (2004) Intellectual property strategy in the context of inter-organizational relations: the case of international agricultural research. In: Evenson R. E. and Santaniello V., editors, *The Regulation of Agricultural Biotechnology*, Oxford: CABI Publishing.

Binenbaum E., Nottenburg C., Pardey P., Wright B. and Zambrano P. (2000) South–north trade, intellectual property jurisdictions, and freedom to operate in agricultural research on staple crops, *Environment and Production Technology Division of the International Food Policy Research Institute, Discussion Paper No. 70, December 2000*.

Bingham J. and Payne P. I. (1993) Wheat for the UK – a plant breeder's perspective, *Journal of the Royal Agricultural Society of England* **154**, 29–44.

Bingham J., Law C. N. and Miller T. (1991) *Wheat Yesterday, Today and Tomorrow*, Cambridge: Plant Breeding International and AFRC Institute of Plant Science Research.

Biotechnology Industry Organization (2002) *Reference Document for Confinement and Development of Plant-Made Pharmaceuticals in the United States*, available online at: www.bio.org/healthcare/pmp/PMPConfinementPaper.pdf.

Blackwell E. (2004) *Hunger*, London: Arrow/Random House.

Blakesley D. and Marks T. (2003) Clonal forestry. In: Thomas B., Murphy D. J. and Murray B. G., editors, *Encyclopedia of Applied Plant Sciences*, Oxford: Elsevier/Academic Press.

Blome R. (1686) *The Gentleman's Recreation*, London.

Blouet B. W. and Luebke F., editors (1979) *Cultural Heritage of the Plains*, Lincoln, Nebraska: University of Nebraska Press.

Boden R., Cox D., Nedeva M. and Barker K. (2004) *Scrutinising Science: The Changing UK Government of Science*, London: Palgrave.

Bolen C. D. (1993) New crops from a seed company perspective. In: Janick J. and Simon J. E., editors, *New Crops*, pp. 658–660, New York: Wiley.

Borlaug N. E. (1954) Mexican wheat production and its role in the epidemiology of stem rust in North America, *Phytopathology* **44**, 398–404.

Borlaug N. E. (1958) The impact of agricultural research on Mexican wheat production, *Transactions of the New York Academy of Science* **20**, 278–295.

Borlaug N. (1970) The Green Revolution, peace and humanity, *Nobel Lecture*, available online at: http://www.nobel.se/cgi-bin/print.

Borlaug N. and Dowswell C. (2001) The unfinished Green Revolution: the future role of science and technology in feeding the developing world, *Proceedings of the Seeds of Opportunity, London*, pp. 1–12.

Bouvier L., Fillon F. R. and Lespinasse Y. (1994) Oryzalin as an efficient agent for chromosome doubling of haploid apple shoots in vitro, *Plant Breeding* **113**, 343–344.

Boyer J. S. (1982) Plant productivity and environment, *Science* **218**, 443–448.

Brasher P. (2006) Drug corn could have promising Iowa future, *Des Moines Register*, 13 April 2006, available online at: http://desmoinesregister.com/apps/pbcs.dll/article?AID = /20060413/BUSINESS01/604130410/1023/SPORTS13.

Bratspies R. (2004) Consuming (f)ears of corn: public health and biopharming, *American Journal of Law and Medicine* **30**, 371–404.

Brennan M., Pray C. E., Naseem A. and Oehmke J. F. (2005) An innovation market approach to analyzing impacts of mergers and acquisitions in the plant biotechnology industry, *AgBioForum* **8**, 89–99, available online at: http://www.agbioforum.org/v8n23/v8n23a05-pray.htm.

Brewster A. L., Chapman A. R. and Hansen S. A. (2005) Facilitating humanitarian access to pharmaceutical and agricultural research, *Innovation Strategy Today* **1**, 203–216, available online at: www.biodevelopments.org/innovation/index.htm.

Briggs S. P. and Koziel M. (1998), Engineering new plant strains for commercial markets, *Current Opinion in Biotechnology* **9**, 233–235.

Britt A. B. and May G. D. (2003) Re-engineering plant gene targeting, *Trends in Plant Science* **8**, 90–95.

Brock R. (2003) Brazil: is the threat for real? *The Corn and Soybean Digest*, 1 March 2003, available online at: http://www.cornandsoybeandigest.com/mag/soybean_brazil_threat_real/.

Broecker W. S. (2003) Does the trigger for abrupt climate change reside in the oceans or in the atmosphere? *Science* **300**, 1519–1522.

Brookfield H. and Stocking M. (1999) Agrodiversity: definition, description and design, *Global Environmental Change* **9**, 77–80.

Broothaerts W., Mitchell H. J., Weir B., Kaines S., Smith L. M., Yang W., Mayer J. E., Roa-Rodriguez C. and Jefferson R. A. (2005) Gene transfer to plants by diverse species of bacteria, *Nature* **433**, 629–633.

Brown J. (2000) Better bananas with biotechnology? *Banana Link*, available online at: www.bananalink.org.uk/documents/Biotechnology_by_James_Brown.doc.

Brown M., Harris C. and Jackson C. J. (2004) *The Ration Book Diet*, Stroud: Sutton Publishing.

Brush S. B., editor (2000) *Genes in the Field. On-Farm Conservation of Crop Diversity*, Boca Raton, Florida: Lewis Publishers.

Brush S. B. (2003) The demise of 'common heritage' and protection for traditional agricultural knowledge. In: *Conference on Biodiversity, Biotechnology and the Protection of Traditional Knowledge, 4–6 April 2003, Washington University, St. Louis*, available online at: http://law.wustl.edu/centeris/Confpapers/PDFWrdDoc/StLouis1.pdf.

Bryant J. and Leather S. (1992) Removal of selectable marker genes from transgenic plants: needless sophistication or social necessity? *Trends in Biotechnology* **10**, 274–275.

Buchanan-Wollaston V. and Ainsworth C. (1997), Leaf senescence in *Brassica napus*: cloning of senescence-related genes by subtractive hybridisation, *Plant Molecular Biology* **33**, 821–834.

Buckmann (1857) Untitled, Report to the British Association, p. 207, cited in
    Darwin C. (1883) *The Variation of Animals and Plants Under Domestication*, 2nd
    edition, New York: Appleton & Co., p. 330.

Buerkert A., Oryakhail M., Filatenko A. and Hammer K. (2006) Cultivation and
    taxonomic classification of wheat landraces in the Upper Panjsher Valley of
    Afghanistan after 23 years of war, *Genetic Resources and Crop Evolution* **53**, 91–97.

Buhler W., Morse S., Arthur E., Bolton S. and Mann J. (2002) *Science, Agriculture
    and Research, a Compromised Participation?* London: Earthscan Publications.

Burdick A. (2005) *Out of Eden. An Odyssey of Ecological Invasion*, New York: Farrar,
    Straus and Giroux.

Burke H. (2004) Monsanto exits Argentina soy biz despite soy boom, *Reuters News
    Service*, 18 January 2004, available online at: http://www.forbes.com/business/
    newswire/2004/01/18/rtr1216109.html%2019jan04.

Busch L. (2004) Lessons unlearned: how biotechnology is changing society. In:
    Biotechnology: Science and Society at a Crossroad, pp. 27–38, Ithaca, New York:,
    National Agricultural Biotechnology Council, available online at: http://nabc.
    cals.cornell.edu/pubs/pubs_reports.html.

Busch L. and Lacy W. B. (1983) *Science, Agriculture, and the Politics of Research*,
    Boulder, Colorado: Westview Press.

Busch L., Lacy W. B., Burkhardt J., Hemken D., Moraga-Rojel J., Koponen T. and
    de Souza Silva J. (1995) *Making Nature, Shaping Culture: Plant Biodiversity in
    Global Context*, Lincoln, Nebraska: University of Nebraska Press.

Busch L., Allison R., Harris C., Rudy A., Ten Eyck T., Coppin D., Shaw B.,
    Fairweather J., Konefal J. and Oliver C. (2004) *External Review of the
    Collaborative Research Agreement between Novartis Agricultural Discovery
    Institute, Inc., and The Regents of the University of California*, University of
    California, available online at: http://www.berkeley.edu/news/media/releases/
    2004/07/external_novartis_review.pdf.

Butler L. J. and Marion B. W. (1985) The impacts of patent protection on the US
    seed industry and public plant breeding, *North Central Regional Research
    Publication 304*, Madison, Wisconsin: College of Agricultural and Life Sciences,
    University of Wisconsin.

Butt A., Mousley C., Morris K., Beynon J., Can C., Holub E., Greenberg J. T. and
    Buchanan-Wollaston V. (1998), Differential expression of a senescence-
    enhanced metallothionein gene in *Arabidopsis* in response to isolates of
    Peronospora parasitica and *Pseudomonas syringae*, *Plant Journal* **16**, 209–221.

Cahoon E. B., Hall S. E., Ripp K. G., Ganzke T. S., Hitz W. D. and Coughlan S. J. (2003)
    Metabolic redesign of vitamin E biosynthesis in plants for tocotrienol production
    and increased antioxidant content, *Nature Biotechnology* **21**, 1082–1087.

Calgene Inc. (1990) Request for advisory opinion on *kanr* gene: safety and use in the
    production of genetically engineered plants, *FDA Docket Number: 90A-041*,
    Washington, DC: US Food and Drug Administration.

Camerarius (1694) Epistola ad M. B. Valentini de sexu plantarum. In: *Ostwald's
    Klassiker der exakten Naturwissenschaften*, No. 105, 1899, Leipzig: Verlag von
    Wilhelm Engelmann.

Campbell G. R. (1996) Can learning become the center of Land Grant Universities and Cooperative Extensions? *Staff Paper 96.6*, Madison, Wisconsin: University of Wisconsin-Extension.

Canter P. H., Thomas H. and Ernst E. (2005) Bringing medicinal plants into cultivation: opportunities and challenges for biotechnology, *Trends in Biotechnology* **23**, 180–185.

Caplan A., Herrera-Estrella L., Inzé D., van Heute E., van Montagu M., Schell J. and Zambyrski P. (1983) Introduction of genetic material into plant cells, *Science* **222**, 815–821.

Carew R. (2000) Intellectual property rights: implications for the canola sector and publicly funded research, *Canadian Journal of Agricultural Economics* **48**, 175–194.

Carew R. and Devadoss S. (2003) Quantifying the contribution of plant breeders' rights and transgenic varieties to canola yields: evidence from Manitoba. *Canadian Journal of Agricultural Economics* **51**, 371–395.

Carlson P. S., Smith H. H., and Dearing R. D. (1972) Parasexual interspecific plant hybridization, *Proceedings of the National Academy of Sciences USA* **69**, 2292–2294.

Cassman K. G., Dobermann A., Walters D. T., and Yang H. S. (2003) Meeting cereal demand while protecting natural resources and improving environmental quality, *Annual Review of Environment and Resources* **28**, 315–358.

Cauvin J. (2000) *The Birth of the Gods and the Origins of Agriculture*, translated by Watkins T., Cambridge: Cambridge University Press.

Ceccarelli S., Bailey E., Grando S. and Tutwiler R. (1997) Decentralized-participatory plant breeding: a link between formal plant breeding and small farmers. In: *New Frontiers in Participatory Research and Gender Analysis, Proceedings of an International Seminar on 'Participatory Research and Gender Analysis for Technology Development', Colombia CA*, pp. 65–74, available online at: http://www.icarda.cgiar.org/Farmer_Participation/PDF/Papers/3CALI96.pdf.

Ceccarelli S., Grando S., Tutwiler R., Baha J., Martinil A. M, Salahieh H., Goodchild A. and Michael M. (2000) A methodological study on participatory barley breeding I. Selection phase, *Euphytica* **111**, 91–104.

CGIAR (2005a) *Healing Wounds; How the International Research Centers of the CGIAR Help Rebuild Agriculture in Countries Affected by Conflicts and Natural Disasters, Consultative Group on International Agricultural Research, January 2005*, available online at: http://www.cgiar.org/publications/HealingWounds/index.htm.

CGIAR (2005b) CGIAR system research priorities 2005–2015, *Report of the Science Council, June 2005*, available online at: http://www.sciencecouncil.cgiar.org/activities/spps/pubs/RP0515.pdf.

CGIAR (2006) Intensified research effort yields climate-resilient agriculture to blunt impact of global warming, prevent widespread hunger, *News Release*, 4 December 2006, available online at: http://www.cgiar.org/newsroom/releases/news.asp?idnews = 521.

Chan C. K., Jalani B. S. and Araffin D. (2003) Refining plantation technologies for sustainable production: the path to eco-economy. In: *Proceedings of the PIPOC 2003 International Palm Oil Congress*, pp. 283–303, Kuala Lumpur: Malaysian Palm Oil Board.

Chapman R. and Tripp R. (2003) Case studies of agricultural extension services using privatized service provision, *Agricultural Research and Extension Network (AgREN), Case Studies Series*, available online at: http://www.rimisp.cl/agren03/casestudies.pdf.

Charles D. (2001) *Lords of the Harvest*, Cambridge, Massachusetts: Perseus.

Chase-Dunn C. (1980), The development of core capitalism in the antebellum United States: tariff politics and class struggle in an upwardly mobile semi-periphery. In: Bergesen A. J., editor, *Studies of the Modern World-System*, pp. 189–230, New York:, Academic Press, chapter available online at: http://www.irows.ucr.edu/cd/papers/ustariffpol.htm.

Check E. (2004) David versus Goliath, *Nature* **432**, 546–548.

Chevé M. (2000) Irreversibility of pollution accumulation: new implications for sustainable endogenous growth, *Environmental and Resource Economics* **16**, 93–104.

Chi H. S. (2003) The efficiencies of various embryo rescue methods in interspecific crosses of *Lilium, Botanical Bulletin of the Academica Sinica* **43**, 139–146.

Chilton M. D. (1988) Plant genetic engineering: progress and promise, *Journal of Agricultural and Food Chemistry* **36**, 3–5.

Chilton M. D. (2001) *Agrobacterium*. A memoir, *Plant Physiology* **125**, 9–14.

Choi D., Kim H. N., Yun H. K., Park J. A., Kim T. W. and Bok S. H. (1996), Molecular cloning of a metallothionein-like gene from *Nicotiana glutinosa* L. and its induction by wounding and tobacco mosaic virus infection, *Plant Physiology* **112**, 353–359.

Chung S. M, Vaidya M. and Tzfira T. (2006) *Agrobacterium* is not alone: gene transfer to plants by viruses and other bacteria, *Trends in Plant Science* **11**, 1–4.

CIMMYT (2005) *Fifth External Programme and Management review 2004/2005, April 2005.*

Cisar G. and Cooper D. B. (2004) *Hybrid Wheat, FAO Review*, available online at: http://www.fao.org/DOCREP/006/Y4011E/y4011e0c.htm.

Clark R. (2004) Agricultural enclosures: the major phase, 1760 onwards, *The Literary Encyclopedia*, available online at: www.LitEncyc.com/php/stopics.php?rec=ture&UID=1472.

Cleveland D. A. and Soleri D. (2000) Knowledge, action, and the environment: what can theories of knowledge and crop genetics contribute to collaboration between scientific plant breeders and local farmers? *ASA 2000 Conference, Participating in Development: Approaches to Indigenous Knowledge April 2000, School of Oriental and African Studies, London*, available online at: http://www.asa2000.anthropology.ac.uk/cleveland/cleveland.html.

Cockburn A. (2004) Commercial plant breeding: What is in the biotech pipeline? *Journal of Commercial Biotechnology* **10**, 209–223.

Cohen J. I. (2005) Poorer nations turn to publicly developed GM crops, *Nature Biotechnology* **23**, 27–33.

Comis D. (1999) Tracking manure-borne pathogens, *Agricultural Research Magazine* **47** (11), 22, available online at: http://www.ars.usda.gov/is/AR/archive/nov99/track1199.pdf.

Commandeur U., Twyman R. M. and Fischer R. (2003) The biosafety of molecular farming in plants, *AgBiotechNet* **5**, 1–9.

Commission on Intellectual Property Rights (2002) *Integrating Intellectual Property Rights and Development Policy, Report of the Commission on Intellectual Property Rights (UK), Agriculture and Genetic Resources*, pp. 57–7, London: Commission on Intellectual Property Rights, c/o Department for International Development, available online at: http://www.iprcommission.org/graphic/documents/final_report.htm.

Conway G. (1997) *The Doubly Green Revolution*, London: Penguin.

Conway G. (2003) *From the Green Revolution to the Biotechnology Revolution: Food for Poor People in the 21st Century*, The Rockefeller Foundation, available online at: http://www.rockfound.org/documents/566/Conway.pdf.

Conway G. and Toenniesson G. (1999) Feeding the world in the twenty-first century, *Nature* **402**, C55–C58.

Cook G. (2004) *Genetically Modified Language*, London: Routledge.

Cook R. (1937) A chronology of genetics, *Yearbook of Agriculture, 1937*.

Coors J. G. (2002) Changing role of plant breeding in the public sector. In: *Proceedings of the 56th Annual Corn and Sorghum Research Conference, 5–7, December 2001*, pp. 48–66, Chicago, Illinois: American Seed Trade Association, available online at: www.public.iastate.edu/~brummer/agron521/papers/CoorsASTA.pdf.

Corley H. (2000) *New Technologies for Plantation Crop Improvement*, Silsoe: Cranfield University, available online at: http://www.taa.org.uk/WestCountry/corley.html.

Corley R. H. V. and Tinker P. B. (2003) *The Oil Palm*, 4th edition, Oxford: Blackwell.

Cotsafis O. and Guiderdoni E. (2005) Enhancing gene targeting efficiency in higher plants: rice is on the move, *Transgenic Research* **14**, 1–14.

Cox G. W. (2004) *Alien Species and Evolution: The Evolutionary Ecology of Exotic Plants, Animals, Microbes, and Interacting Native Species*, Washington, DC: Island Press.

Cox T. S. (2002) The mirage of genetic engineering, *American Journal of Alternative Agriculture* **17**, 41–44.

Crabb A. R. (1947) *The Hybrid-Corn Makers: Prophets of Plenty*, New Brunswick, New Jersey: Rutgers University Press.

Cramer C. L., Boothe J. G. and Oishi K. K. (1999) Transgenic plants for therapeutic proteins: linking upstream and downstream strategies, *Current Topics in Microbiology and Immunology* **240**, 95–118.

Crawford M. A., Costeloe K., Ghebremeskel K., Phylactos A., Skirvin L. and Stacey F. (1997) Are deficits of arachidonic and docosahexaenoic acids responsible for the neural and vascular complications of preterm babies? *American Journal of Clinical Nutrition* **66**, 1032S–1041S.

Crawley M. J., Hails R. S., Rees M., Kohn D. and Buxton J. (1993) Ecology of transgenic oilseed rape in natural habitats, *Nature* **363**, 620–623.

Crystal D. (2005) *The Stories of English*, pp. 99–100, London: Penguin.

Cunfer G. (2002) Causes of the dust bowl. In: Knowles A. K., editor, *Past Time, Past Place: GIS for History*, pp. 93–104, Redlands, California: ESRI Press.

Cyranoski D. (2003) Rice genome: a recipe for revolution? *Nature* **422**, 796–798.

Dale E. C. and Ow D. W. (1991) Gene transfer with subsequent removal of the selection gene from the host genome, *Proceedings of the National Academy of Sciences USA* **88**, 1055–1056.

Dale P. J. (2004) Public-good plant breeding: what should be done next? *Journal of Commercial Biotechnology* **10**, 199–208.

Dalton T. (2004) A household hedonic model of rice traits: economic values from farmers in West Africa, *Agricultural Economics* **31**, 149–159.

Dalton T. and Guei R. (2003) Productivity gains from rice genetic enhancements in West Africa: countries and ecologies, *World Development* **31**, 359–374.

Daly M. (2006) Clinton's GM recipe to feed armies of the hungry, *Sydney Morning Herald*, 13 April 2006, available online at: http://smh.com.au/news/world/clintons-gm-recipe-to-feed-armies-of-the-hungry/2006/04/12/1144521401084.html.

Daniell H. (2002) Molecular strategies for gene containment in transgenic crops, *Nature Biotechnology* **20**, 581–586.

Daniell H., Kumar S. and Dufourmantel N. (2005) Breakthrough in chloroplast genetic engineering of agronomically important crop, *Trends in Biotechnology* **23**, 238–245

Darwin C. (1859) The Origin of Species by means of Natural Selection, London: Penguin., This book is a reproduction of the first edition of Darwin's famous work. However, a more definitive version is contained in the 6th edition, from 1868, which is available for free online at: http://www.literature.org/authors/darwin-charles/the-origin-of-species-6th-edition/.

Darwin C. (1883) *The Variation of Animals and Plants Under Domestication*, 2nd edition, New York: Appleton & Co, available online at: http://www.esp.org/books/darwin/variation/facsimile/title3.html.

Davies W. P. (2003) An historical perspective from the Green Revolution to the Gene Revolution, *Nutrition Reviews* **61**, S124–S134.

Davis C. C. and Wurdack K. J. (2004) Host-to-parasite gene transfer in flowering plants: phylogenetic evidence from Malpighiales, *Science* **305**, 676–678.

Davis C. C., Anderson W. R. and Wurdack K. J. (2005) Gene transfer from a parasitic flowering plant to a fern. *Proceedings of the Royal Society B, Biological Sciences* **272**, 2237–2242.

Davoudi S. (2006) Monsanto: giant of the $6.15 GM market, *Financial Times*, 16 November 2006, available online at: http://search.ft.com/ftArticle?queryText=%22AventistSA%22&javascriptEnabled=true &id=061115010778.

Day Rubenstein K. and Heisey P. W. (2005) Can technology transfer help public-sector researchers do more with less? The case of the USDA's agricultural research service, *AgBioForum* **8**, 134–142, available online at: http://www.agbioforum.org/v8n23/v8n23a10-heisey.htm.

Dedijer S. (1963) Underdeveloped science in underdeveloped countries, *Minerva* 2, 61–81.

DEFRA (2003) *Review of Non-native Species Policy: Report of the Working Group*, available online at: http://www.defra.gov.uk/wildlife-countryside/resprog/findings/non-native/.

Delmer D. P. (2005) Agriculture in the developing world: connecting innovations in plant research to downstream applications, *Proceedings of the National Academy of Sciences USA* 102, 15739–15746.

Delseney M. (2004) Re-evaluating the relevance of ancestral shared synteny as a tool for crop improvement, *Current Opinion in Plant Biology* 7, 126–131.

Devos K. M. and Gale M. D. (2000) Genome relationships: the grass model in current research, *Plant Cell* 12, 637–646.

Dhlamini Z. (2006) The role of non-GM biotechnology in developing world agriculture, *Policy Brief*, February 2006, Science and Development Network, available online at: http://www.scidev.net.

Diao X., Freeling M. and Lisch D. (2006) Horizontal transfer of a plant transposon, *public library of science, biology* 4, e5, available online at: http://biology. plosjournals.org/perlserv/?request = get-document&doi = 10.1371/journal. pbio.0040005.

Diaz K. (2004) Brazil: the new breadbasket, *Star Tribune*, 7 March 2004, available online at: http://www.startribune.com/stories/462/4647358.html.

Dierker D. and Phillips P. W. B. (2003) The search for the holy grail? maximizing social welfare under Canadian biotechnology patent policy, *IP Strategy Today*, Number. 6–2003, pp. 45–62, available online at: http://biodevelopments.org/ip/index.htm.

Diez M. C. F. (2002) The impact of plant varieties rights on research: The case of Spain. *Food Policy* 27, 171–183.

Dirnberger D., Steinkellner H., Abdennebi L., Remy J. J and van de Wiel D. (2001) Secretion of biologically active glycoforms of bovine follicle stimulating hormone in plants, *European Journal of Biochemistry* 268, 4570–4579.

Djurfeldt G., Holmen H., Jirstrom M. and Larsson R., editors (2005) *The African Food Crisis: Lessons from the Asian Green Revolution*, Wallingford: CABI Publishing.

Doggett H. (1988) *Sorghum*, Harlow: Longman.

Downey R. K. and Craig B. M. (1964) Genetic control of fatty acid biosynthesis in rapeseed (*Brassica napus* L.), *Journal of the American Oil Chemists Society* 41, 475–478.

Downing J. (2006) Kansas opens arms to Ventria's biotech rice, *Sacramento Bee*, 13 June 2006, available online at: http://www.sacbee.com/content/business/story/14267204p-15078948c.html.

Doyle J. (1985) *Altered Harvest: Agriculture, Genetics, and the Fate of the World's Food Supply*, London: Penguin.

Dreher K., Khairallah M., Ribaut J. M. and Morris M. (2003) Money matters (I): costs of field and laboratory procedures associated with conventional and marker-assisted selection at CIMMYT, *Molecular Breeding* 11, 221–234.

Driscoll M. (2003) Can bio-crops really end world hunger? George Bush says GM food will be Africa's salvation, *The Times*, 29 June 2003.

Dubcovsky J. (2004) Marker-assisted selection in public breeding programs: the wheat experience, *Crop Science* **44**, 1895–1897.

Dunwell J. M. (2005) Intellectual property aspects of plant transformation, *Plant Biotechnology Journal* **3**, 371–384.

Durant J., Bauer M. and Gaskell G., editors (1998) *Biotechnology in the Public Sphere: A European sourcebook*, London: Science Museum Publications.

Durrant A. (1962) The environmental induction of heritable changes in *Linum*, *Heredity* **17**, 27–61.

Duvick D. N. (1996) Plant breeding, an evolutionary concept, *Plant Science* **36**, 539–548.

Duvick D. N. (1999) Heterosis: feeding people and protecting natural resources. In: Coors J. G. and Pandey S., editors, *The Genetics and Exploitation of Heterosis in Crops*, Madison, Wisconsin: American Society of Agronomy Inc., available online at: http://www.biotech-info.net/heterosis_duvick1.pdf.

Duvick D. (2003) The current state of plant breeding: how did we get here? In: Sligh M. and Lauffer L., editors, *Summit Proceedings: Summit on Seeds and Breeds for 21st Century Agriculture*, pp. 71–92, Pittsboro, North Carolina: Rural Advancement Foundation International, available online at: http://www.rafiusa.org/pubs/Seeds%20and%20Breeds.pdf.

Duvick D. (2004) New technologies for sustained productivity growth: plant breeding, *USDA Agricultural Outlook Forum 2004, Ensuring a Healthy Food Supply, Arlington, Virginia*, available online at: www.usda.gov/oce/forum/Archives/2004/speeches/Duvick%20feb%2011.doc.

Economist (2005a) The triumph of the commons – can open source revolutionise biotech? *Economist*, February 2005, p. 75.

Economist (2005b) Short cut to China, *Economist*, 26 March 2005, p. 67.

Economist (2005c) Economics focus: punch-up over handouts, *Economist*, 26 March 2005, p. 102.

Economist (2005d) Global or national? The perils facing big oil, *Economist*, 30 April 2005, supplement pp. 8–12.

Economist (2005e) Prescription for change. A survey of pharmaceuticals, *Economist*, 18 June 2005, supplement p. 4.

Economist (2005f) Peru. Blooming desert: an agricultural revolution, *Economist*, 9 July 2005, p. 45.

Economist (2005g) Brazilian agriculture: the harnessing of nature's bounty, *Economist*, 5 November 2005, pp. 95–98.

Economist (2005h) Changing gear: decent reforms are enriching the Vietnamese; the future looks fine, *Economist*, 28 November 2005, pp. 71–72.

Economist (2006) A seedbed of revolution, *Economist*, 16 September 2006, p. 102.

Egelyng H. (2005) Evolution of capacity for institutionalized management of intellectual property at international agricultural research centers: a strategic case study, *AgBioForum* **8**, 7–17.

Ehrlich P. (1968) *The Population Bomb*, New York: Ballantine Books.

Eicher C. (1999) Institutions and the African farmer, *Third Distinguished Economist Lecture, CIMMYT Economics Program, 15 January 1999, International Maize and Wheat Improvement Center, El Batan (CIMMYT)*.

Einsiedel E. F. and Medlock J. (2005) A public consultation on plant molecular farming, *AgBioForum* **8**, 26–32, available online at: www.agbioforum.org/v8n1/v8n1a04-einsiedel.htm.

Elbehri A. (2005), Biopharming and the food system: examining the potential benefits and risks, *AgBioForum* **8**, 18–25.

Ellstrand N. C. (2003a) *Dangerous Liaisons? When Cultivated Plants Mate with their Wild Relatives*, Baltimore, Maryland: Johns Hopkins University Press.

Ellstrand N. C. (2003b) Going to 'great lengths" to prevent the escape of genes that produce speciality chemicals, *Plant Physiology* **132**, 1770–1774.

Engels J. M. M. and Visser L. (2003) *A Guide to Effective Management of Germplasm Collections*, Rome: International Plant Genetic Resources Institute (IPGRI).

Epinat J. C., Arnould S., Chames P., Rochaix P., Desfontaines D., Puzin C., Patin A., Zanghellini A., Paques F. and Lacroix E. (2003) A novel engineered meganuclease induces homologous recombination in yeast and mammalian cells, *Nucleic Acids Research* **31**, 2952–2962.

Erikson O., Hertzberg M. and Nasholm T. (2004) A conditional marker gene allowing both positive and negative selection in plants, *Nature Biotechnology* **22**, 455–458.

Evans L. T. (2005) The changing context for agricultural science, *Journal of Agricultural Science* **143**, 7–10.

Evanson R. E. and Gollin D. (2003a) Assessing the impact of the Green Revolution, 1960 to 2000, *Science* **300**, 758–762.

Evanson R. E. and Gollin D., editors (2003b) *Crop Variety Improvement and its Effect on Productivity: The Impact of International Agricultural Research*, Wallingford: CAB International.

Ewen S. W. B. & Pusztai A. (1999) Effect of diets containing genetically modified potatoes expressing *Galanthus nivalis* lectin on rat small intestine, *Lancet* **354**, 1353–1354.

Fazackerley A. (2004) Staff at risk in RAE run-up, *Times Higher Education Supplement*, 20 May 2004, p. 1.

Fazackerley A. (2005) Revamped RAE fails to win over medics and engineers, *Times Higher Education Supplement*, 21 January 2005, p. 4.

Fearnside P. M. (2000) Soybean cultivation as a threat to the environment in Brazil, *Environmental Conservation* **28**, 23–38, available online at: http://philip.inpa.gov.br/publ_livres/Preprints/2001/SOY-sw.pdf.

Featherstone A. and Conforte D. (2002) *The Future of South American Agriculture, 2002 Risk & Profit Conference, 15–16 August 2002 Manhattan, Kansas*, available online at: www.agmanager.info/events/risk_profit/2002/Featherstone.pdf.

Fedoroff N. V. and Brown N. M. (2004) *Mendel in the Kitchen: Scientist's View of Genetically Modified Food*, Washington, DC: National Academy of Sciences, available online at: http://www.nap.edu/books/0309092051/html/.

Feil B. and Schmid J. E. (2002) *Dispersal of Maize, Wheat and Rye Pollen. A Contribution to Determining the Necessary Isolation Distances for the Cultivation of Transgenic Crops*, Aachen: Shaker Verlag.

Fernandez M. and Smith C. (2005) Impacts of biotech regulation on small business and university research: possible barriers and potential solutions, *Conference of Pew Initiative on Food and Biotechnology and USDA Animal and Plant Health Inspection Service (APHIS), 2–3 June 2004, Washington DC*, available online at: http://pewagbiotech.org/events/0602/.

Fernandez-Cornejo *et al.* (2004) *The Seed Industry in US Agriculture*, Economic Research Service/USDA, AIB-786.

Finkel E. (1999) Australian center develops tools for developing world, *Science* **285**, 1481–1483.

Fischer K. S. and Cordova V. G. (1998) Impact of IRRI on rice science and production. In: Pingali P. L. and Hossain M., editors, *The Impact of Rice Research*, pp. 27–51, Manila: IRRI

Fischer R., Emans N. J., Twyman R. M. and Schillberg S. (2004) Molecular farming in plants: technology platforms. In: Goodman R. B., editor, *Encyclopedia of Plant and Crop Science*, New York: Marcel Dekker.

Fisher R. A. (1918) The correlation between relatives on the supposition of Mendelian inheritance, *Philosophical Transactions of the Royal Society of Edinburgh* **52**, 399–433.

Fishman A. N., Gandal N. and Shy O. (1993) Planned obsolescence as an engine of technological progress, *Journal of Industrial Economics* **41**, 361–370.

Fitzpatrick M. (2004) *MMR and Autism: What Patients Need to Know*, London: Routledge.

Flavell R. B., Dart E., Fuchs R. L. and Fraley R. T. (1992) Selectable marker genes: safe for plants? *Bio/Technology* **10**, 141–144.

Flint J. (1998) Agricultural industry giants moving towards genetic monopolism, *Farm Journal*, see source online at: http://archive.greenpeace.org/geneng/reports/food/intrfo09.htm.

Flowers T. J. (2004) Improving crop salt tolerance, *Journal of Experimental Botany* **55**, 307–319.

Folger K. R., Wong E. A., Wahl G. and Capecchi M. (1982) Patterns of integration of DNA microinjected into cultured mammalian cells: evidence for homologous recombination between injected plasmid DNA molecules, *Molecular Cell Biology* **2**, 1372–1387.

Food and Agriculture Organisation (2003) *The State of Food Insecurity in the World*, 5th Edition, Rome: Food and Agriculture Organisation of the United Nations, available online at: http://www.fao.org/documents/show_cdr.asp?url_file=/docrep/006/j0083e/j0083e00.htm.

Food and Agriculture Organisation (2004a) The state of food and agriculture: agricultural biotechnology – meeting the needs of the poor? *FAO Agriculture Series No. 35*, Rome: Food and Agriculture Organisation of the United Nations, available online at: http://www.fao.org/documents/show_cdr.asp?url_file=/docrep/007/y5650e/y5650e00.htm.

Food and Agriculture Organisation (2004b) *World Agriculture: Towards 2015/2030*, New York: Earthscan, available online at: http://www.fao.org/es/ESD/gstudies. htm.

Forbes, J. C. and Watson R. D. (1992) *Plants in Agriculture*, Cambridge: Cambridge University Press.

Forster B. P., Ellis R. P., Thomas W. T., Newton A. C., Tuberosa R., This D., el-Enein R. A., Bahri M. H., Ben Salem (2000) The development and application of molecular markers for abiotic stress tolerance in barley, *Journal of Experimental Botany* **51**, 19–27.

Fowler C. (2003) An international perspective on trends and needs in public agricultural research. In: Sligh M. and Lauffer L., editors, *Summit Proceedings: Summit on Seeds and Breeds for 21st Century Agriculture*, pp. 1–11, Pittsboro, North Carolina: Rural Advancement Foundation International, available online at: http://www.rafiusa.org/pubs/Seeds%20and%20Breeds.pdf.

Fowler C. and Mooney P. (1990) *Shattering: Food, Politics, and the Loss of Genetic Diversity*, Tucson, Arizona: University of Arizona Press.

Fox J. L. (2003) Puzzling industry response to ProdiGene fiasco, *Nature Biotechnology* **21**, 3–4.

Freese B. (2002) Manufacturing drugs and chemicals in crops, *Genetically Engineered Food Alert*, available online at: http://www.gefoodalert.org/pages/home.cfm.

Frey K. J. (1996) National plant breeding study, volume I. Human and financial resources devoted to plant breeding research and Development in the United States in 1994, *Agriculture and Home Economics Experiment Station (special report 98)*, Ames, Iowa.

Frey K. J. (1997) National plant breeding study, volume II. National plan for promoting breeding programs for minor crops in the United States, *Agriculture and Home Economics Experiment Station (special report 100)*, Ames, Iowa.

Frey K. J. (1998) The national plant breeding study, volume III. National plan for genepool enrichment of U.S. crops, *Agriculture and Home Economics Experiment Station (special report 101)*, Ames, Iowa.

Fry J. (1998) Implication of recent developments in Asian economies and in the global economy for the oil palm industry. In: Angga Jatmita *et al.*, editors, *Proceedings of the 1998 International Oil Palm Conference*, pp. 28–35, Medan: Indonesian Oil Palm Research Institute.

Fuglie K. O. and Schimmelpfennig D. E., editors (2000) *Public Private Collaboration in Agricultural Research: New Institutional Arrangements and Economic Implications*, Ames, Iowa: Iowa State University Press.

Fukuda-Parr S. (2006) Introduction: global actors, markets and rules driving the diffusion of genetically modified (GM) crops in developing countries, *International Journal of Science, Technology, and Globalization* **2**, 1–11, available online at: http://bcsia.ksg.harvard.edu/research.cfm?program = STPP&project = STG&pb_id = 537.

Funk C. D. (2001) Prostaglandins and leukotrienes: advances in eicosanoid biology, *Science* **294**, 1871–1875.

Fussell B. (1992) *The Story of Corn*, New York: Knopf.

Gager C. S. and Blakeslee A. F. (1927) Chromosome and gene mutations in *Datura* following exposure to radium rays, *Proceedings of the National Academy of Sciences USA* **13**, 75–79.

Gao A. G., Hakimi S. M., Mittanck C. A., Wu Y., Woerner B. M., Stark D. M., Shah D. M., Liang J. and Rommens C. M. (2000) Fungal pathogen protection in potato by expression of a plant defensin peptide, *Nature Biotechnology* **18**, 1307–1310.

Gärtner F. C. (1849) *Versuche und Beobachtungen über die Bastarderzeugung im Pflanzenreiche*, Stuttgart: KF Herring.

Gaskell G. and Bauer M. W., editors (2001) *Biotechnology 1996–2000, the Years of Controversy*, London: Science Museum.

Gatehouse J. (1999) Letter to the editor, *Lancet* **354**, 7 October 1999.

Gay P. B. and Gillespie S. H. (2005) Antibiotic resistance markers in genetically modified plants: a risk to human health? *Lancet Infectious Diseases* **5**, 637–646.

Gepts P. (2004) Who owns biodiversity, and how should the owners be compensated? *Plant Physiology* **134**, 1295–1307.

Gerlach V. L. and Dyer T. A. (1980) Sequence organization of the repeat units in the nucleus of wheat which contain 5S rRNA genes, *Nucleic Acids Research* **8**, 4851–486.

Gerpacio R. V. (2003) The roles of public sector versus private sector in R&D and technology generation: the case of maize in Asia, *Agricultural Economics* **29**, 319–331.

Ghatnekar L., Jaarola, M. and Bengtsson B. O. (2006) The introgression of a functional nuclear gene from *Poa* to *Festuca ovina*, *Proceedings of the Royal Society, B: Biological Sciences* **273**, 395–399.

Gherbi H., Gallego M. E., Jalut N., Lucht J. M., Hohn B. and White C. I. (2001) Homologous recombination *in planta* is stimulated in the absence of Rad50, *EMBO Reports* **2**, 287–291.

Gill P., Jeffreys A. J. and Werrett D. J. (1985) Forensic application of DNA 'finger prints', *Nature* **318**, 577–579.

Glass B. (1955) Genetics in the service of man, *Johns Hopkins Magazine* **6** (5), 2–8 February 1955.

Gleba Y. Y. and Hoffman F. (1979) '*Arabidobrassisa*': Plant-genome engineering by protoplast fusion, *Naturwissenschaften* **66**, 47–554.

Gleba Y., Klimyuk V. and Marillonnet S. (2005) Magnifection – a new platform for expressing recombinant vaccines in plants, *Vaccine* **23**, 2042–2048.

Goeschl T. and Swanson T. (2003) The development impact of genetic use restriction, technologies: a forecast based on the hybrid crop experience *Environment and Development Economics* **8**, 149–165.

Goff S. and Salmeron J. M. (2004) Back to the future of cereals, *Scientific American* **291**, August 2004, 42–49.

Goldberger J., Merrill J. and Hurley T. (2005) Bt corn farmer compliance with insect resistance management requirements in Minnesota and Wisconsin, *AgBioForum* **8**, 151–160, available online at: http://www.agbioforum.org/v8n23/v8n23a12-hurley.htm.

Goodman D., Sorj B. and Wilkinson J. (1987) *From Farming to Biotechnology*, Oxford: Blackwell.

Goodman M. M. (2004) Plant breeding requirements for applied molecular biology, *Crop Science* **44**, 1913–1914.

Graff G. and Zilberman D. (2004) Explaining Europe's resistance to agricultural biotechnology, *Agricultural and Resource Economics Update* 7, No.5, May/June 2004, pp. 1–4, available online at: http://are.berkeley.edu/˜ggraff.

Graff G. D., Cullen S. E., Bradford K. J., Zilberman D. and Bennett A. B. (2003) The public-private structure of intellectual property ownership in agricultural biotechnology, *Nature Biotechnology* **21**, 989–995.

Graham R. D. (2003) Biofortification: a global challenge program, *IRRI mini review* **28**, 1–5, available online at: http://www.irri.org/publications/irrn/pdfs/vol28no1/irrn28-1mini1.pdf.

Grant-Downton R. T. and Dickinson H. G. (2005) Epigenetics and its implications for plant biology 1. The epigenetic network in plants, *Annals of Botany (London)* **97**, 1143–1164.

Grant-Downton R. T. and Dickinson H. G. (2006) Epigenetics and its implications for plant biology 2. The 'epigenetic epiphany': epigenetics, evolution and beyond, *Annals of Botany (London)* **97**, 11–27.

Gressel J. (1992) Indiscriminate use of selectable markers – sowing wild oats, *Trends in Biotechnology* **10**, 382.

Gressel J. and Al Ahmad H. (2005) Assessing and managing biological risks of plants used for bioremediation, including risks of transgene flow, *Zeitschrift für Naturforschung* **60c**, 154–165.

Guha S. and Maheshwari S. C. (1964) *In vitro* production of embryos from anthers of *Datura, Nature* **204**, 497.

Guha S. and Maheshwari S. C. (1966) Cell division and differentiation of embryos in the pollen grains of *Datura* in vitro, *Nature* **212**, 97–98.

Gurr S. J. and Rushton P. J. (2005a) Engineering plants with increased disease resistance: what are we going to express? *Trends in Biotechnology* **23**, 275–282.

Gurr S. J. and Rushton P. J. (2005b) Engineering plants with increased disease resistance: how are we going to express it? *Trends in Biotechnology* **23**, 283–290.

Haas J. D., Beard J. L., Murray-Kolb L. E., del Mundo A. M., Felix A. and Gregorio G. B. (2005) Iron-biofortified rice improves the iron stores of nonanemic filipino women, *Journal of Nutrition* **135**, 2823–2830, available online at: http://www.nutrition.org/current.shtml.

Habetha M. and Bosch T. C. (2005) Symbiotic Hydra express a plant-like peroxidase gene during oogenesis, *Journal of Experimental Biology* **208**, 2157–6.

Halfter U., Morris P. C. and Willmitzer L. (1992) Gene targeting in *Arabidopsis thaliana, Molecular and General Genetics* **231**, 186–193.

Halpin C. (2005) Gene stacking in transgenic plants – the challenge for 21st century plant biotechnology, *Plant Biotechnology Journal* **3**, 141–155.

Hambridge G. and Bressman E. M. (1936) Foreword and summary, *Yearbook of Agriculture 1936*, Washington, DC: US Government Printing Office.

Hammer K. (2003) A paradigm shift in the discipline of plant genetic resources, *Genetic Resources and Crop Evolution* **50**, 3–10.

Han N. M., May C. Y., Ngan M. A., Hock C. C. and Ali Hashim M. (2004) Isolation of palm tocols using supercritical fluid chromatography, *Journal of Chromatographic Science* **42**, 536–539.

Hanin M. and Paszkowski J. (2003) Plant genome modification by homologous recombination, *Current Opinion in Plant Biology* **6**, 157–162.

Hare B., Plyusnina I., Ignacio N. *et al.* (2005) Social cognitive evolution in captive foxes is a correlated by-product of experimental domestication, *Current Biology* **15**, 226–230, available online at: http://www.current-biolgy.com/content/article/abstract?uid = PIIS0960982205000928.

Hare P. D. and Chua N. H. (2002) Excision of selectable marker genes from transgenic plants, *Nature Biotechnology* **20**, 575–580.

Harl N. E., Ginder R. G., Hurgurgh C. R. and Moline S. (2003) *The StarLink™ Situation*, Ames, Iowa: Iowa State University Extension, available online at: http://www.extension.iastate.edu/grain/resources/publications/buspub/0010star.PDF.

Harlan H. V. and Martini M. L. (1936) Problems and results in barley breeding. In: *Yearbook of Agriculture 1936*, Washington, DC: US Government Printing Office.

Harrison P. and Pearce F. (2001) *AAAS Atlas of Population and Environment*, Washington, DC: American Association for the Advancement of Science, available online at: http://atlas.aaas.org/.

Harsch E. (2004) Farmers embrace African 'miracle' rice: high-yielding 'Nerica' varieties to combat hunger and rural poverty, *Africa Recovery – United Nations Department of Public Information* **17**, 10–15.

van Harten A. M. (1997) *Mutation Breeding*, Cambridge. Cambridge University Press.

Hartwich F, Janssen W. and Tola J. (2003) Public–private partnerships for agroindustrial research: recommendations from an expert consultation, *ISNAR Briefing Paper 61*, available online at: http://www.isnar.cgiar.org/publications/briefing/bp61.htm.

Haworth P. L. (1915) *George Washington: Farmer. Being an account of his home life and agricultural activities*, republished in 2004 by Whitefish, Montana: Kessinger Publishing Co., available as an ebook online at: http://www2.cddc.vt.edu/gutenberg/1/1/8/5/11858/11858-8.txt.

Hayashi M. A., Kano E. and Goto, editors (1992) numerous articles from the International Symposium on Transplant Production Systems, in *Acta Horticulturae* **319**, available online at: http://www.actahort.org/books/319/.

Heady E. O. (1982) The adequacy of agricultural land: a demand-supply perspective. In: Crosson P. R., editor, *The Cropland Crisis: Myth or Reality?* Baltimore, Manyland: Resources for the Future Inc.

Hedden P. (2002) The genes of the Green Revolution, *Trends in Genetics* **19**, 5–9.

Heisey P. W., Srinivasan C. S. and Thirtle C. G. (2001) Public sector plant breeding in a privatizing world, USDA Economic Research Service, *Agriculture Information Bulletin No. 772*, pp. 1–22, available online at: http://www.ers.usda.gov/publications/aib772/.

Heisey P. W., Lantican M. A. and Dubin H. J. (2002) *Assessing the Benefits of International Wheat Breeding Research in the Developing World: The Global Wheat Impacts Study, 1966–1997*, Mexico, DF: CIMMYT.

Heisey P. W., King J. L. and Rubenstein K. D. (2005) Patterns of public-sector and private-sector patenting in agricultural biotechnology, *AgBioForum* **8**, 73–82, available online at: http://www.agbioforum.org/v8n23/v8n23a03-heisey.htm.

Heller M. A. and Eisenberg R. S. (1998) Can patents deter innovation? The anticommons in biomedical research, *Science* **280**, 698–701.

Herrera S. (2005) Syngenta's gaff embarrasses industry and White House, *Nature Biotechnology* **23**, 514.

Hiatt A. C., Cafferkey R. and Bowdish K. (1989) Production of antibodies in transgenic plants, *Nature* **342**, 76–78.

Hides D. and Humphreys M. (2000) From WPBS to IGER, *IGER Innovations 2000*, pp. 6–13, available online at: www.iger.bbsrc.ac.uk/Publications/Innovations/In2000/ch1.pdf.

Hightower J. (1973) *Hard Tomatoes, Hard Times*, Cambridge, Massachusetts: Schenkman Publishing Co.

Hileman B. (2003) ProdiGene and StarLink incidents provide ammunition to critics, *Chemical and Engineering News* **81**, 25–33, available online at: http://pubs.acs.org/cen/coverstory/8123/8123biotechnologyb.html.

Hillman G., Hedges R., Moore A., Colledge S. and Pettitt P. (2001) New evidence of Lateglacial cereal cultivation at Abu Hureyra on the Euphrates, *Holocene* **11**, 383–395.

Hills M. J. and Murphy D. J. (1990) The biotechnology of oilseeds, *Biotechnology and Genetic Engineering Reviews* **9**, 1–46.

Ho C. T., Chen Q. and Zhou R. (1996) Flavor compounds in fats and oils. In: Hui Y. H., editor, *Bailey's Industrial Oil and Fat Products*, Fifth Edition, Volume I: Edible Oil and Fat Products: General Applications, Chapter 4, pp. 83–104, New York: Wiley.

Hoffert E. (2005) Agragen sows a crop of false assertions, *Grand Forks Herald*, editorial, 26 June 2005, available online at: http://www.grandforks.com/mld/grandforks/news/editorial/11987962.htm.

Holmes F. L. (1973) Liebig, Justus von, *Dictionary of Scientific Biography*, volume 8, pp. 329–335, New York: Charles Scribner's Sons.

Hong S., Gronert K., Devchand P. R., Moussignac R. L. and Serhan C. N. (2003) Docosatrienes and 17S-resolvins generated from docosahexaenoic acid in murine brain, human blood, and glial cells. Autacoids in anti-inflammation, *Journal of Biological Chemistry* **278**, 14677–14687.

Hooper W. D. (1999) *Cato and Varro: On Agriculture*, reprint of 1935 edition, Cambridge, Massachusetts: Harvard University Press.

Horsch R. and Montgomery J. (2004) Why we partner: collaborations between the private and public sectors for food security and poverty alleviation through agricultural biotechnology, *AgBioForum* **7**, 80–83, available online at: http://www.agbioforum.org/v7n12/v7n12a15-horsch.htm.

Houghton J. (1681–1683) *A Collection of Letters for the Improvement of Husbandry and Trade (1681–1683)*, London: Houghton.

Huang J., Pray C. and Rozelle S. (2002) Enhancing the crops to feed the poor, *Nature* **418**, 678–684.

Huang Z., Dry I., Webster D., Strugnell R. and Wesselingh S. (2001) Plant-derived measles virus hemagglutinin protein induces neutralizing antibodies in mice, *Vaccine* **19**, 2163–2167.

Huffman W. E. (2004) Production, identity preservation, and labeling in a marketplace with genetically modified and non-genetically modified foods, *Plant Physiology* **134**, 3–10.

Hughes S. (2006) Genomics and crop plant science in Europe, *Plant Biotechnology Journal* **4**, 3–5.

Hunzicker G. M., Elmar E. W. and Kubigsteltig I. (2004) Exploring a new detergent-inducible promoter active in higher plants and its potential biotechnological application. In: *Proceedings of the 4th International Crop Science Congress, 26 September – 1 October 2004, Brisbane, Australia*, available online at: www.cropscience.org.au.

Huppatz J. L., Llewellyn D. J., Last D. L., Higgins T. J. and Peacock W. J. (1995) *Development of herbicide-resistant crops – strategies, benefits and risks in herbicide-resistant crops and pastures*. In: McLean G. D. and Evans G., editors, *Australian farming systems*, pp. 15–24, Canberra: Bureau of Resource Sciences.

Innes N. L. (1982) Patents and plant breeding, *Nature* **298**, 786.

Innes N. L. (1984) Public and private plant breeding of horticultural food crops in Western Europe, *Horticultural Science* **19**, 803–808.

Innes N. L. (1992) *Plant Breeding and Intellectual Property Rights*, University of Kiel, available online at: http://www.agric-econ.uni-kiel.de/Abteilungen/II/forschung/file5.pdf.

Jacobs P. (2003) Contaminated grain still showing up in corn supply, *Centre Daily Times*, Pennsylvania State University, 1 December 2003, available online at: http://www.centredaily.com/mld/centredaily/news/7386628.htm.

Jain H. K. (1991) Plant genetic resources and the efficiency factor in agriculture. In: Getupig I. P. and Swaminathan M. S., editors, *Biotechnology for Asian Agriculture: Public Policy Implications*, Kuala Lumpur: Asian and Pacific Development Centre.

Jain H. K. (2000) *Agriculture in the 21ˢᵗ Century: A New Global Order for Research*, ISNAR-CGIAR publications, available online at: www.isnar.cgiar.org/publications/pdf/vision/jain.pdf.

Jain H. K. and Kharkwal (2004) *Plant Breeding: Mendelian to Molecular Approaches*, Dordrecht: Kluwer.

James C. (2006) *Global Status of Commercialized Biotech/GM Crops: 2005, ISAAA Briefs No. 34*, Ithaca, New York: ISAAA, available online at: http://www.isaaa.org.

Jana A., Brewster A., Stephen A., Hansen S. A. and Kisielewski M. (2006) *The Effects of Patenting in the AAAS Scientific Community*, Washington DC: American Association for the Advancement of Science, available online at: http://sippi.aaas.org/survey/AAAs_IP_Survey_Report.pdf.

Janis M. D. (2002) Intellectual property issues in plant breeding and plant biotechnology. In: Martin M. A., editor, *Biotechnology, Gene Flow, and Intellectual Property Rights: An Agricultural Summit*, Proceeding of a conference held in Indianapolis, Indiana, 13 September 2002, Purdue University, Research Bulletin RB-995, Lafayette, Indiana.

Janis M. D. and Kesan J. P. (2002) U.S. plant variety protection: sound and fury...? *Houston Law Review* **39**, 727–778.

Jansen K. (1938) Some west Baltic pollen diagrams, *Quartar* **1**, 124–139.

Jefferson-Moore K. Y. and Traxler G. (2005) Second-generation GMOs: where to from here? *AgBioForum* **8**, 143–150, available online at: http://www.agbioforum. org/v8n23/v8n23a11-jefferson.htm.

Jensen K. and Murray F. (2005) Intellectual property landscape of the human genome, *Science*, **310** 239–240.

John Innes Centre (2002) John Innes Centre loses research partner as company restructures, *Press Release*, 18 September 2002, available online at: http://www.jic. bbsrc.ac.uk/corporate/Media_and_Public/Releases/020918.htm.

John Innes Centre (2005) Re-structuring at the John Innes Centre, *Press Release*, 8 April 2005, available online at: http://www.jic.ac.uk/corporate/media-and-public/current-releases/050408.htm.

Johnson V. A. (1986) Future prospects in genetic improvement in yield of wheat. In: *Genetic Improvement in Yield of Wheat*, Madison Wisconsin: Crop Science Society of America and American Society of Agronomy.

Jones C. C. (1957) The Burlington Railroad and agricultural policy in the 1920s, *Agricultural History* **31**, 67–74.

Jones C. W., Mastrangelo I. A., Smith H. H., Liu H. Z. and Meck R. A. (1976) Interkingdom fusion between human (HeLa) cells and tobacco hybrid (GGLL) protoplasts, *Science* **193**, 401–403.

Jones M. D. and Brooks J. S. (1950) Effectiveness of distance and border rows in preventing outcrossing in corn, *Oklahoma Agricultural Experimental Station, Bulletin no. T-38*.

Jones S. (2003) A system out of balance – the privatization of the Land Grant University breeding programs. In: Sligh M. and Lauffer L., editors, *Summit Proceedings: Summit on Seeds and Breeds for 21st Century Agriculture*, pp. 109–110, Pittsboro, North Carolina: Rural Advancement Foundation International, available online at: http://www.rafiusa.org/pubs/Seeds%20and%20Breeds.pdf.

Joshi L., Shah M. M., Flynn C. R. and Panitch A. (2004) Plant-produced recombinant therapeutics. In: Goodman R. B., editor, *Encyclopedia of Plant and Crop Science*, New York: Marcel Dekker.

Joshi S. P., Gupta V. S., Aggarwal R. K., Ranjekar P. K. and Brar D. S. (2000) Genetic diversity and phylogenetic relationship as revealed by inter simple sequence repeat (ISSR) polymorphism in the genus *Oryza*, *Theoretical and Applied Genetics* **100**, 1311–1320.

Jouhikainen K., Lindgren L., Jokelainen T., Hiltunen R., Teeri, T. H. and Oksman-Caldentey K. M. (1999) Enhanced scopolamine production in *Hyoscyamus muticus* L. hairy root cultures by genetic engineering, *Planta* **208**, 545–551.

Kalaitzandonakes N. and Bijman J. (2003) Who is driving biotechnology acceptance? *Nature Biotechnology* **21**, 366–369, available online at: http://www. nature.com/nbt/journal/v21/n4/full/nbt0403-366.html.

Kang M. S., editor (2002) *Quantitative Genetics, Genomics and Plant Breeding*,
    Wallingford: CABI Publishing.

Kaplan J. K. (1998) Conserving the world's plants, *Agricultural Research*, September
    1998, available online at: http://www.ars.usda.gov/is/timeline/germplasm.htm.

Karpenstein-Machan M. and Heyn J. (1992) Yield and yield structure of the winter
    cereals triticale and wheat in the middle mountain areas of northern Hessen,
    Agribiological Research **45**, 88–96.

Kasha K. J., Hu T. C., Oro R., Simion E. and Shim Y. S. (2001) Nuclear fusion leads
    to chromosome doubling during mannitol pretreatment of barley (*Hordeum
    vulgare* L.) microspores, *Journal of Experimental Botany* **52**, 1227–1238.

Kempin S. A., Liljegren S. J., Block L. M., Rounsley S. D., Yanofsky M. F. and Lam E.
    (1997) Targeted disruption in *Arabidopsis, Nature* **389**, 802–803.

Kenney M., Kloppenburg J., Buttell F. H. and Cowan J. T. (1983) *Genetic
    Engineering and Agriculture: Socioeconomic Aspects of Biotechnology R and D in
    developed and Developing Countries*, BIOTECH 83, Northwood: Online
    Publications

Ketchum R. E. B., Gibson D. M., Croteau R. and Shuler M. L. (1999) The influence
    of methyl jasmonate on the production of paclitaxel and related taxanes in
    suspension cell cultures of *Taxus canadensis, Biotechnology and Bioengineering*
    62, 97–105.

Khush G. S. (2005) What it will take to Feed 5.0 Billion Rice consumers in 2030,
    Plant Molecular Biology. **59**, 1–6.

Khush G. S., Coffman W. R. and Beachell H. M. (2001) The history of rice breeding:
    IRRI's contribution. In: Rockwood W. G., editor, *Rice Research and Production
    in the 21st Century: Symposium Honoring Robert F. Chandler, Jr*, pp. 117–135, Los
    Baños: International Rice Research Institute.

Kikkert J. R., Vidal J. R. and Reisch B. I. (2005) Stable transformation of plant cells
    by particle bombardment/biolistics, *Methods in Molecular Biology* **286**, 61–78.

King J. L. and Schimmelpfennig D. E. (2005) Mergers, acquisitions, and stocks of
    agricultural biotechnology intellectual property, *AgBioForum* **8**, 83–88, available
    online at: http://www.agbioforum.org/v8n23/v8n23a04-king.htm.

Kirk D. D., McIntosh K., Walmsley A. M. and Peterson R. D. K. (2005) Risk
    analysis for plant-made vaccines, *Transgenic Research* **14**, 449–462.

Kleingartner L. W. (2002) NuSun sunflower oil: redirection of an industry. In: Janick J.
    and Whipkey A. editors, *Trends in New Crops and New Uses*, pp. 135–138,
    Alexandria, Virginia: ASHS Press.

Kloppenburg J. R. (1988) *First the Seed: The Political Economy of Plant
    Biotechnology, 1492–2000*, 1st Edition, Cambridge: Cambridge University Press.

Kloppenburg J. R. (2004) *First the Seed: The Political Economy of Plant
    Biotechnology, 1492–2000*, 2nd Edition, Madison, Wisconsin: University of
    Wisconsin Press.

Klose N. (1950) *America's Crop Heritage: The History of Foreign Plant Introduction by
    the Federal Government*, Ames, Iowa: Iowa State College Press.

Knapp S. J. (1990) New temperate oilseed crops. In: Janick J. and Simon J. E.,
    editors, *Advances in New Crops*, pp. 203–210, Portland, Oregon: Timber Press,

available online at: http://www.hort.purdue.edu/newcrop/proceedings1990/v1-203.html.

Knapp S. J. (1998) Breakthroughs towards the domestication of *Cuphea*. In: Janick J. and Simon J. E., editors, *New crops*, pp. 372–379, New York: Wiley, available online at: http://www.hort.purdue.edu/newcrop/proceedings1993/v2-372.html.

Knapp S. J. and Crane J. M. (1999) Breeding advances and germplasm resources in meadowfoam: a very long chain oilseed. In: Janick J., editor, *Perspectives on New Crops and New Uses*, Alexandria, Virginia: ASHS Press.

Kneen B. (1992) *The Rape of Canola*, Toronto: EC Press.

Knight J. (2003) A dying breed, *Nature* **421**, 568–570, available online at: http://cuke.hort.ncsu.edu/cucurbit/breedingnature.html.

Kodym A. and Afza R. (2003) Physical and chemical mutagenesis, *Methods in Molecular Biology* **236**, 189–20.

Koebner R. M. D. and Summers R. W. (2003) 21st century wheat breeding: plot selection or plate detection, *Trends in Biotechnology* **21**, 59–63.

Kohli A., Twyman R. M., Abranches R., Wegel E., Stoger E. and Christou P. (2003) Transgene integration, organization and interaction in plants, *Plant Molecular Biology* **52**, 247–258.

Kölreuter D. J. G. (1761) *Vorläfige Nachricht von einigen das Geschlecht der Pflanzen betreffenden Versuchen und Beobachtungen, nebst Fortsetzung 1, 2, 3*, Leipzig: Gledisch.

Koo B., Pardey P. G. and Wright B. D., editors (2004) *Saving Seeds: The Economics of Conserving Crop Genetic Resources Ex Situ in the Future Harvest Centres of CGIAR*, Wallingford: CABI Publishing, available online at: http://www.ifpri.org/pubs/otherpubs/savingseeds.htm.

Kothari A. (1994) Agricultural biodiversity: Luxury or necessity, *Seminar 418*, quoted in Rangnekar (2000c)

Kranthi K. R. (2005) Bollworm resistance to Bt cotton in India, *Nature Biotechnology* **23**, 1476–1477.

Krattiger A. F. (2000) Food biotechnology: promising havoc or hope for the poor? *Proteus* **17**, 38–44.

Kropotkin P. (1880) *An Appeal to the Young*, reprinted in 1889 in *Revolutionary Classics*, New York: Charles H. Kerr Publishing Company, available online at: http://dwardmac.pitzer.edu/Anarchist_Archives/kropotkin/appealtoyoung.html.

Kryder R. D., Kowalski S. P., Krattiger A. F. (2000) The intellectual and technical property components of pro-vitamin A rice (GoldenRice ™): a preliminary freedom-to-operate review, *International Service for the Acquisition of Agri-Biotech Applications*, available online at: http://www.isaaa.org/Publications/briefs/briefs_20.htm.

Kuchel H., Ye G., Fox R. and Jeffries S. (2005) Genetic and economic analysis of a targeted marker-assisted wheat breeding strategy, *Molecular Breeding* **16**, 67–78.

Kuiper H. A., Noteborn P. J. M. and Peijnenburg A. C. M. (1999) Adequacy of methods for testing the safety of genetically modified foods, *Lancet* **354**, 1315–1316.

Laibach F. (1943) *Arabidopsis thaliana* (L.) Heynh. als object fur genetische und entwicklungsphysiologische untersuchungen, *Botanisches Archiv* **44**, 439–455.

390   *References*

Lambrecht B. (2001) *Dinner at the New Gene Café*, New York: St Martin's Press.
Lamkey K. R. (2003) Plant breeding: research and education agenda. In: Sligh M.
    and Lauffer L., editors, *Summit Proceedings: Summit on Seeds and Breeds for 21st*
    Century Agriculture, pp. 129–141, Pittsboro, North Carolina: Rural
    Advancement Foundation International.
Lantican M. A., Dubin H. J. and Morris M. L. (2005) *Impacts of International Wheat*
    Breeding Research in the Developing World, 1988–2002, D. F. Mexico:
    CIMMYT, available online at: http://www.cimmyt.org/english/wps/publs/
    catalogdb/index.cfm.
Lapan H. E. and Moschini G. (2004) Identity preservation and labeling of genetically
    modified products: system design and enforcement issues, *Working Paper*
    04-WP 375 October 2004, Ames, Iowa: Center for Agricultural and Rural
    Development Iowa State University, available online at: http://www.agmrc.org/
    NR/rdonlyres/C5C73C31–0C9B-4A37–9209–48DAC7DFF044/0/
    CARD04WP_375.pdf.
Lappé M. (1998) *Against the Grain*, London: Earthscan.
Latado R. R., Derbyshire M. T. V., Tsai S. M. and Tulmana Neto A. (2000)
    Obtenção de híbridos somáticos de limão 'Cravo' e tangerina 'Cleópatra',
    Pesquisa Agropecuária Brasileira **37**, 1735–1741.
Latham J. R., Wilson A. K. and Steinbrecher R. A. (2006) The mutational
    consequences of plant transformation, *Journal of Biomedicine and Biotechnology*
    2006, Article ID 25376, pp. 1–7, available online at: http://www.hindawi.com/
    GetArticle.aspx?pii = S1110724306253762.
Lawler A. (2003) University–industry collaboration: last of the big-time spenders?
    Science **299**, 330–333.
Lee L. S. (1988) Citrus polyploidy – origins and potential for cultivar improvement,
    Australian Journal of Agricultural Research **39**, 735–747.
Lee M. and Lau E. (2004) A new field of rice, *The Sacramento Bee*, 25 January 2004,
    available online at: http://www.checkbiotech.org/blocks/dsp_document.cfm?
    doc_id = 7009.
Leng T. (2004) New system to beat the heart, *Global Oils and Fats Business Magazine*
    1, 41–42.
Leonard K. J. (1987) The host population as a selective factor. In: Wolfe MS and
    Caten C. E., editors, *Populations of Plant Pathogens: their Dynamics and*
    Genetics, pp. 163–179.
Lesser W. (1998) Intellectual property rights and concentration in agricultural
    biotechnology, *AgBioForum* **1**, 56–61.
Lesser W. (2005) Intellectual property rights in a changing political environment:
    perspectives on the types and administration of protection, *AgBioForum* **8**, 64–72,
    available online at: http://www.agbioforum.org/v8n23/v8n23a02-lesser.htm.
Lev-Yadun S., Abbo S. and Doebley J. (2002) Wheat, rye, and barley on the cob?
    Nature Biotechnology **20**, 337–338.
Lewandowski I. (1997) Micropropagation of *Miscanthus* × *giganteus*, In Bajaj
    Y. P. S., editor, *High-Tech and Micropropagation V, Biotechnology*, volume **39**,
    pp. 239–255 Berlin: Springer.

Lewis O. (1938) *The Story of Huntingdon, Stanford, Hopkins and Crocker*, New York: Alfred A Knopf.

Lewontin R. (1982) Agricultural research and the penetration of capital, *Science for the People* **14**, 12–17.

von Liebig J. (1840) *Organic Chemistry and Its Applications in Agriculture and Physiology*, London: Taylor and Walton.

Lilja N. and Dalton T. (1997) Developing African public goods: rice varietal selection in Côte d'Ivoire. A discussion paper presented at the *American Agricultural Economics meeting, 1997, Salt Lake City, Utah*, cited in Ashby and Lilja (2004).

Lindner R. (2004a) Privatised provision of essential plant breeding infrastructure, *Australian Journal of Agricultural and Resource Economics* **48**, 301–321.

Lindner R. (2004b) Economic issues for plant breeding – public funding and private ownership, *Australasian Agribusiness Review* **12**, Paper 6, available online at: http://www.agribusiness.asn.au/Review/2004V12/Lindner.htm.

Locke J. (1690) *Second Treatise of Government*, edited commentaries with an Introduction by C. B. McPherson, Indianapolis, Minnesota: Hackett Publishing Company, 1980, original full text available online at: http://www.american.edu/dgolash/locke-treatise.htm.

Louwaars N. and Minderhoud M. (2001) When a law is not enough: biotechnology patents in practice, *Biotechnology and Development Monitor* **46**, 16–19.

Low J., Walker T. and Hijmans R. (2001) The potential impact of orange-fleshed sweetpotatoes on vitamin A intake in Sub-Saharan Africa, Paper presented at a regional workshop on food-based approaches to human nutritional deficiencies, *The VITAA Project, Vitamin A and Orange-Fleshed Sweetpotatoes in Sub-Saharan Africa, 9–11 May 2001, Nairobi, Kenya.*

Lucca P., Hurrell R. and Potrykus I. (2002) Fighting iron deficiency anemia with iron-rich rice, *Journal of the American College of Nutrition* **21**, 184S–190S.

Lurquin P. L. (2001) *The Green Phoenix*, New York: Columbia University Press.

Lurquin P. L. (2002) *High Tech Harvest*, Boulder, Colorado: Westview Press.

Lutz W., Sanderson W. and Scherbov S. (2001) The end of world population growth, *Nature* **421**, 543–545.

Ma J. K. C., Chikwamba R., Sparrow P., Fischer R., Mahoney R. and Twyman R. M. (2005) Plant-derived pharmaceuticals – the road forward, *Trends in Plant Science* **10**, 580–585.

Macilwain C. (2005) Stray seeds had antibiotic-resistance genes. Accidental release of genetically modified crops sparks new worries, *Nature* **434**, 548.

Mackey M. and Montgomery J. (2004) Plant biotechnology can enhance food security and nutrition in the developing world, *Nutrition Today* **34**, 52–58.

MacLeod D. (2006) Research exercise to be scrapped, *The Guardian*, 22 March 2006, available online at: http://education.guardian.co.uk/RAE/story/0,,1737082,00.html.

Mahoney R. J. (2005) Genetically modified crops, *Issues in Science and Technology Online*, Spring 2005, available online at: http://search.nap.edu/issues/21.3/21.3preview.html.

Mahoney R. *et al.* (2005) *Blueprint for the Development of Plant-derived Vaccines for Diseases of the Poor in Developing Countries, A Discussion Document*, Biodesign Institute at Arizona State University, Tempe, Arizona, available online at: http://www.biodesign.asu.edu/assets/pdfs/idv-provacs/WHO-Blueprint.pdf.

Maliga P. (2002) Engineering the plastid genome of higher plants, *Current Opinion in Plant Biology* **5**, 164–172.

Maliga P. (2004) Plastid transformation in higher plants, *Annual Review of Plant Biology* **55**, 289–313.

Maluszynski M., Nichterlei K., van Zanten L. and Ahloowalia B. S. (2000) Officially released mutant varieties – the FAO/IAEA database, *Mutation Breeding Review* **12**, 1–84.

Maluszynski M., Szarejko I. and Maluszynska J. (2003) Mutation techniques. In: Thomas B., Murphy D. J. and Murray B., editors, *Encyclopedia of Applied Plant Sciences*, pp. 186–201, Oxford: Elsevier/Academic Press.

Manelsdorf P. (1951) Hybrid corn, *Scientific American* **185**, 39–47.

Marcus G. J. (1980) *The Conquest of the North Atlantic*, Woodbridge: The Boydell Press.

Maredia M. K. and Byerlee D. (2000) Efficiency of research investments in the presence of international spillovers: wheat research in developing countries, *Agricultural Economics* **22**, 1–16.

Marillonnet S., Thoeringer C., Kandzia R., Klimyuk V. and Gleba Y. (2005) Systemic *Agrobacterium tumefaciens*-mediated transfection of viral replicons for efficient transient expression in plants, *Nature Biotechnology* **23**, 718–723.

Martin E. W. (1873) *A Complete and Graphic Account of the Crédit Mobilier Investigation from 'Behind the Scenes in Washington'*, New York: The Continental Publishing Company and National Publishing Co.

Martin W. (1999) Mosaic bacterial chromosomes: a challenge *en route* to a tree of genomes, *BioEssays* **21**, 99–104.

Martineau B. (2001) *First Fruit*, New York: McGraw-Hill.

Mascia P. N. and Flavell R. B. (2004) Safe and acceptable strategies for producing foreign molecules in plants, *Current Opinion in Plant Biology* **7**, 189–195.

Masle J., Gilmore S. R. and Farquhar G. D. (2005) The ERECTA gene regulates plant transpiration efficiency in *Arabidopsis*, *Nature* **436**, 866–870.

McCallum C. M., Comai L., Greene E. A. and Henikoff S. (2000a) Targeted screening for induced mutations, *Nature Biotechnology* **18**, 455–457.

McCallum C. M., Comai L., Greene E. A. and Henikoff S. (2000b) Targeting induced local lesions in genomes (TILLING) for plant functional genomics, *Plant Physiology* **123**, 439–442.

McGuire S. (1997) The effects of privatization on winter-wheat breeding in the UK, *Biotechnology and Development Monitor* **33**, 8–11.

McHughen, A. (2000) *Pandora's Picnic Basket*, Oxford: Oxford University Press.

McVey M. J., Baumel C. P. and Wisner R. (2000) Is Brazilian soybean production a threat to US exports? *Iowa Farm Outlook*, 19 June 2000, Iowa State University,

Ames, Iowa, available online at: www.econ.iastate.edu/outreach/agriculture/periodicals/ifo/PRDCTION3baumel.pdf.

Meadows D. H., Meadows D. L., Randers J. and Behrens W. W. (1972) *The Limits to Growth*, London: Pan.

Meager L. (1697) *The Mystery of Husbandry*, London: W. Onley.

Medina-Bolivar F., Wright R., Funk V., Sentz D., Barroso L., Wilkins T. D., Petri W. and Cramer C. L. (2003) A non-toxic lectin for antigen delivery of plant-based mucosal vaccines, *Vaccine* **21**, 997–1005.

Mendes-da-Glória F. J., Mourão Filho F. A. A., Camargo L. E. A. and Mendes B. M. J. (2000) Caipira sweet orange + Rangpur lime: a somatic hybrid with potential for use as rootstock in the Brazilian citrus industry, *Genetics and Molecular Biology* **23**, 661–665.

Mengiste T., Revenkova E., Bechtold N. and Paszkowski J. (1999) An SMC-like protein is required for efficient homologous recombination in, *Arabidopsis EMBO Journal* **18** 4505–4512.

Mentewab A. and Stewart C. N. (2005) Overexpression of an *Arabidopsis thaliana* ABC transporter confers kanamycin resistance to transgenic plants, *Nature Biotechnology* **23**, 1177–1180.

Meyer P. (2005) *Plant Epigenetics*, Oxford: Blackwell.

Miao Z. H. and Lam E. (1995) Targeted disruption of the TGA3 locus in *Arabidopsis thaliana*, *Plant Journal* **7**, 359–65.

Mill J. S. (1867) *Inaugural Address delivered to the University of St Andrews*, 1 February 1867, London: Longmans, Green, Reader, and Dyer.

Miller H. I. and Conko G. (2004) *The Frankenfood Myth*, Westport, Connecticut: Praeger.

Mira C. (2006) Detienen un embarque de soja argentina en Liverpool, *La Nacion*, 20 February 2006 available online at: http://buscador.lanacion.com.ar/Nota.asp?nota_id = 778880&high = Liverpool.

Mithen R. F. (2001) Glucosinolates and their degradation products, *Advances in Botanical Research* **35**, 213–262.

Mithen R. F., Faulkner K., Magrath R., Rose P., Williamson G. and Marquez G. (2003) Development of isothiocyanates-enriched broccoli and its enhanced ability to induce phase 2 detoxification enzymes in mammalian cells, *Theoretical and Applied Genetics* **106**, 727–734.

Mittler R. (2006) Abiotic stress, the field environment and stress combination, *Trends in Plant Science* **11**, 15–19.

Moehs C. P. (2005) TILLING: Harvesting functional genomics for crop improvement, *Information Systems for Biotechnology*, March 2005, available online at: http://www.isb.vt.edu/news/2005/news05.Mar.html.

Mohan Jain S. (2005) Major mutation-assisted plant breeding programs supported by FAO/IAEA, *Plant Cell, Tissue and Organ Culture* **82**, 133–123.

Mohan Jain S., Brar D. S. and Ahloowalia B. S., editors (1998) *Somaclonal Variation and Induced Mutations in Crop Improvement*, Dordrecht: Kluwer.

Monbiot G. (2004) Feeding cars, not people, *The Guardian*, 22 November 2004, available online at: http://www.monbiot.com/archives/2004/11/23/feeding-cars-not-people.

Mooney P. R. (1983) The law of the seed: another development and plant genetic resources, *Development Dialogue* **1–2**, 1–172.

Moore A. M. T., Hillman G. C. and Legge A. J. (2000) *Village on the Euphrates: From Foraging to Farming at Abu Hureyra*, Oxford: Oxford University Press.

Moore G., Devos K., Wang Z. and Gale M. (1995) Grasses, line up and form a circle, *Current Biology* **5**, 737–739.

Mor T. S., Sternfeld M., Soreq H., Arntzen C. J. and Mason H. S. (2001) Expression of recombinant human acetylcholinesterase in transgenic tomato plants, *Biotechnology and Bioengineering* **75**, 259–266.

Morais R. C. (2005) The great brazilian land grab, *Forbes*, 25 July 2005, available online at: http://www.forbes.com/business/global/2005/0725/052.html.

Moreau L., Lemarié S., Charcosset A. and Gallais A. (2000) Economic efficiency of one cycle of marker-assisted selection, *Crop Science* **40**, 329–337.

Moriguchi T., Kita M., Hisada S., Endo-Ingaki T. and Omura M. (1998), Characterisation of gene repertoires at mature stage of citrus fruits through random sequencing of redundant metallothionein-like genes expressed during fruit development, *Gene* **211**, 221–227.

Morris M. L. and Bellon M. R. (2004) Participatory plant breeding research: opportunities and challenges for the international crop improvement system, *Euphytica* **136**, 21–35.

Morris M., Dreher K., Ribaut J. M. and Khairallah M. (2003) Money matters (II): costs of maize inbred conversion schemes at CIMMYT using conventional and marker-assisted selection, *Molecular Breeding* **11**, 235–247.

Mower J. P Stefanovic S., Young G. J and Palmer J. D. (2004) Gene transfer from parasitic to host plants, *Nature* **432**, 165–166.

Muntzing A. (1979) Triticale: result and problems, *Zeitschrift für Pflanzenzuchtung, supplement,* 10, 1–103.

Murphy D. J. (1991) Designer oilseed crops. Genetic engineering of new oilseed crops for edible and non-edible applications, *Agro-Industry Hi-Tech* **5**, 5–10.

Murphy D. J., editor (1994) *Designer Oil Crops*, Weinheim: VCH Press.

Murphy D. J. (1995) The use of conventional and molecular genetics to produce new diversity in seed oil composition for the use of plant breeders – progress, problems and future prospects, *Euphytica* **85**, 433–440.

Murphy D. J. (1996) Engineering oil production in rapeseed and other oil crops, *Trends in Biotechnology* **14**, 206–213.

Murphy D. J. (1998) Impact of genomics on improving the quality of agricultural products. In: Dixon G. K., Copping L. C. and Livingstone D., editors, *Genomics: Commercial Opportunities from a Scientific Revolution*, pp. 199–210, Oxford: BIOS Scientific Publishing.

Murphy D. J. (1999a) Le colza carbure aux transgènes (Energy production from rapeseed oil), *Biofutur* **195**, 22–23.

Murphy D. J. (1999b) Production of novel oils in plants, *Current Opinion in Biotechnology* **10**, 175–180.

Murphy D. J. (1999c) The future of new and genetically modified oil crops. In: Janick J. editor, *Perspectives on New Crops and New Uses*, Alexandria,

Virginia: ASHS Press, available online at: http://www.hort.purdue.edu/newcrop/proceedings1999/v4-216.html.

Murphy D. J. (2002) Novel oils from plants – genes, dreams and realities, *Phytochemistry Reviews* **1**, 67–77, available online at: http://reo.nii.ac.jp/journal/HtmlIndicate/Contents/SUP0000001000/JOU0001000496/ISS0000008263/ART0000088455/ART0000088455_abstract.html.

Murphy D. J. (2003a) Agricultural biotechnology and oil crops – current uncertainties and future potential, *Applied Biotechnology and Food Science Policy* **1**, 25–38, available online at: http://www.extenza-eps.com:443/extenza/loadPDF?objectIDValue = 19503.

Murphy D. J. (2003b) Biofuels from crop plants. In: Thomas B., Murphy D. J. and Murray B., editors, *Encyclopedia of Applied Plant Sciences*, pp. 263–266, Oxford: Elsevier/Academic Press.

Murphy D. J. (2003c) Escaping from our boxes, *Times Education Supplement, Cymru*, 25 June 2003.

Murphy D. J. (2004a) Public sector R&D – does it have a future? *AgBioForum*, 11 March, 2004, available online at: http://www.agbioworld.org/.

Murphy, D. J. (2004b) Biotechnology, its impact and future prospects. In: Archer M. A. and Barber J., editors, *Molecular to Global Photosynthesis*, London: Imperial College Press. available online at: http://www.icpress.co.uk/books/lifesci/p218.html.

Murphy D. J. (2005) The study and utilisation of plant lipids: from lipid rafts to margarine. In: Murphy D. J. editor, *Plant Lipids: Biology, Utilisation and Manipulation*, Oxford: Blackwell.

Murphy D. J. (2006a) Plant manipulation to change lipid composition. In: Gunstone F., editor, *Modifying Lipids for Use in Food*, pp. 273–305, Oxford: Woodhead Press.

Murphy D. J. (2006b) Monster, marvel, or just misunderstood? *Public Service Review* **12**, 114–118, available online at: http://www.publicservice.co.uk/pdf/devolved/issue12/DG12%20Denis%20Murphy%20ATL.pdf.

Murphy D. J. (2007) *People, Plants, and Genes*, Oxford: Oxford University Press.

Murphy D. J., Fairbairn D. J. and Slocombe S. P. (1994) Genes for altering plant metabolism, *International Patent Application WO 94/01565*, European Patent Office.

Murphy D. J., Fairbairn D. J. and Bowra D. (1999) Expression of unusual fatty acids in transgenic rapeseed causes induction of glyoxylate cycle genes, *John Innes & Sainsbury Laboratory Annual Report 1998/99*, pp. 44–46, Norwich: John Innes Centre.

Naseem A., Oehmke J. F. and Schimmelpfennig D. E. (2005) Does plant variety intellectual property protection improve farm productivity? Evidence from cotton varieties, *AgBioForum* **8**, 100–107, available online at: http://www.agbioforum.org/v8n23/v8n23a06-oehmke.htm.

NASULGC (1995) *The Land Grant Tradition*, National Association of State Universities and Land-Grant Colleges, available online at: http://www.nasulgc.org/publications/Land_Grant/Land_ Grant_Main.htm.

National Academy of Sciences (2000) *Transgenic Plants and World Agriculture*, available online at: http://www.nap.edu/catalog/9889.html.

Naudin C. (1863) Nouvelles recherches sur l'hybridité dans les végétaux, *Annales Sciences Naturales, 4ème série, Botanique* **19**, 180–203.

Nehru J. (1961) Science and India, *Proceedings of the National Institute of Sciences of India* **27A**, 564.

Newman J. H. (1852) Discourses on the Scope and Nature of University Education, Dublin, republished as: Ker I. T., editor, *The Critical Edition of The Idea of a University by J. H. Newman*, Oxford: Oxford University Press, 1976.

Nielsen N. and Richie R. (2005) TILLING genes to improve soybeans, *Agricultural Research* **53**, July 2005, USDA-ARS, available online at: http://www.ars.usda.gov/is/AR/archive/jul05/genes0705.htm.

Normile D. (2004) CGIAR: lab network eyes closer ties for tackling world hunger, *Science* **303**, 1281–1283.

Normile D. (2005) centers embrace an alliance but remain wary of a merger, *Science* **307**, 498.

Nourse T. (1700) *Campania Felix or a Discourse of the Benefits and Improvements of Husbandry*, London.

NRC (National Research Council) (1972) *Genetic Vulnerability of Major Crops*, Washington DC: National Academy of Sciences.

NRC (National Research Council, Committee on Managing Global Genetic Resources) (1993) *Managing Global Genetic Resources – Agricultural Crop Issues and Policies*, Washington DC: National Academy Press.

Nuccio M. L., Rhodes D., McNeil S. D. and Hanson A. D. (1999) Metabolic engineering for osmotic stress resistance, *Current Opinion in Plant Biology* **2**, 128–134.

Nuffield (1999) *Genetically Modified Crops: the Ethical and Social Issues*, London: Nuffield Council on Bioethics, available online at: http://www.nuffieldbioethics.org/go/ourwork/gmcrops/publication_301.html.

O'Connor E. (2003) Biomass production, In: Thomas B., Murphy D. J. and Murray B., editors, *Encyclopedia of Applied Plant Sciences*, pp. 266–273, Oxford: Elsevier/Academic Press.

Oehmke J. F., Pray C. E. and Naseem A. (2005) Preface: innovation and dynamic efficiency in agricultural biotechnology (2005), *AgBioForum* **8**, 50–51, available online at: http://www.agbioforum.org/v8n23/v8n23a01-oehmke.htm.

Oil World (2005) Vol **48**, 1 April 2005, Hamburg: ISTA Mielke.

Oksman-Caldentey K. M. (2000) Metabolic engineering of plants for production of pharmaceuticals. In: *Proceedings of the International Conference on Biotechnology: Practice in Non-Food Products*, pp. 93–96, Europoint, Netherlands.

Osmani S. R. (1998) Did the green revolution hurt the poor? A re-examination of the early critique. In: Pingali P. L. and Hossain M., editors, *The Impact of Rice Research*, pp. 193–213, Manila: IRRI.

Overton M. (1996) *Agricultural Revolution in England. The Transformation of the Agrarian Economy 1500–1850*, Cambridge Studies in Historical Geography 23, Cambridge: Cambridge University Press.

Packer K. and Webster A. (1996) Patenting culture in science: Reinventing the wheel of scientific credibility, *Science, Technology and Human Values* **21**, 427–453.

Padidam M. (2003) Chemically regulated gene expression in plants, *Current Opinion in Plant Biology* **6**, 169–177.

Paine J. A., Shipton C. A., Chaggar S., Howells R. M., Kennedy M. J., Vernon G., Wright S. Y., Hinchliffe E., Adams J. L., Silverstone A. L. and Drake R. (2005) Improving the nutritional value of Golden Rice through increased pro-vitamin A content, *Nature Biotechnology* **23**, 482–487.

Palladino P. (1990) The political economy of applied research: plant breeding in Great Britain, 1910–1940, *Minerva* **28**, 446–468.

Palladino P. (1996) Science, technology and the economy: plant breeding in Great Britain, 1920–1970, *Economic History Review* **49**, 116–136.

Palladino P. (2002) *Plants, Patients and the Historian. (Re)membering in the Age of Genetic Engineering*, Manchester: Manchester University Press.

Panayotou T. (1993) *Empirical Tests and Policy Analysis of Environmental Degradation at Different Stages of Economic Development, World Employment Program. Working Paper*, Geneva: International Labour Office.

Panayotou T. (1997) Demystifying the Environmental Kuznets curve: turning a black box into a policy tool, *Environment and Development Economics* **2**, 465–484.

Paradise J., Andrews L. and Holbrook T. (2005) Intellectual property: patents on human genes: an analysis of scope and claims, *Science* **307**, 1566–1567.

Pardey P. G. and Beintema N. M. (2005) *Slow Magic, Agricultural R&D a Century After Mendel*, Agricultural Science and Technology Indicators Initiative, Washington DC: International Food Policy Research Institute.

Perl A., Galili S., Shaul O., Ben-Tzvi I. and Galili G. (1993) Bacterial dihydrodipicolinate synthase and desensitized asparate kinase: two novel selectable markers for plant transformation, *Bio/Technology* **11**, 715–718.

Perrin R. K., Hunnings, K. A. and Ihnen L. A. (1983) Some effects of the US plant variety protection act of 1970, *Economics Research Report No 46*, North Carolina: North Carolina State University Department of Economics and Business, Raleigh.

Peterson M. D. editor (1984) *The Writings of Thomas Jefferson*, New York: Literary Classics of the United States Inc..

Peterson R. K. D. and Arntzen C. J. (2004) On risk and plant-based biopharmaceuticals, *Trends in Biotechnology* **22**, 64–66.

Petrov D. A., Sangster K. A., Spencer Johnston J., Hartl D. L. and Shaw T. L. (2000) Evidence of DNA loss as a determinant of genome size, *Science* **287**, 1060–1062.

Pew Initiative (2002) *Pharming the Field*, available online at: http://pewagbiotech.org/.

Pew Initiative (2004) *Feeding the World: A Look at Biotechnology and World Hunger, Issue brief, March 2004*, available online at: www.pewbiotech.org.

Phillips L., Bridgeman J. and Ferguson-Smith M. (2000) *The BSE Enquiry. Report, Evidence and Supporting Papers of the Enquiry into the Emergence and Identification of Bovine Spongiform Encephalopathy (BSE) and Variant Creutzfeldt–Jakob Disease (vCJD) and the Action Taken in Response to it up to 20 March 1996*, London: The Stationery Office.

Phillips P. W. B. (1999) IRPs, Canola and public research in Canada. In: Santaniello V., Evenson R. E., Zilberman D. and Carlson G. A., editors, *Agriculture and*

Intellectual Property Rights Economic, Institutional and Implementation Issues in Biotechnology, Wallingford: CAB International, available online at: www.ag. usask.ca/departments/agec/cmtc/pdfs/IPRs_Canola_Phillips_new.pdf.

Phillips P. (2003a) The economic impact of herbicide tolerant canola in Canada. In: Kalaitzanondakes N., editor, *Economic and Environmental Impacts of First Generation Biotechnologies*, New York: Kluwer Academic.

Phillips P. W. B. (2003b) IRPs and the industrial structure of the North American seed industry. In: IPRs *in Agriculture: Implications for Seed Producers and Users*, Denver, Colorado: American Society of Agronomy, Crop Science Society of America, Soil Science Society of America Annual Meetings, available online at: http://www.iprsonline.org/resources/food.htm.

Phillips P. and Dierker D. (2001) Public good and private greed: strategies for realizing public benefits from a privatized global agri-food research effort. In: Pardey P., editor, *The Future of Food: Biotechnology Markets and Policies in an International Setting*, Baltimore, Maryland: Johns Hopkins University Press.

Phillips R. L. (1993) Plant genetics: out with the old, in with the new? *American Journal of Clinical Nutrition* **58**, 259S–263S.

Pickett A. A. (1993) Hybrid wheat: results and problems, advances in Plant Breeding, *Supplement to Journal of Plant Breeding* **15**, 1–259.

Pimentel D. (2002) *Biological Invasions*, Florida: CRC Press, Boca Raton.

Pimentel D. and Patzek T. W. (2005) Ethanol production using corn, switchgrass, and wood; biodiesel production using soybean and sunflower, *Natural Resources Research* **14**, 65–76.

Pingali P. and Traxler G. (2002) Changing locus of agricultural research: will the poor benefit from biotechnology and privatization trends? *Food Policy* **27**, 223–238.

Pinstrup-Andersen P. and Schiøler E. (2000) *Seeds of Contention*, Baltimore, Maryland: Johns Hopkins University Press.

Piperno D. R., Weiss E., Holst I. and Nabel D. (2004) Processing of wild cereal grains in the Upper Palaeolithic revealed by starch grain analysis, *Nature* **430**, 670–673.

Pirie M. (1988) *Privatization: Theory, Practice and Choice*, Aldershot: Avebury.

Pirie M.(2002) Privatization. In: Henderson, D. R., editor, *The Concise Encyclopedia of Economics*, Indianapolis, Minnesota: Liberty Fund, Inc., available online at: http://www.econlib.org/library/Enc/Privatization.htm.

Pollack A. (2006) DuPont and Syngenta join in modified-seed venture, *New York Times*, 11 April 2006, available online at: http://www.nytimes.com/2006/04/11/business/11place.html?_r=2&oref=slogin&oref=slogin.

Pollan M. (1998) Playing God in the garden, *New York Times Sunday Magazine*, 25 October, 1998.

Popovsky M. (1984) *The Vavilov Affair*, Hamden, Connecticut: Archon Books.

Porsch P., Janhke A. and During K. (1998) A plant transformation vector with minimal T-DNA II. Irregular patterns of the T-DNA in the plant genome, *Plant Molecular Biology* **37**, 581–585.

Prasad M. N. V. and de Oliveira Freitas H. M. (1999) Feasible biotechnological and bioremediation strategies for serpentine soils and mine spoils, *Electronic Journal of Biotechnology* **2**(1), available online at: http://www.ejb.org/content/vol2/issue1/full/5/index.html.

Pray C. E. (1991) Plant breeders' rights legislation, enforcement and R&D: lessons for developing countries. In: *Sustainable Agricultural Development: The Role of International Competition, Proceedings of the 21ˢᵗ International Conference of Agricultural Economists, Tokyo.*

Pray C. E. (1996) The impact of privatizing agricultural research in Great Britain: an interim report on PBI and ADAS, *Food Policy* **21**, 305–318.

Pray C. E. (1998) Private funding for public research. In: Tabor S. R., Janssen W. and Bruneau H., editors, *Financing Agricultural Research: A Sourcebook*, available online at: www.isnar.cgiar.org/publications/pdf/FSB/fsb-11c.pdf.

Pray C. E. and Naseem A. (2005) Intellectual property rights on research tools: incentives or barriers to innovation? Case studies of rice genomics and plant transformation technologies, *AgBioForum* **8**, 108–117, available online at: http://www.agbioforum.org/v8n23/v8n23a07-pray.htm.

Pray C. E., Oehmke J. F., and Naseem A. (2005) Innovation and dynamic efficiency in plant biotechnology: an introduction to the researchable issues, *AgBioForum* **8**, 52–63, available online at: http://www.agbioforum.org/v8n23/v8n23a01-oehmke.htm.

Price S. C. (1999) Public and private plant breeding, *Nature Biotechnology* **17**, 938–939, available online at: http://www.grain.org/publications/technology-example-wisonsin.cfm.

Priest S. H. (2001) *A Grain of Truth. The Media, the Public and Biotechnology*, Lanham, Maryland: Rowman & Littlefield.

Pringle P. (2003) *Food, Inc. Mendel to Monsanto – the Promises and Perils of the Biotech Harvest*, New York: Simon and Schuster.

Puchta H. (2003a) Marker-free transgenic plants, *Plant Cell, Tissue and Organ Culture* **74**, 123–134.

Puchta H. (2003b) Towards the ideal GMP: homologous recombination and marker gene excision, *Journal of Plant Physiology* **160**, 743–754.

Qi B., Fraser T., Mugford S., Dobson G., Sayanova O., Butler J., Napier J. A., Stobart A. K. and Lazarus C. M. (2004) Production of very long chain polyunsaturated omega-3 and omega-6 fatty acids in plants, *Nature Biotechnology* **22**, 739–745.

Qualset C. O. and Shands H. L. (2005) *Safeguarding the Future of U.S. Agriculture: The Need to Conserve Threatened Collections of Crop Diversity Worldwide*, report from: University of California, Division of Agriculture and Natural Resources, Genetic Resources Conservation Program, Davis, CA, available online at: http://www.grcp.ucdavis.edu/publications/index.htm.

Raboin L. M., Carreel F., Noyer J. L., Baurens F. C., Horry J. P., Bakry F., du Montcel H. T., Ganry J., Lanaud C., Lagoda P. J. L. (2005) Diploid ancestors of triploid export banana cultivars: molecular identification of $2n$ restitution gamete donors and n gamete donors, *Molecular Breeding* **16**, 333–341.

Rahman S. M., Kinoshita T., Anai T. and Takagi Y. (2001) Combining ability in loci for high oleic and low linolenic acids in Soybean, *Crop Science* **41**, 26–29.

Rahman S. M., Anai T., Kinoshita T. and Takagi Y. (2003) A novel soybean germplasm with elevated saturated fatty acids, *Crop Science* **43**, 527–531.

Ramirez R. and Quarry W. (2004) Communication strategies in the age of decentralisation and privatisation of rural services: lessons from two African experiences, *Agricultural Research & Extension Network, Paper No. 136, July 2004*, London: Overseas Development Institute, available online at: http://www.odi.org.uk/agren/.

Raney T. and Pingali P. (2004) Private research and public goods: implications of biotechnology for biodiversity, *ESA Working Paper No. 04–07*, Food and Agriculture Organisation of the United Nations, available online at: www.fao.org/es/esa.

Rangnekar D. (2000a) *Intellectual Property Rights and Agriculture: an Analysis of the Economic Impact of Plant Breeders' Rights*, document prepared for Action Aid, available online at: www.actionaid.org.uk/content_document.asp?doc_id = 241.

Rangnekar D. (2000b) Planned obsolescence and plant breeding: empirical evidence from wheat breeding in the UK (1965–1995), *Kingston University, Economics Discussion Paper 00/08*, available online at: http://www.kingston.ac.uk/~ku00386/school/research/0008.pdf.

Rangnekar D. (2000c) Plant breeding, biodiversity loss and intellectual property rights, Kingston University, *Economics Discussion Paper 00/05*, available online at: www.kingston.ac.uk/~ku00386/school/research/0005.pdf.

Rasmussen W. D. (1967) Technological change in western beet production, *Agricultural History* **41**, 31–35.

Raybould A. F. and Gray A. J. (1993) Genetically modified crops and hybridization with wild relatives: a UK perspective, *Journal of Applied Ecology* **30**, 199–219.

Rea P. A. (2005) A farewell to bacterial ARMs. A plant gene that confers antibiotic resistance provides a 'cleaner' selectable marker for plant transgenesis, *Nature Biotechnology* **23**, 1005–1006.

Reif J. C., Zhang P., Dreisigacker S., Warburton M. L., van Ginkel M., Hoisington D., Bohn M. and Melchinger A. E. (2005) Wheat genetic diversity trends during domestication and breeding, *Theoretical and Applied Genetics* **110**, 859–6.

Reinholz E. (1947) *Field Information Agency Technical Report No. 106*, pp. 1–70.

Rieseberg L. H., Wood T. E. and Baack E. J. (2006) The nature of plant species, *Nature* **440**, 524–527.

Rifkin J. (1999) *The Biotech Century*, London: Phoenix.

Roberts G. (2003) Review of research assessment, *Report by Sir Gareth Roberts to the UK Funding Bodies*, available online at: http://www.ra-review.ac.uk/reports/roberts.asp.

Rodgers K. K. (2003) *The Potential of Plant-Made Pharmaceuticals*, available online at: www.plantpharma.org/ials/fileadmin/The_Potential_of_Plant-Made_Pharmaceuticals.pdf.

Romeis J., Meissle M. and Bigler F. (2006) Transgenic crops expressing *Bacillus thuringiensis* toxins and biological control, *Nature Biotechnology* **24**, 63–71.

Rommens C. M. (2004) All Native DNA transformation: a new approach to plant genetic engineering, *Trends in Plant Science* **9**, 457–464.

Rosegrant M. W. and Cline S. A. (2003) Global food security: challenges and policies, *Science* **302**, 1917–1919.

Rowell A. (2003) *Don't Worry. It's Safe to Eat. The True Story of GM Food, BSE & Foot and Mouth*, London: Earthscan.

Ruse M. and Castle D., editors (2002) *Genetically Modified Foods. Debating Biotechnology*, Amherst, New York: Prometheus Books.

Sala F. and Labra M. (2003) Somaclonal variation. In: Thomas B., Murphy D. J. and Murray B., editors, *Encyclopedia of Applied Plant Sciences*, pp. 1417–1422, Oxford: Elsevier/Academic Press.

Salz P. A. (2005) Sharing your innovations is potentially profitable, *Wall Street Journal, Europe*, 24 March 2005.

Sampson A. (1975) *The Seven Sisters, The Great Oil Companies and the World They Made*, London: Hodder and Stoughton.

Sanderson M. (1972) *The Universities and British Industry, 1850–1970*, pp. 31–60, 184–213, London: Routledge & Kegan Paul.

San Miguel P., Gaut B. S., Tikhonov A., Nakajima Y. and Bennetzen J. L. (1998) The paleontology of retrotransposons in maize: dating the strata, *Nature Genetics* **20**, 43–45.

Scarano M. T., Abbate L., Ferrante S., Lucretti S., Nardi S. L. and Tusa N. (2002) Molecular characterization of Citrus symmetric and asymmetric somatic hybrids by means of ISSR-PCR and PCR-RFLP. In: Vasil I. K., editor, *Plant Biotechnology 2002 and Beyond*, pp. 549–550, Dordrecht: Kluwer.

Schnepf R. D., Dohlman E. and Bolling C. (2001) *Agriculture in Brazil and Argentina: Developments and Prospects for Major Field Crops*, Economic Research Service, USDA, Agriculture and Trade Report No. WRS013, available online at: www.ers.usda.gov/publications/wrs013/.

Schumpeter J. (1954) *Capitalism, Socialism and Democracy*, London: George Allen and Unwin.

Sechley K. A. and Schroeder H. (2002) Intellectual property protection of plant biotechnology inventions, *Trends in Biotechnology* **20**, 455–461.

Shah D. M., Rommens C. M. T. and Beachy R. N. (1995) Resistance to diseases and insects in transgenic plants: progress and applications to agriculture, *Trends in Biotechnology* **13**, 362–368.

Shanahan M. and Cockburn K. (2005) Rice researchers forced to flee Côte d'Ivoire violence, *Science and Development Network*, 2 March 2005, available online at: http://www.scidev.net/news/index.cfm?fuseaction = readnews&itemid = 1963 &language = 1.

Shands H. L. (1986) *A Synopsis of the U.S. National Germplasm System*, unpublished ms. July 1986, Washington DC: USDA, cited on p. 272 of Kloppenburg (2004).

Shands H. L. and Stoner A. K. (1997) Agricultural germplasm and global contributions. In: Hoaglund K. E. and Rossman A. Y., editors, *Beltsville Symposia in Agricultural Research: Global Genetic Resources: Access, Ownership, and Intellectual Property Rights*, pp. 97–106, Washington, DC: Association of Systematics Collections.

Shapira G., Stachelek J. L., Lestsou A., Soodak L. K. and Liskay R. M. (1983) Novel use of synthetic oligonucleotide insertion mutants for the study of homologous recombination in mammalian cells, *Proceedings of the National Academy of Sciences USA* **80**, 4827–4831.

Shapiro C. (2001) Navigating the patent thicket: cross licences, patent tools and standard setting, in innovation policy and the economy. In: Jaffe A., Lerner J. and Stern S., editors, *Innovation Policy and the Economy*, Volume I, Cambridge, Massachusetts: National Bureau of Economic Research.

Sharp M. (2005) Cavalier, how Britain discarded many of its strongest scientific institutions – and why it needs to rebuild, book review, *ResearchResearch*, available online at: http://www.researchresearch.com/.

Shepard S. (2004) Bioscience in Jonesboro, Ark. ASU institute studies biotech agriculture, *Memphis Business Journal*, 8 October 2004, available online at: http://www.bizjournals.com/memphis/stories/2004/10/11/story2.html.

Shorto R. (2005) *The Island at the Center of the World: The Epic Story of Dutch Manhattan and the Forgotten Colony that Shaped America*, London: Vintage Books.

Shull (1946) Hybrid seed corn, *Science* **103**, 547–550.

Simmonds I. and Tooley M., editors (1981) *The Environment in British Prehistory*, Ithaca, New York: Cornell University press.

Simmonds N. (1979) *Principles of Crop Improvement*, 1st edition, New York: Longman.

Simmonds N. W. (1990) The social context of plant breeding, *Plant Breeding Abstracts* **60**, 337–341.

Simmonds N. and Smartt J. (1995) *Principles of Crop Improvement*, 2nd edition, Oxford: Blackwell.

Singh R. B. (1995) Agricultural biotechnology in the Asia–Pacific region. In: *Agricultural Biotechnology in the Developing World*, Food and Agriculture Organization of the United Nations, available online at: http://www.fao.org/docrep/V4845E/V4845E07.htm.

Slade A. J. and Knauf V. C. (2005) TILLING moves beyond functional genomics into crop improvement, *Transgenic Research* **14**, 109–114.

Slade A. J., Fuerstenberg S. I., Loeffler D., Steine M. N. and Facciotti D. (2005) A reverse genetic, nontransgenic approach to wheat crop improvement by TILLING, *Nature Biotechnology* **23**, 75–78.

Smith A. (1776) *An Inquiry into the Nature and Causes of the Wealth of Nations*, 2003 edition, New York: Bantam Books, available online at: http://www.adamsmith.org/smith/won-intro.htm.

Smith N., Kilpatrick J. B. and Whitelam G. C. (2001) Superfluous transgene integration in plants, *Critical Reviews in Plant Science* **20**, 215–249.

Snowdon R. J. and Friedt W. (2004) Molecular markers in Brassica oilseed breeding: current status and future possibilities, *Plant Breeding* **123**, 1–8.

Sobel D. (1995) *Longitude: The True Story of a Lone Genius who Solved the Greatest Scientific Problem of his Time*, London: Fourth Estate.

Spector A. (1999) Essentiality of fatty acids, *Lipids* **34**, 1–3.

Spillane C., Curtis M. D. and Grossniklaus U. (2004) Apomixis technology development – virgin births in farmers' fields? *Nature Biotechnology* **22**, 687–991.

Srinivasan C. S. (2004) Plant variety protection, innovation, and transferability: some empirical evidence. *Review of Agricultural Economics* **26**, 445–471.

Srinivasan C. S., Thirtle C. G. and Palladino P. (2003) Winter wheat in England and Wales, 1923–1995: what do indices of genetic diversity reveal? *Plant Genetic Resources* **1**, 43–57.

Stadler L. J. (1928) Genetic effects of X-rays in maize, *Proceedings of the National Academy of Sciences USA* **14**, 69–75.

Steele K. A., Virk D. S., Prasad S. C., Kumar P., Singh D. N., Gangwar J. S. and Witcombe J. R. (2002) Combining PPB and marker-assisted selection: strategies and experiences with rice, extended abstract for *The Quality of Science in Participatory Plant Breeding Workshop, September 30–October 4, 2002, IPGRI, Rome*, available online at: http://www.cazs.bangor.ac.uk/ricemas/downloads. htm.

Steele K. A., Singh D. N., Kumar R., Prasad S. C., Billore M., Virk D. S., Shashidhar H. E. and Witcombe J. R. (2004) Marker-assisted selection of four root QTL and aroma to improve Kalinga III: evaluation of pyramid lines and advanced bulks from backcrossing, presentation at the *Indian National Conference on Increasing Rice Production under Water Limited Environment, December 2004, BAU, Ranchi*, available online at: http://www.cazs.bangor.ac.uk/ricemas/downloads. htm.

Stiglitz J. E. (1993) *Economics*, New York: Norton.

Stoger E., Ma J. K., Fischer R. and Christou P. (2005) Sowing the seeds of success: pharmaceutical proteins from plants, *Current Opinion in Biotechnology* **16**, 167–173

Stone R. (1995) Sweeping patents put biotech companies on the warpath, *Science* **268**, 656–658.

Stoskopf N. C. (1985) *Cereal Grain Crops*, Chapter 20, Reston, Virginia: Reston Publishing Co., Inc.

Streatfield S. J. *et al.* (2003) Corn as a production system for human and animal vaccines, *Vaccine* **21**, 812–815.

van Stuyvenberg J. H. (1969) *Margarine: An Economic, Social and Scientific History, 1869–1969*, Liverpool: Liverpool University Press.

Sullivan S. N. (2004) Plant genetic resources and the law: past, present, and future, *Plant Physiology* **135**, 10–15.

Sun J. and Zuo J. (2003) Excision of selectable marker genes from transgenic plants by site-specific DNA recombination, *AgBiotechNet* **5**, 1–4.

Sundstrom F. J., Williams J., Van Deynze A. and Bradford K. J. (2002) Identity preservation of agricultural commodities, University of California, Division of Agriculture and Natural Resources, *Agricultural Biotechnology in California Series, Publication 8077*, available online at: http://sbc.ucdavis.edu/Publications/ ABC_Series.htm.

Swift J. (1725), A proposal that all the ladies and women of ireland should appear constantly in irish manufactures. In: Davies H., editor (1955) *The Prose Works of Jonathan Swift*, volume 12.

Syvanen M. and Kado C. I., editors (2002) *Horizontal Gene Transfer*, London: Academic Press.

Tabashnik B. E., Carriere Y., Dennehy T. J., Morin S., Sisterson M. S., Roush R. T., Shelton A. M. and Zhao J. (2003) Insect resistance to transgenic Bt crops: lessons from the laboratory and field, *Journal of Economics and Entomology* **96**, 1031–1036.

Takahashi R., Kurosaki H., Yumoto S., Han O. K., and Abe J. (2001) Genetic and linkage analysis of cleistogamy in soybean, *Journal of Heredity* **92**, 89–9.

Tarczynski M. C., Jensen R. G. and Bonhert H. J. (1992) Expression of a bacterial *mtlD* gene in transgenic tobacco leads to production and accumulation of mannitol, *Proceedings of the National Academy of Sciences USA* **89**, 2600–2604.

Tarrant (2001) High Profile: Norman Borlaug, *The Dallas Morning News*, 21 January, 2001, available online at: http://www.normanborlaug.org/267339_borlaug_21liv..html.

Taxler G. (2006) The GMO experience in North and South America, *International Journal of Science, Technology, and Globalization* **2**, 46–64, available online at: http://bcsia.ksg.harvard.edu/research.cfm?program = STPP&project = STG&pb_id = 537.

Taylor M. R. and Tick J. S. (2002) The StarLink case: issues for the Future, *Pew Initiative on Biotechnology*, available online at: http://pewagbiotech.org/agtopics/index.php?TopicID = 7.

Taylor N. L., Anderson M. K., Wuesenberry K. H. and Watson L. (1976) Doubling the chromosome number of *Trifolium* species using nitrous oxide, *Crop Science* **16**, 516–518.

Teich A. H., editor (1986) *Biotechnology and the Environment*, Discussion section, p. 108, Washington: American Association for the Advancement of Science.

Tencalla F. (2005) Science, politics, and the GM debate in Europe, *Regulatory Toxicology and Pharmacology* **44**, 43–48.

Terada R., Urawa H., Inagaki V., Tsugane K. and Lida S. (2002) Efficient gene targeting by homologous recombination in rice, *Nature Biotechnology* **20**, 1030–1033.

Thatcher A. L. (2004) Continued losses put pressure on Monsanto product launch, *Information System for Biotechnology News Report*, November 2004, available online at: http://www.isb.vt.edu/news/2004/news04.nov.html#nov0405.

Thirtle C. G., Beck H. S., Palladino P., Upton M. and Wise W. S. (1991) Agriculture and Food. In: Nicholson R., Cunningham C. M. and Gummett P., editors, *Science and Technology in the United Kingdom*, Harlow: Longman.

Thirtle C. G., Bottomley P., Palladino P., Schimmelpfennig D. and Townsend R. (1998) The rise and fall of public sector plant breeding in the United Kingdom: a causal chain model of basic and applied research and diffusion, *Agricultural Economics* **19**, 127–143.

Thomas B., Murphy D. J. and Murray B. G., editors, (2003) *Encyclopedia of Applied Plant Sciences*, Oxford: Elsevier/Academic Press.

Thro A. M. (2004) Europe on transgenic crops: how public plant breeding and eco-transgenics can help in the transatlantic debate, *AgBioForum* **7**, 142–148, available online at: http://www.agbioforum.org.

Thrupp L. A., editor (1996) *New Partnerships for Sustainable Agriculture*, Washington: World Resources Institute.

Tilman D., Cassman K. G., Matson P. A., Naylor R. and Polasky S. (2002) Agricultural sustainability and intensive production practices, *Nature* **418**, 671–677.

Tinker P. B. (2000) The future research requirements for the oil palm plantation. In: Pushparajah E., editor, *Plantation Tree Crops in the New Millennium: the Way Ahead, Proceedings of the International Planters Conference*, pp. 3–40, Kuala Lumpur: The Incorporated Society of Planters.

Toomey G. (1999) *Farmers as Researchers: The Rise of Participatory Plant Breeding, IDRC Project no. 950019*, Ottawa: International Development Research Centre.

Tracy W. F. (2003) What is plant breeding? In: Sligh M. and Lauffer L., editors, *Summit Proceedings: Summit on Seeds and Breeds for 21st Century Agriculture*, pp. 35–42, Pittsboro, North Carolina: Rural Advancement Foundation International, available online at: http://www.rafiusa.org/pubs/Seeds%20and%20Breeds.pdf.

Tran D. V. (2001) Closing the rice yield gap for food security, rice research for food security and poverty alleviation. In: *Proceedings of the International Rice Research Conference, 31 March–3 April 2000, Los Baños*, pp. 27–41, Los Baños: IRRI, available online at: http://ressources.ciheam.org/om/pdf/c58/03400070.pdf.

Trewavas A. (2002) Malthus foiled again and again, *Nature* **418**, 668–670.

Tripp R. and Byerlee D. (2000) Public plant breeding in an era of privatisation, *Natural Resource Perspectives* **57**, June 2000, available online at: http://www.odi.org.uk/nrp/57.pdf.

Trut L. (1999) Early canid domestication: the farm-fox experiment, *American Scientist* **87**, 160–169.

Turner M. K. (1993) Pharmaceuticals from agriculture: manufacture or discovery? *Industrial Crops and Products* **1**, 125–131.

Turner M., Kugbei S. and Bishaw Z. (1999) Privatization of the seed sector in the Near East and North Africa. In: *Seed Policy and Programmes for Sub-Saharan Africa, Proceedings of the Regional Technical Meeting on Seed Policy and Programmes in the Near East and North Africa, June 1999, Larnaca, Cyprus*, available online at: http://www.fao.org/ag/AGP/AGPS/Cyprus/Paper4.htm.

Turuspekov Y., Mano Y., Honda I., Kawada N., Watanabe Y. and Komatsuda T. (2004) Identification and mapping of cleistogamy genes in barley, *Theoretical and Applied Genetics* **109**, 480–487.

Tusser T. (1560) *Five Hundred Points of Good Husbandry – Comparing Good Husbandry*, reprinted in 1984, Oxford: Oxford University Press.

van Tuyl J. M., Meijer B. and van Diën M. P. (1992) The use of oryzalin as an alternative for colchicine in in-vitro chromosome doubling of *Lilium* and *Nerine*, *Acta Horticulturae* **325**, 625–629.

Twain M. (1883) *Mark Twain: Mississippi Writings: Tom Sawyer, Life on the Mississippi, Huckleberry Finn, Pudd'nhead Wilson*, New York: Library of America, available online at: http://mark-twain.classic-literature.co.uk/life-on-the-mississippi/index.asp.

UK Government (1993) *White Paper Realising Our Potential – a Strategy for Science, Engineering and Technology, Cm 2250*, London: HMSO,

UK Parliament (1998) Research and development, *Hansard Debates*, Written Answers to Questions, 1 December 1988, column 322: 31, available online at: http://www.parliament.the-stationery-office.co.uk/pa/cm198889/cmhansrd/1988-12-01/Writtens-1.html.

UK Parliament (2004a) The work of the biotechnology and biological sciences research council, *Report of House of Commons Science and Technology Committee, 29 April 2004 (HC 526)*, London: Stationery Office, available online at: http://www.publications.parliament.uk/pa/cm200304/cmselect/cmsctech/6/602.htm.

UK Parliament (2004b) Research assessment exercise: a re-assessment, *Report of House of Commons Science and Technology Committee, 23 September 2004 (HC 586)*, available online at: www.parliament.uk/parliamentary_committees/science_and_technology_committee.cfm.

UK Parliament (2006) Research council support for knowledge transfer, *House of Commons Science and Technology Committee, Third Report, 15 June 2006, (HC 995-I)*, available online at: http://www.publications.parliament.uk/pa/cm200304/cmselect/cmsctech/6/602.htm

United States Court of Appeals for the Federal Circuit (2005) In: *re. Dane K Fisher and Raghunath V Lalgudi, 04–1465, Serial No. 09/619,643, 7 September 2005*, available online at: http://fedcir.gov/opinions/04-1465.pdf.

USDA (2005) Audit report animal and plant health inspection service, controls over issuance of genetically engineered organism release permits, Office of Inspector General, *Audit Report 50601–8-Te*, available online at: http://www.usda.gov/oig/webdocs/50601-08-TE.pdf.

US–EC Task Force on Biotechnology Research (2005) *Future of Plant Biotechnology, 21–22 June, 2005*, available online at: http://europa.eu.int/comm/research/biotechnology/ec-us/docs/ec-us_tfws_2005_june_arlington_abstracts.pdf.

Vain P. (2005) Plant transgenic science knowledge, *Nature Biotechnology* **23**, 1348–1349.

Vandekerckhove J., van Damme J., van Lijsebettens M., Botterman J., De Block M., Vanderweile M., de Clerq A., Leemans J., Van Montagu M. and Krebbers E. (1989) Enkephalins produced in transgenic plants using modified 2S seed storage proteins, *Bio/Technology* **7**, 929–932.

Various (2002) *Abrupt Climate Change: Inevitable Surprises*, Washington DC: National Academies Press, available online at: http://www.nap.edu/books/0309074347/html/.

Various (2004) *Biological Confinement of Genetically Engineered Organisms*, Washington DC: National Academies Press, available online at: http://www.nap.edu/openbook/0309090857/html/.

Varshney R. K., Graner A. and Sorrells M. E. (2005) Genomics-assisted breeding for crop improvement, *Trends in Plant Science* **10**, 621–630.

Vavilov N. I. (1926) Centres of origin of cultivated plants, *Trudi po Prikladnoi Botanike Genetikei Seleksii* [Bulletin of Applied Botany and Genetics] **16**, 139–248 [in Russian].

Vavilov N. I. (1935) The phytogeographical basis for plant breeding, Volume I, Moscow, Origin and geography of cultivated plants. In: *The Phytogeographical Basis for Plant Breeding*, translated by Love D., pp. 316–366, Cambridge: Cambridge University Press.

Vavilov N. I. (1992) *Origin and Geography of Cultivated Plants*, Dorofeev V. F., editor, Cambridge: Cambridge University Press.

Venkat K. (2001) Natural medicines from plants. In: Kuokka A., Nuutila A. M. and Oksman-Caldentey K. M., editors, *Plant Biotechnology – Better Products from Better Plants*, Abstracts, p. 4,. Helsinki: VTT Biotechnology.

Vernooy R. (2003) *Seeds That Give: Participatory Plant Breeding*, Ottawa: International Development Research Centre.

de Vetten N., Wolters A. M., Raemakers K., van der Meer I., ter Stege R., Heeres E., Heeres P. and Visser R. (2003) A transformation method for obtaining marker-free plants of a cross-pollinating and vegetatively propagated crop, *Nature Biotechnology* **21**, 439–442.

Vickers J. and Yarrow G. (1988) *Privatization: An Economic Analysis*, Cambridge, Massachusetts: MIT Press.

Vidal J. (2002) Hunger in a world of plenty, *The Guardian*, 22 August 2002, available online at: http://www.guardian.co.uk/worldsummit2002/earth/story/0,,777671,00.html

Vietmeyer N. D., editor (1989) *Triticale – a Promising Addition to the World's Cereal Grains, Report of an Ad Hoc Panel*, Board on Science and Technology for International Development, National Research Council, Washington DC: National Academy Press.

Villareal R. L., Varughese G. and Abdalla O. S. (1990) Advances in spring triticale breeding, *Plant Breeding Reviews* **8**, 43–90.

Vinocur B. and Altman A. (2005) Recent advances in engineering plant tolerance to abiotic stress: achievements and limitations, *Current Opinion in Biotechnology* **16**, 123–132.

Virk D. S., Singh D. N., Prasad S. C., Gangwar J. S. and Witcombe J. R. (2003) Collaborative and consultative participatory plant breeding of rice for the rainfed uplands of eastern India, *Euphytica* **132**, 95–108.

Virmani S. S. (1994) *Hybrid Rice Technology: New Developments and Future Prospects*, Selected papers from the *International Rice Research Conference*, Virmani S. S., editor, Manila: International Rice Research Institute.

Visser M. (1986) *Much Depends on Dinner: The Extraordinary History and Mythology, Allure and Obsessions, Perils and Taboos of an Ordinary Meal*, New York: Grove Press.

Voltaire (1759) *Candide* (translated by Norman Cameron), London: Penguin, available online at: http://www.online-literature.com/voltaire/candide/.

de Vries (1901) *The Mutation Theory*, originally published as: *Die Mutationstheorie*, Leipzig: Veit & Co. Germany

de Vries J. and Toenniessen G. (2001) *Securing the Harvest: Biotechnology, Breeding and Seed Systems for African Crops*, pp. 99–166, Wallingford: CABI.

Waage J., editor (2002) *Crop Diversity at Risk: The Case for Sustaining Crop Collections*, Wye: Department of Agricultural Sciences, Imperial College.

Waines J. G. (1998) *In situ* conservation of wild relatives of crop plants in relation to their history. In: Damania A. B., Valkoun J., Willcox G. and Qualset C. O., editors, *The Origins of Agriculture and Crop Domestication*, Aleppo: ICARDA, available online at: http://www.ipgri.cgiar.org/publications/pubfile.asp?ID.PUB=47.

Wallace H. A. and Bressman E. N. (1949) *Corn and Corn Growing*, New York: John Wiley and Sons.

Wallace H. A. and Brown W. L. (1956) *Corn and its Early Fathers*, East Lansing, Michigan: Michigan State University Press.

Walsh B. (2001) Quantitative genetics in the age of genomics, *Theoretical and Population Biology* **59**, 175–178.

Wambugu F. (1999) Why Africa needs agricultural biotech, *Nature* **400**, 15–16.

Wang W., Vinocur B. and Altman A. (2003a) Plant responses to drought, salinity and extreme temperatures: towards genetic engineering for stress tolerance, *Planta* **218**, 1–14.

Wang X., Chen C., Wang L., Chen D., Guang W. and French J. (2003b) Conception, early pregnancy loss, and time to clinical pregnancy: a population-based prospective study, *Fertility and Sterility* **79**, 577–584.

Warnken P. F. (1999) *The Development and Growth of the Soybean Industry in Brazil*, Ames, Iowa: Iowa State University Press.

Webster A. J. (1989) Privatisation of public sector research: the case of a plant breeding institute, *Science and Public Policy* **16**, 224–232.

Webster A. and Packer K. (1996) Patenting in public sector research: tensions between theory and practice. In: Kirkland J., editor, *Barriers to International Technology Transfer*, Dordrecht: Kluwer.

Weil C. (2005) Single base hits score a home run in wheat, *Trends in Biotechnology* **23**, 220–222.

Weiss H., Wetterstrom W., Nadel D. and Bar-Yosef O. (2004) The broad spectrum revisited: evidence from plant remains, *Proceedings of the National Academy of Sciences USA* **101**, 9551–9555.

Weston R. (1645) *Discours of Husbandrie*.

Whitney (2005) Study: seed diversity is at risk; losing genetic traits could lead to disease, pestilence problem, *The Modesto Bee*, 5 February 2005, available online at: http://www.modbee.com/business/story/10071538p-10898904c.html.

Wiegmann A. F. (1828) *Über die Bastarderzeugung im Pflanzenreiche: eine von den königiche Akademie der Wissenschaften zu Berlin gekrönte Preisschrift*, Braunschweig: F. Vieweg.

van Wijk J. (1995) Broad biotechnology patents hamper innovation, *Biotechnology and Development Monitor* **25**, 15–17.

van Wijk J. (2002) Food insecurity: prevalence, causes, and the potential of transgenic 'Golden Rice', *Phytochemistry Reviews* **1**, 141–151.

Wilkes G. (1983) Current status of crop plant germplasm: CAC critical review, *Plant Science* **127**, 137–181.

Wilkie T. (1989) The Thatcher effect in science. In: *The Thatcher Effect*, Kavanagh D. and Seldon A., editors, Oxford: Clarendon Press.

Wilkie T. (1991) *British Science and Politics Since 1945*, Oxford: Blackwell.

Wilson C. H. (1954) *The History of Unilever: a Study in Economic Growth and Social Change*, London: Cassell.

Wilson J. D., Whittingham M. J. and Bradbury R. B. (2005) The management of crop structure: a general approach to reversing the impacts of agricultural intensification on birds? *Ibis* **147**, 453–463.

Wilson W. W. and Dahl B. L. (2002) Costs and risks of testing and segregating GM wheat, *Agricultural Experiment Station, Agribusiness and Applied Economics Report No. 501*. Fargo, North Dakota: North Dakota State University, available online at: http://www.agecon.lib.umn.edu/cgi-bin/pdf_view.pl?paperid=6149&ftype=.pdf.

Winston M. L. (2002) *Travels in the Genetically Modified Zone*, Cambridge, Massachusetts: Harvard University Press.

Wise T. A. (2004) The paradox of agricultural Subsidies: measurement issues, agricultural dumping, and policy reform, *GDAE Working Paper No. 04–02*, Medford, Massachusetts: Global Development and Environment Institute, Tufts University.

Witcombe J. R., Joshi A. and Goyal S. N. (2003) Participatory plant breeding in maize: a case study from Gujarat, India, *Euphytica* **130**, 413–422.

White P. J. and Broadley M. R. (2005) Biofortifying crops with essential mineral elements, *Trends in Plant Science* **10**, 586–593.

Wolpert L. (1993) *The Unnatural Nature of Science*, London: Faber and Faber.

Won H. and Renner S. S. (2003) Horizontal gene transfer from flowering plants to Gnetum, *Proceedings of the National Academy of Sciences USA* **100**, 10824–10829.

Wong G., Tan C. C. and Soh A. C. (1997) Large scale propagation of oil palm clones – experiences to date, *Acta Horticulturae* **447**, 649–658.

Worlidge J. (1669) *Systema Agriculturae. The Mystery of Husbandry Discovered*, London: printed by T. Johnson for S. Speed.

Wright B. D. and Pardey P. G. (2006) Changing intellectual property regimes: implications for developing country agriculture, *International Journal of Science, Technology, and Globalization* **2**, 93–114, available online at: http://bcsia.ksg.harvard.edu/research.cfm?program=STPP&project=STG&pb_id=537.

Wrigley E. A. (2004) *Poverty, Progress and Population*, Cambridge: Cambridge University Press.

WTO (2006) *World Trade Report 2006: Exploring the links between subsidies, trade and the WTO*, Geneva: WTO, available online at: http://www.wto.org/english/news_e/pres06_e/pr447_e.htm.

Wu G., Truksa M., Datla N., Vrinten P., Bauer J., Zank T., Cirpus P., Heinz E. and Qiu X. (2005) Stepwise engineering to produce high yields of very long-chain polyunsaturated fatty acids in plants, *Nature Biotechnology* **23**, 1013–1017.

Yamaguchi T. and Blumwald E. (2005) Developing salt-tolerant crop plants: challenges and opportunities, *Trends in Plant Science* **10**, 615–620.

Yamaguchi-Shinozaki K. and Shinozaki K. (2001) Improving plant drought, salt and freezing tolerance by gene transfer of a single stress-inducible transcription factor, *Novartis Foundation Symposium* **236**, 176–186.

Yamamoto Y. T., Rajbhandari N., Lin X., Bergmann B. A., Nishimura Y. and Stomp A. (2001) Genetic transformation of duckweed *Lemna gibba* and *Lemna minor*, *In Vitro Cellular and Developmental Biology – Plant* **37**, 349–353.

Ye X., Al-Babili S., Kloti A., Zhang J., Lucca P., Beyer P. and Potrykus I. (2000) Engineering the provitamin A (β-carotene) biosynthetic pathway into (carotenoid-free) rice endosperm, *Science* **287**, 303–305.

Young N. D. (1999) A cautiously optimistic vision for marker-assisted breeding, *Molecular Breeding* **5**, 505–510.

Yukimune Y., Tabata H., Higashi Y. and Hara Y. (1996) Methyl jasmonate-induced overproduction of paclitaxel and baccatin III in *Taxus* cell suspension cultures, *Nature Biotechnology* **14**, 1129–1132.

Zeleny M. (1987) Management support systems: towards integrated knowledge management, *Human Systems Management* **7**, 59–70.

Zhai W., Li X., Tian W., Zhou Y., Pan X., Cao S., Zhao Y., Zhao B., Zhang Q. and Zhu L. (2000) Introduction of a blight-resistance gene, Xa21, into five Chinese rice varieties through an *Agrobacterium*-mediated system, *Science in China, Series C* **43**, 361–368.

Zhang H. X., Hodson J. N., Williams J. P. and Blumwald E. (2001) Engineering salt-tolerant Brassica plants: characterization of yield and seed oil quality in transgenic plants with increased vacuolar sodium accumulation, *Proceedings of the National Academy of Sciences USA* **98**, 12832–12836.

Zimmermann M. B. and Hurrell R. F. (2002) Improving iron, zinc and vitamin A nutrition through plant biotechnology, *Current Opinion in Biotechnology* **13**, 142–145.

Zohary D. (2004) Unconscious selection and the evolution of domesticated plants, *Economic Botany* **58**, 5–10.

Zohary D. and Hopf M. (2000) *Domestication of Plants in the Old World*, 3rd edition, Oxford: Oxford University Press.

Index

411

Printed in the United States
by Baker & Taylor Publisher Services